16th International Summer School on Vacuum, Electron, and Ion Technologies 2009

(VEIT 2009)

Journal of Physics: Conference Series 223

Sunny Beach, Bulgaria
28 September-2 October 2009

ISBN: 978-1-61738-481-3
ISSN: 1742-6588

Printed from e-media with permission by:

Curran Associates, Inc.
57 Morehouse Lane
Red Hook, NY 12571

Some format issues inherent in the e-media version may also appear in this print version.

Printed by Curran Associates, Inc. (2010)

For permission requests, please contact the Institute of Physics
at the address below.

Institute of Physics
Dirac House, Temple Back
Bristol BS1 6BE UK

Phone: 44 1 17 929 7481
Fax: 44 1 17 920 0979

techtracking@iop.org

Additional copies of this publication are available from:

Curran Associates, Inc.
57 Morehouse Lane
Red Hook, NY 12571 USA
Phone: 845-758-0400
Fax: 845-758-2634
Email: curran@proceedings.com
Web: www.proceedings.com

TABLE OF CONTENTS

GAS DISCHARGE AND PLASMA MODELING AND PROCESSING

THIN FILMS DEPOSITION, PROPERTIES AND APPLICATIONS

ELECTRON BEAM, ION BEAM AND LASER PROCESSING

ANALYSES, METHODS, APPLICATIONS

PREFACE

The International Summer School on Vacuum, Electron and Ion Technologies (VEIT) has been organized biennially since 1977 when the series of VEIT Schools was launched by the Institute of Electronics, Bulgarian Academy of Sciences with the aim to act as a forum for interchange and dissemination of knowledge and ideas on the latest developments in electron-, ion-, and plasma-assisted technologies. Beginning from 2001, the school has been jointly organized with the Institute of Ion Beam Physics and Materials Research, Forschungszentrum Dresden-Rossendorf, Germany. Whereas, the school initially provided a meeting place for researchers mainly from Eastern and Central European countries, its importance grew issue by issue. The school is now a major scientific event and a meeting place for young scientists from Eastern and Western Europe involved in research and development associated with high-tech industries. Many former school participants have gone on to become leading scientists in research establishments and companies throughout the world. Leading international companies, such as High Voltage Engineering, Balzers, Varian, and Hauzer have used the VEIT forum to present their products through oral presentations, poster contributions or exhibits. The School Proceedings have been published in special issues of the international journals *Vacuum, Plasma Processes and Polymers, Journal of Physics: Conference Series.*

The Sixteenth VEIT school was held in the Black Sea resort Sunny Beach, Bulgaria on 28 September to 2 October 2009. It was attended by close to 110 participants from 13 countries: Belgium, Bulgaria, Czech Republic, France, Germany, The Netherlands, Romania, Slovak Republic, Spain, Sweden, Ukraine, UK and USA.

Following the tradition of publishing the VEIT Proceedings, a selection of papers presented at the event is published in this volume of *Journal of Physics: Conference Series*, under the originality and quality criteria of acceptance by the journal, including peer reviewing.

The school comprised nine oral and three poster sessions. Seventeen invited talks of general interest and six progress reports were presented orally. In total 63 contributed papers were presented during the poster sessions. There were several scientific highlights covering fundamentals of interaction of fast particles with solids and challenging practical applications ranging from novel techniques for creating hard coatings, optical/protective layers, biocompatible materials to nanosized structures produced by evaporation, sputtering or external irradiation. Latest results were presented on ion-beam synthesis and modification in both low-energy (deposition and film growth) and high-energy (sputtering, implantation) regimes, and processing of solid materials aiming at patterning the surface or to create nanophase systems towards for electronic or tribological/wear resistant applications.

Despite the busy scientific program, the atmosphere was relaxed and informal. The early afternoons of most conference days were free to stimulate both scientific and social interaction between participants, which often took place on the beach. The social program included a welcome reception, a conference banquet, and an outing to historical landmarks in Bulgaria.

VEIT 2009 owes its success to many people. The International Advisory Committee shaped the scientific program and ensured high-quality plenary presentations by careful selection of invited speakers. The Local Committee bore the brunt of the organization both at the conference site and in dealing with correspondence, abstracts, and manuscripts for these proceedings. We are grateful to our sponsors Forschungszentrum Dresden-Rossendorf, Dresden, Germany and Bulgarian Academy of Sciences, Sofia, Bulgaria for their generosity that enabled us to support the attendance of students and provided support to deal with mailing, printing, renting the conference site, etc. We would like to thank also all authors for their valuable contributions to these proceedings and to the school, as well as all reviewers for their important work.

The next conference in the series will be held in September 2011.

Wolfhard Möller, Nikolay Guerassimov, Chavdar Ghelev
Guest Editors

All papers published in this volume of *Journal of Physics: Conference Series* have been peer reviewed through processes administered by the proceedings Editors. Reviews were conducted by expert referees to the professional and scientific standards expected of a proceedings journal published by IOP Publishing.

Journal of Physics: Conference Series **223** (2010) 012001 doi:10.1088/1742-6596/223/1/012001

Negative ion density measurements in an inductively driven hydrogen discharge

St Lishev, A Shivarova and Ts Tsankov[1]

Faculty of Physics, St. Kl. Ohridski University of Sofia
5 J. Bouchier Blvd., 1164 Sofia, Bulgaria

E-mail: tsankov@phys.uni-sofia.bg

Abstract. The study presents experiments on the determination of the electronegativity and its axial variation in an inductively driven hydrogen discharge, performed by using the laser photodetachment technique. The results showing high electronegativity with nonmonotonic axial variations and a maximum next to the rf power deposition region provide an experimental indication that a design of an efficient source of negative hydrogen ions based on a single-chamber discharge might be possible.

1. Introduction

The plasma sources of negative hydrogen ions have formed a subject of extended research owing to their applications in the accelerator physics [1], ion implantation technology [2] and fusion plasma heating [3]. The volume production of the negative ions results from dissociative attachment of electrons to (highly) vibrationally excited molecules. In search for conditions for efficient production of negative hydrogen ions the idea for the tandem type of a design of the source has been reached [3]. In its modification into an inductively-driven rf plasma source [4] this concept has emerged to a design of the source as a two-chamber source with rf power deposition into the first – small-size – chamber and ion extraction from the second – large-size – chamber. However, recent theoretical studies [5, 6] show that the region of the rf power deposition – the driver region – sustains high concentration of negative ions due to their accumulation in the on-axis discharge region, and trace out a concept for a design of the source as a matrix of small radius discharges with a single hole extraction from each of them.

This study, being in the scope of recent work on small-size inductively-driven two-chamber negative-hydrogen-ion sources, extends previous experiments [7] on registration of the concentration of the negative hydrogen ions H^- in the second chamber of the source towards registration of the plasma electronegativity, i.e. the ratio of the concentrations of the negative ions and of the electrons, in the first chamber. The obtained axial variation of the electronegativity in the plasma region of the rf power deposition and in its vicinity shows nonmonotony with a maximum just outside the power deposition region and comes as an experimental support of the concept [6] for a rf-driven negative hydrogen ion source based on single-chamber discharges.

[1] To whom any correspondence should be addressed.

2. Experimental arrangements

The experimental set-up (figure 1), described in more details in [7], is a two-chamber rf plasma source inductively driven by a cylindrical coil (a 9-turn coil). The first chamber of the source, a quartz tube, has a diameter of $d_1 = 4.5$ cm. The modifications, made here, concern the experimental arrangements for the laser photodetachment technique. In general, this technique [8] in its combination with a probe is a method for determination of the electronegativity $\alpha = n_-/n_\mathrm{e}$ based on measurements of the ratio of the plateau amplitude I_ph of the pulse in the electron saturation current to the probe to its steady state value $I_{\mathrm{e(dc)}}$; in α, n_- and n_e are, respectively, the density of the negative ions and the electron density. The pulse is due to the extra electrons detached – by the laser beam photons ($\mathrm{H}^- + h\nu \rightarrow \mathrm{H} + \mathrm{e}$) – from the negative ions and collected by the probe. As it has been done before [7], the second harmonics of a Surelite III-10 Nd:YAG laser is used. However, here, in accordance with the aim of the study – determination of the axial variation of α – the laser beam path is aligned axially, centred at the discharge axis, as shown in figure 1. Respectively, the probe is axially movable over the entire length of the discharge, also centred at the discharge axis. The laser beam diameter is fixed at 1 cm by a diaphragm and the laser pulse energy is measured with a Gentec joulemeter. The relation of α to the measured current ratio $(I_\mathrm{ph}/I_{\mathrm{e(dc)}})$ is $\alpha = \left(I_\mathrm{ph}/I_{\mathrm{e(dc)}}\right) / (\delta n_-/n_-)$, where the photodetachment fraction $(\delta n_-/n_-)$ is obtained for the measured values of the laser pulse energy density W, by using the well-established dependence of $(\delta n_-/n_-)$ on W [7, 8].

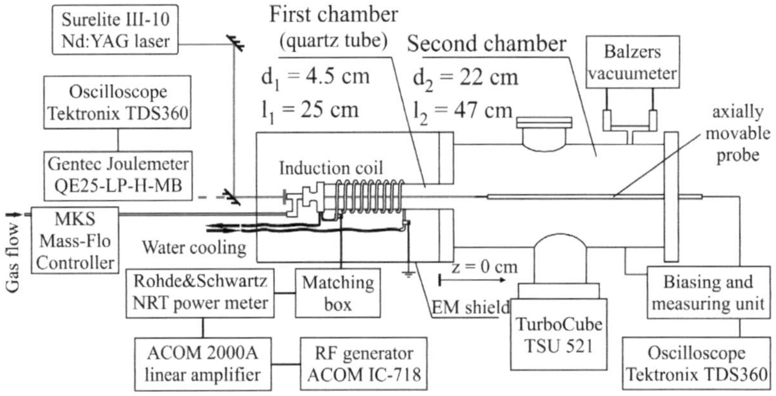

Figure 1. The experimental set-up.

The probe is 8 mm in length, with a radius of 0.25 mm. In the measurements it is moved from the position $z = 4$ cm inside the second chamber till almost the middle of the coil ($z = -12$ cm). A second probe (a SmartProbe^{TM} probe system) is used for measuring the ion saturation current which provides indications for the axial variation of the negative ion density n_-. The results are in arbitrary units, since – at the current stage – the electron temperature (necessary for the determination of the plasma density) cannot be reliably determined.

3. Results and discussions

The results (figures 2 and 3) presented here are for the axial variation of the electronegativity α and of the negative ion density n_- in a discharge sustained in the gas pressure range $p = (4$–$10)$ mTorr in flowing gas (a gas flow of 20 sccm) by applying a rf power $P = (700$–$1000)$ W at frequency of $f = 27$ MHz. In the figures, the $(z = 0)$-position is at the transition between the two chambers and, thus, the negative z-values are inside the first chamber where the driver is located. The vertical lines mark the position of the middle and the end of the coil.

The obtained axial variations of the electronegativity α for different gas-pressure values are in figure 2. Nonmonotony of the axial profiles of α is the main conclusion: starting with high

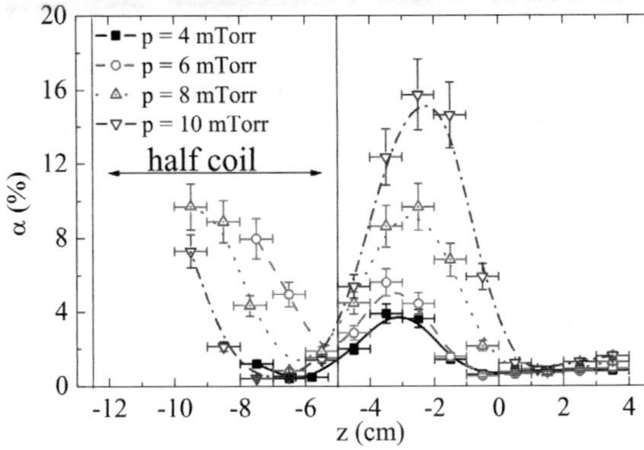

Figure 2. Axial variation of the electronegativity $\alpha = n_-/n_e$ for different gas-pressure values p and applied power of $P = 700$ W.

values in the plasma region under the coil, α decreases forming a minimum close to the end of the coil and then increases again reaching a maximum in the region between the coil end and the transition between the two chambers, followed by a decrease towards the second chamber. With the measured axial profile of the ion saturation current shown in figure 3(a), the axial profile of the negative ion density, though obtained only in arbitrary units (figure 3(b)), shows the same behaviour: high values of n_- in the plasma region under the coil, minimum close to its end and a maximum towards the transition between the two chambers.

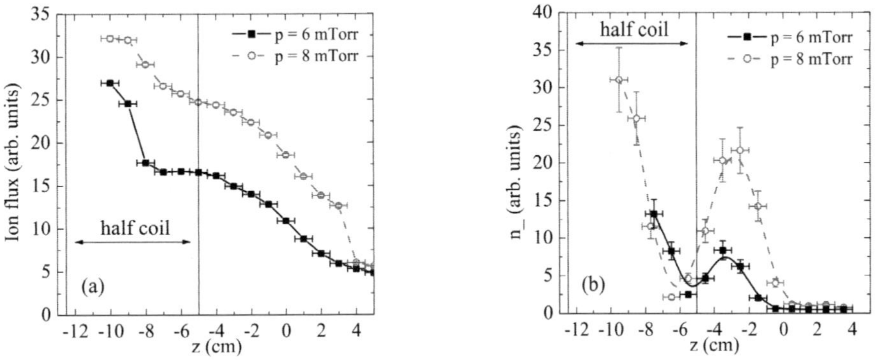

Figure 3. Axial variation of the ion saturation current to the probe in (a) and of the negative ion density, in arbitrary units, in (b) for $p = 6$ and 8 mTorr and $P = 700$ W.

Clearly, the explanation of such a peculiar behaviour of the profiles of α and n_- requires detailed theoretical and experimental knowledge of the plasma parameter distribution, still not available for two-chamber plasma sources. However, even at this stage the results give indications for locality in the H^--balance established under the conditions of nonlocality in the formation of the discharge structure. A similar conclusion has been reached also before [7], in the experiments carried out in the second chamber of the source. Such a conclusion is supported also by the results from the model [9] of hydrogen discharges in a tandem source, with the same configuration and size as that in the experiment here, obtained, however, at higher gas pressures. The theoretical results show a strong decrease of the electron temperature T_e and density n_e from the driver of the discharge towards the second chamber of the source. However, whereas the axial n_e-decrease starts from the centre of the region of the rf power deposition, that of T_e starts in the vicinity of the transition between the two chambers. As it is widely accepted, an effective H^- production by electron attachment to vibrationally excited molecules ($e + H_2 \rightarrow H_2(v) + e$,

$e + H_2(v) \rightarrow H^- + H$) requires a combination of high T_e with high n_e and N_2 (the concentration of the hydrogen molecules), for the vibrational excitation of the molecules, and a combination of low T_e with high n_e, for the dissociative attachment of the electrons. Rovibrational excitation involving surface processes can also contribute to the H^--production [10]. The high values of α and n_- (figures 2 and 3(b)) in the plasma region under the coil could be related to the high values of n_e and $H_2(v)$ there, the minimum close to the end of the coil could be due to lower n_e under the condition of still high T_e and the maximum outside the plasma region covered by the coil could be related to lower T_e there. High N_2 and lowering of the concentrations of the positive ions and of the hydrogen atoms outside the power deposition region [9] also favour sustaining high n_- since the former supports the H^--production and the latter reduces the H^--losses. The shift of the position of the maximum of α with p calls for influence of effects of nonlocality, i.e. for contribution by H^--ions produced in the vicinity of the transition between the two chambers and accelerated to the position of the maximum in the dc potential drop towards the second chamber, which should be a pressure dependent process. However, the shift of the maximum with p is very slight and, thus, this effect is weak. The increase in the value of the maximum of α and n_- with p is in conformity with the gas-pressure dependence of n_-, known from former experiments [11]. Increasing the rf power (up to 1 kW) preserves the structuring of the α-profile.

4. Conclusions

In general, the experimental results obtained showing high values of α and n_- in the first chamber, moreover, quite higher than in the second one, come in conformity with the suggestion [6], stemming from theoretical results, for a design of a source based on the driver region and performed as a matrix of small-radius discharges with a single hole extraction from each of them. Moreover, with the registration of maxima of α and n_- next to the region of the rf power deposition, the experimental results specify better the design of the source: The extraction of the H^- ions from the single discharge of the matrix should be from the position of the maxima of α and n_-, which is constructively better suited compared to extraction from the rf power deposition region itself.

Acknowledgments

The authors thank Prof. Dr. Kiss'ovski and Dr. Kolev for contributions to the experimental arrangements. The work is within projects DO02-267 and BUL/1026323 supported, respectively, by the National Science Fund of Bulgaria and by the Alexander von Humboldt Foundation and it is part of the programme of the Bulgarian Association EURATOM/INRNE (task 2). Support to the experimental equipment by the National Science Fund (Bulgaria) and Sofia University through project D01-413 is also highly acknowledged.

References

[1] Peters J 2000 *Rev. Sci. Instrum.* **71** 1069–74
[2] Ishikawa J 1994 *Rev. Sci. Instrum.* **65** 1290–4
[3] Bacal M 2006 *Nucl. Fusion* **46** S250–9
[4] Speth E *et al* 2006 *Nucl. Fusion* **46** S220–38
[5] Paunska Ts, Schlüter H, Shivarova A and Tarnev Kh 2006 *Phys. Plasmas* **13** 023504
[6] Paunska Ts, Shivarova A and Tarnev Kh 2009 *J. Appl. Phys.*, submitted
[7] Lishev St, Shivarova A and Tsankov Ts 2009 *AIP Conf. Proc.* **1097** 127–36
[8] Bacal M 2000 *Rev. Sci. Instrum.* **71** 3981–4006
[9] Paunska Ts, Shivarova A, Tarnev Kh and Tsankov Ts 2009 *AIP Conf. Proc.* **1097** 12–21
[10] Gabriel O, Schram D C and Engeln R 2008 *Phys. Rev.* E **78** 016407
[11] Bacal M, Bruneteau A M, Graham W G, Hamilton G W and Nachman M 1981 *J. Appl. Phys.* **52** 1247–54

16th International Summer School on Vacuum, Electron, and Ion Technologies (VEIT 2009) IOP Publishing
Journal of Physics: Conference Series **223** (2010) 012002 doi:10.1088/1742-6596/223/1/012002

2D model of a hydrogen inductive discharge with a planar coil

Ts Paunska[1], A Shivarova[1], Kh Tarnev[2] and Ts Tsankov[1,3]
[1]Faculty of Physics, St. Kl. Ohridski University of Sofia
5 J. Bouchier Blvd., 1164 Sofia, Bulgaria
[2]Department of Applied Physics, Technical University of Sofia,
8 Kl. Ohridski Blvd., 1000 Sofia, Bulgaria

E-mail: tsankov@phys.uni-sofia.bg

Abstract. Based on a two-dimensional (2D) fluid-plasma model of low-pressure hydrogen discharges sustained by rf power deposition of the type of inductive discharges with a planar coil, the study presents the spatial distribution of the plasma parameters and of the fluxes in the discharge and discussions outlining plasma maintenance in regions outside the rf power input.

1. Introduction

The use of inductively coupled plasmas with planar coils as reactors for processing in the microelectronics industry has stimulated since the 90's of the last century active theoretical research on the source operation within 2D and 3D discharge modelling [1, 2]. Inductive discharges with hydrogen plasmas are currently also actively studied as sources of negative ions for the neutral beam injection heating in large Tokamak devices. However, in the latter case the source, performed as large scale experimental arrangements of a two-chamber source, has an inductive discharge with a cylindrical coil as a driver [3]. On the other hand, recent results for high negative ion density in smaller radius discharges [4] has traced out a suggestion for a design of the source as a matrix of small radius discharges [5]. Since an inductive discharge with a planar coil would be a technically better solution, this study is directed towards description of such small radius discharges, not yet studied in the literature on the microelectronics processing technology.

This study presents results from a 2D-model of small-radius hydrogen discharges with rf power deposition of the type of inductive discharges with a planar coil. Due to the small radius of the discharge compared to its length, remote plasma maintenance clearly shows evidence. The latter requires detailed analysis of the spatial distribution of the particle and electron-energy fluxes as shown by the results presented here. The model is a first step towards self-consistent description of a small-radius negative hydrogen ion source inductively driven by a planar coil and, therefore, rovibrational excitation, negative hydrogen ions and variation of the gas temperature are currently not accounted for.

2. Description of the model

Hydrogen discharge maintenance in a straight cylindrical metal discharge vessel and power deposition with a super-Gaussian shape along its axis located at one of its end plates ($z = 0$ in

[3] To whom any correspondence should be addressed.

the figures further on) is described within the fluid plasma theory. The initial set of equations is based on the fluid-plasma theory. The continuity equations

$$\frac{\partial n_\alpha}{\partial t} + \nabla \cdot \vec{\Gamma}_\alpha = \frac{\delta n_\alpha}{\delta t} \tag{1}$$

are solved for the electrons ($\alpha = \mathrm{e}$), the three types of positive ions (H^+, H_2^+ and H_3^+ denoted by $\alpha = \mathrm{i}j$, with $j = 1, 2, 3$, respectively) and the hydrogen atoms ($\alpha = \mathrm{a}$). Here n_α and $\vec{\Gamma}_\alpha$ are, respectively, the particle densities and their fluxes and $\delta n_\alpha/\delta t$ describes the production and the losses of the corresponding type of particles α. In the electron energy balance

$$\frac{3}{2}\frac{\partial\left(n_\mathrm{e} T_\mathrm{e}\right)}{\partial t} + \nabla \cdot \vec{J}_\mathrm{e} = P_\mathrm{ext} - P_\mathrm{coll} + P_\mathrm{dc} \tag{2}$$

T_e is the electron temperature (in energy units), P_ext is the externally applied rf power density, $P_\mathrm{dc} = -e\vec{\Gamma}_\mathrm{e} \cdot \vec{E}_\mathrm{dc}$ is the Joule losses/input related to the dc electric field $\vec{E}_\mathrm{dc}$, with potential Φ ($\vec{E}_\mathrm{dc} = -\nabla\Phi$) and P_coll summarises the electron energy losses in collisions with neutrals. The electron energy flux $\vec{J}_\mathrm{e}$ is with its two compounds, the conductive $\vec{J}_\mathrm{e}^\mathrm{cond} = -\chi_\mathrm{e}\nabla T_\mathrm{e}$ and convective $\vec{J}_\mathrm{e}^\mathrm{conv} = (5/2)T_\mathrm{e}\vec{\Gamma}_\mathrm{e}$ fluxes; χ_e is the thermal conductivity coefficient. The Poisson equation

$$\Delta\Phi = -\frac{e}{\varepsilon_0}\left(\sum_{j=1}^{3} n_{\mathrm{i}j} - n_\mathrm{e}\right) \tag{3}$$

and the expression for the gas pressure $p = \kappa T_\mathrm{g}\left(n_\mathrm{a} + n_\mathrm{m}\right)$ complete the initial set of equations. Here e and ε_0 are the elementary charge and the vacuum permittivity, T_g is the gas temperature, n_m is the density of the hydrogen molecules and κ is the Boltzmann constant. The boundary conditions [6] are for the fluxes and for the potential at the walls as well as for symmetry on the axis ($r = 0$).

3. Results and discussions

The results discussed here (figures 1–3) are for two values of the total applied rf power $Q = 250$ and 500 W and of the gas pressure $p = 20$ and 50 mTorr. The mean free paths of the charged particles are such that the latter value of p is in the gas pressure range of diffusion-controlled discharges whereas the former is at the transition to the free-fall regime of discharge maintenance.

The contour plots of the plasma-parameter distribution and the arrow plots of the fluxes in figure 1 completed with the axial variations of the contributors to the electron balance and the electron-energy balance in figure 2 outline the following pattern of the discharge behaviour:

(i) The rf power deposition region, extended towards about $z = 6$ cm (figure 2(b)), sustains high density plasma with high T_e and high dc potential (figure 1). The charged particle production there by ionization (atom ionization prevailing over that of the molecules (figure 2(a)) as confirmed also by comparing the density values of the atomic and molecular ions in figures 1(b)–1(d)) compensates the diffusion losses towards the side wall of the vessel (losses in the radial direction (figure 2(a))). An axial electron flux flowing into the on-axis region, due to the formation of a vortex type of an electron flux in the power deposition region, also aids the ionization to compensate the diffusion losses in the radial direction. Upstream axial flux towards the ($z = 0$)-surface shows also evidence as particle losses. Charged particle losses through the fluxes strongly prevail over volume recombination (figure 2). Due to the radially homogeneous rf power deposition and the density drop towards the discharge walls, the highest values of T_e (figure 1(e)) are close to the walls and this drives a conductive flux directed to the on-axis region of the discharge. The latter aids the rf power deposition to compensate the electron energy losses

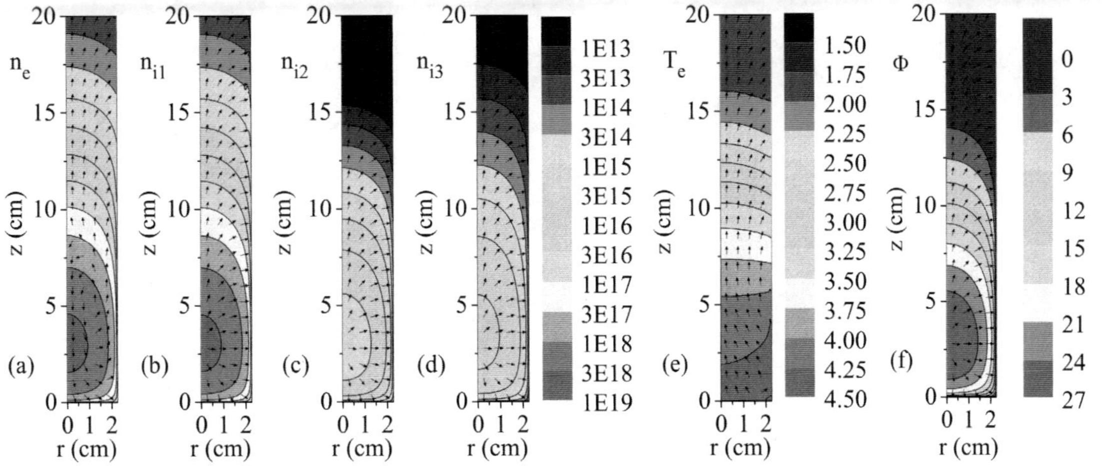

Figure 1. Spatial distributions with contours of constant values of the plasma parameters and arrow plot presentation of the direction of fluxes: n_e [m^{-3}] and $\vec{\Gamma}_e$ in (a), n_{i1} [m^{-3}] and $\vec{\Gamma}_{i1}$ in (b), n_{i2} [m^{-3}] and $\vec{\Gamma}_{i2}$ in (c), n_{i3} [m^{-3}] and $\vec{\Gamma}_{i3}$ in (d), T_e [eV] and $\vec{J}_e$ in (e) and Φ [V] and $\vec{E}_{dc}$ in (f); $p = 20$ mTorr, $Q = 500$ W. Figures (a)–(d) have the same scale, given in (d).

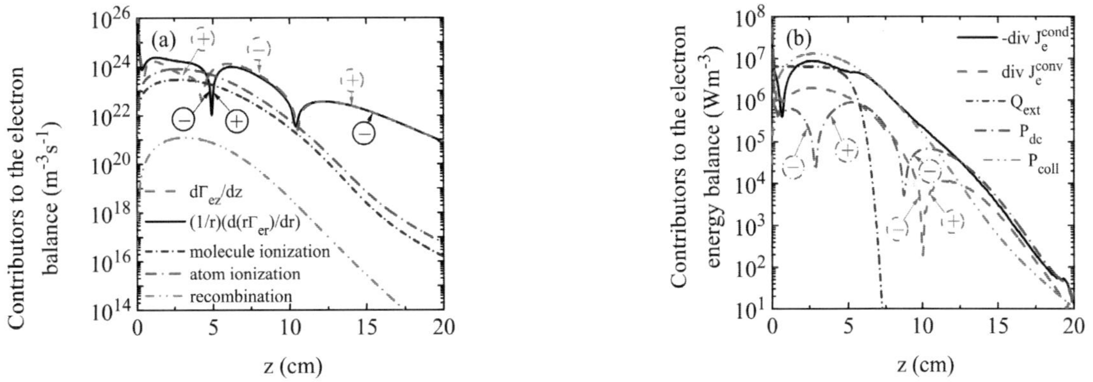

Figure 2. Axial variation of the contributors to the electron balance (a) and to the electron energy balance (b). Regions of electron production/electron energy input and electron losses/electron energy losses via the $(1/r)[\partial(r\Gamma_{er})/\partial r]$, $\partial\Gamma_{ez}/\partial z$, $\nabla\cdot\vec{J}_e^{conv}$ and P_{dc} (the quantities changing – along the discharge length – their role in the electron- and electron-energy balances) are marked, respectively by (+) and (−); $p = 20$ mTorr, $Q = 500$ W.

in collisions (figure 2(b)). The dc electric field directed towards the walls drives the ion fluxes (which are, in fact, drift fluxes) straight to the walls (figures 1(f) and 1(b)–1(d)).

(ii) The region of a remote plasma maintenance (at about $z \gtrsim 10$ cm) is characterized by low plasma density (figures 1(a)–1(d)), low T_e (figure 1(e)) and low dc potential (figure 1(f)). The electron energy input is via the conductive flux which compensates electron energy losses for maintenance of the dc electric field (figure 2(b)). Energy input via the convective flux also adds to the conductive flux. The electron energy losses in collisions, including collisions for ionization, are negligible. The existence of the remote plasma is due to an axial electron flux which compensates the particle losses to the side wall of the discharge vessel, i.e. the radial electron flux. In the remote plasma region the dc electric field is directed to the walls (figure 1(f))

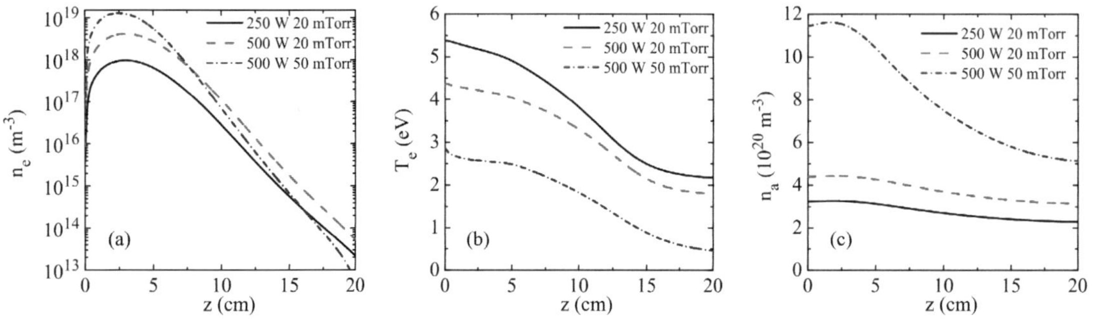

Figure 3. Axial variations at $r = 0$ of n_e (a), T_e (b) and n_a (c) for $p = 20$ and 50 mTorr, $Q = 250$ and 500 W.

like in the power deposition region and the ions drift along the field lines (figures 1(b)–1(d)).

(iii) The region 6 cm$\lesssim z \lesssim$ 10 cm shows the transition between the region of a rf power input and the remote plasma region. In its on-axis region electron energy input via the conductive flux, slightly supported by a Joule heating in the dc field (the P_{dc}-term in (2)), compensates electron energy losses in collisions (figure 2(b)). Input of electrons via the radial flux compensates their losses via their axial flux.

Figure 3 shows the changes in the plasma parameters with varying p and Q. The main trends are similar to those in the first chamber of the tandem source modelled in Refs. [6, 7].

4. Conclusions

The 2D fluid plasma model presented in the study, providing results for the structure of low-pressure rf discharges with inductive driving of the type of a planar coil, outlines remote plasma maintenance in one-chamber (straight tube) gas-discharge vessels: The plasma existence is due to an axial electron flux from the driver that compensates the diffusion losses to the walls and the electron energy input via the conductive flux compensates energy losses for the maintenance of the dc field that forms the fluxes. As it should be expected for sources with remote plasma maintenance, the fluxes are important. The latter shows similarities with the behaviour of the fluxes in two-chamber plasma sources with metal walls and a rf power deposition to the first chamber [6] that determines a discharge regime with a dc current in a rf discharge, due to different electron and ion fluxes.

Acknowledgments

The work is within project DO02-267 supported by the National Science Fund of Bulgaria. Support via project BUL/1026233 of the Alexander von Humboldt Foundation is also acknowledged.

References

[1] Kushner M J, Collison W Z, Grapperhaus M J, Holland J P and Barnes M S 1996 *J. Appl. Phys.* **80** 1337–44

[2] Lymberopoulos D P and Economou D J 1995 *J. Res. Natl. Stand. Technol.* **100** 473–94

[3] Speth E *et al* 2006 *Nucl. Fusion* **46** S220–38

[4] Paunska Ts, Schlüter H, Shivarova A and Tarnev Kh 2006 *Phys. Plasmas* **13** 023504

[5] Paunska Ts, Shivarova A and Tarnev Kh 2009 *J. Appl. Phys.*, submitted

[6] Paunska Ts, Shivarova A, Tarnev Kh and Tsankov Ts 2009 *AIP Conf. Proc.* **1097** 12–21

[7] Paunska Ts, Shivarova A, Tarnev Kh and Tsankov Ts 2009 *AIP Conf. Proc.* **1097** 99–108

3D model of the magnetic filter region in tandem plasma sources

St Lishev[1,3], A Shivarova[1] and Kh Tarnev[2]

[1]Faculty of Physics, St. Kl. Ohridski University of Sofia,
5 J. Bouchier Blvd., 1164 Sofia, Bulgaria
[2]Department of Applied Physics, Technical University of Sofia,
8 Kl. Ohridski Blvd, 1756 Sofia, Bulgaria

E-mail: lishev@phys.uni-sofia.bg

Abstract. Based on a three-dimensional model developed within the fluid-plasma theory, the study presents results on the spatial structure of hydrogen discharges in the magnetic filter region of tandem plasma sources and an analysis on its formation by transport processes – particle and electron-energy fluxes – both across and along the magnetic field.

1. Introduction

The magnetic field is a tool for controlling the plasma parameters in different type of devices used in plasma physics applications. In the sources of negative hydrogen ions developed for nuclear fusion [1], the magnetic field appears as an electron cooler. It separates two regions in the source: regions of high- and of low-energetic electrons. This is considered as necessary for providing proper conditions for the two-step reaction of negative ions production by electron attachment to vibrationally-excited molecules. Because of coupling two regions, the source it known as a tandem source and the magnetic field is called a magnetic filter. The filter field is weak so that the electrons are magnetized and the ions are left unmagnetized. Transport processes across the magnetic field and filtering of the fast electrons based on the temperature dependence of the electron-electron collision frequency have been involved in the understanding of the filter operation since the early years of the modelling of the source [2]. A recent study [3], also developed within the fluid plasma theory and stressing transport processes across the magnetic field, has provided extended analysis based on a two-dimensional (2D) modeling of the filter operation. The model involves only the electron-neutral collision frequency in the charged particle and electron-energy transport coefficients. Reduced nonlocality in the electron heating combined with diffusion and thermal diffusion and completed by ($\mathbf{E}\times\mathbf{B}$)- and diamagnetic- drift action outline the complicated structure of the discharge in the filter region, shown also in experiments [4].

This study extends the analysis in Ref. [3] towards a detailed three-dimensional (3D) description of the discharge structure. The difference in the fluxes across and along the filter field both in the charged-particle and electron-energy balances specifies the formation of different types of distribution of the plasma parameters in the three directions.

[3] To whom any correspondence should be addressed.

2. Model description

Figure 1(a) shows the modeling domain, with the magnetic filter centered at $z = 10$ cm. The discharge is produced by a rf power input P_w with a maximum P_{w0} at $z = 0$ and a super-Gaussian profile in the z-direction (figure 1(b)). The magnetic field is homogeneous in the x- and y-directions and has a Gaussian profile along the z-axis. Thus, the plasma is produced in the region before the filter (i.e., close to $z = 0$) and expands along the discharge vessel (i.e., in the z-direction) across the filter located in the (x-y)-plane at $z = 10$ cm. The species in the hydrogen discharge considered are electrons, the three types of positive ions (H^+, H_2^+ and H_3^+), and hydrogen atoms (H) and molecules (H_2).

The initial set of equations numerically solved by employing COMSOL Multiphysics consists of the continuity equations, the electron-energy balance and the Poisson equation:

$$\mathrm{div}\boldsymbol{\Gamma}_\alpha = \frac{\partial n_\alpha}{\partial t}, \tag{1a}$$

$$\mathrm{div}\mathbf{Q}_e = P_\mathrm{w} - P_\mathrm{coll}, \tag{1b}$$

$$\Delta\phi = \frac{e}{\varepsilon_0}\left(n_\mathrm{e} - \sum_{l=1}^{3} n_{i_l}\right). \tag{1c}$$

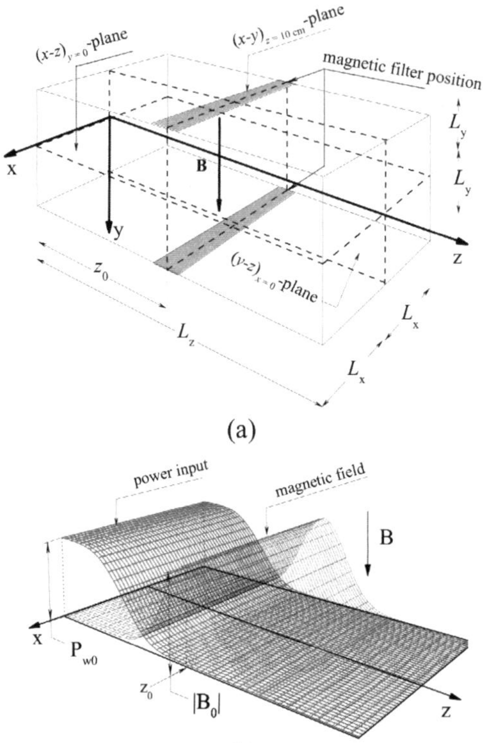

(a)

(b)

Figure 1. Modeling domain (a) and axial profiles of the magnetic field $B(z)$ and of the applied rf power density $P_w(z)$ (b).

Here n_α, with α denoting electrons (α = e), the three types of positive ions ($\alpha = i_l$, l = 1-3) and hydrogen atoms, are the particle densities, e and ε_0 are, respectively, the elementary charge and the vacuum permittivity and is the potential of the dc field in the discharge (E_{dc} = -). The charged particle fluxes $_\alpha$ and the electron energy flux $\mathbf{Q}_e$ are, respectively:

$$\boldsymbol{\Gamma}_\alpha = -Z_\alpha n_\alpha \hat{\mathbf{b}}_\alpha \cdot \mathrm{grad}\phi - \hat{\mathbf{D}}_\alpha \cdot \mathrm{grad}\, n_\alpha - \frac{n_\alpha}{T_\alpha}\hat{\mathbf{D}}_\alpha^T \cdot \mathrm{grad}\, T_\alpha, \tag{2a}$$

$$\mathbf{Q}_\mathrm{e} = -\hat{\chi}_\mathrm{e} \cdot \mathrm{grad}T_\mathrm{e} + \frac{5}{2}T_\mathrm{e}\boldsymbol{\Gamma}_\mathrm{e}. \tag{2b}$$

In (2), $\hat{\mathbf{b}}_\alpha$ and $\hat{\mathbf{D}}_\alpha$ are, respectively, the mobility- and diffusion- tensors, $\hat{\mathbf{D}}_\alpha^T = \hat{\mathbf{D}}_\alpha$ is the thermal diffusion tensor and $\hat{\chi}_e$ is the thermal conductivity tensor [5]. Their components along the magnetic field (i.e., in the y-direction) are, respectively, $b_{\|\alpha} = e/\mu_\alpha \nu_\alpha$, $D_{\|\alpha} = T_\alpha/\mu_\alpha \nu_\alpha$ and $\chi_{\|e} = (5/2)n_e D_{\|e}$ whereas $b_{\ \alpha} = b_{\|\alpha}/[1+(m_\alpha\Omega_\alpha/\mu_\alpha\nu_\alpha)^2]$, $D_{\ \alpha} = D_{\|\alpha}/[1+(m_\alpha\Omega_\alpha/\mu_\alpha\nu_\alpha)^2]$ and $\chi_{\ e} = \chi_{\|e}/[1+(\Omega_e/\nu_e)^2]$ are the corresponding components perpendicularly to the magnetic field (i.e. in the x- and z-directions). In addition, the components $b_{d\alpha} = b_{\ \alpha}(m_\alpha\Omega_\alpha/\mu_\alpha\nu_\alpha)$ and $D_{d\alpha} = D_{\ \alpha}(m_\alpha\Omega_\alpha/\mu_\alpha\nu_\alpha)$ account, respectively, for the (**E**×**B**)- and diamagnetic-drifts; $\chi_{de} = \chi_{\ \alpha}(\Omega_e/\nu_e)$ is the thermal conductivity coefficient related to the diamagnetic drift. The other notations are as follows: T_e is the electron temperature, $\Omega_\alpha = eB/m_\alpha$ are the gyro-frequencies and m_α and μ_α are the mass and the reduced mass in elastic collisions with neutrals with frequency ν_α, the latter specified by $\mu_\alpha\nu_\alpha = \sum_j \mu_{\alpha j}\nu_{\alpha j}$ with j denoting atoms and molecules; $Z_\alpha = \pm 1$ for positive ions and electrons. The flux of the hydrogen atoms is a diffusion flux and the density of the molecular component is determined by the equation of state. In 1(a) $\delta n_\alpha/\delta t$ describes the production and the losses of charged particles and hydrogen atoms, and P_{coll} in 1(b) summarizes the electron energy losses in collisions. The boundary conditions are for the fluxes at the walls where a zero-potential is assumed (metal walls). For more details see [3].

3. Results and discussions

The results presented here are for discharge vessel size $2L_x = 2L_y = L_z = 20$ cm, maximum value of the magnetic field $B_0 = 20$ G, power density applied $P_w = 10^4$ W/m^3, widths $\sigma_B = 1.581$ cm and $\sigma_p = 4.729$ cm of the B- and P_w-profiles and gas temperature $T_g = 300$ K. For the gas-pressure $p = (5\text{-}10)$ mTorr considered, the mean free paths of the charged particles are smaller then the dimensions of the discharge vessel, justifying applicability of the drift-diffusion approximation.

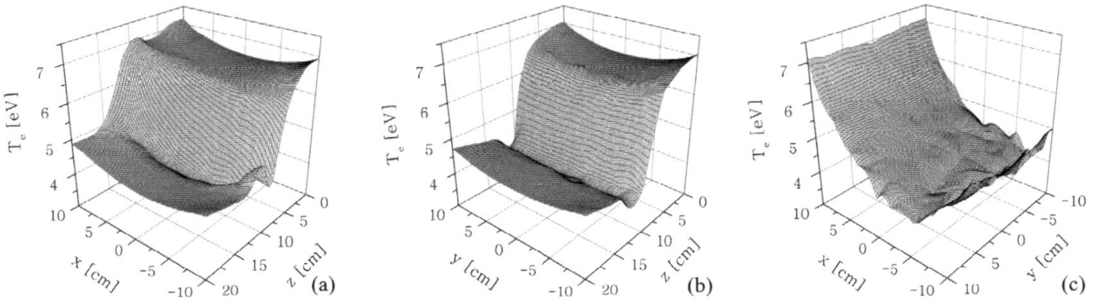

Figure 2. Spatial distribution of the electron temperature in the $(x\text{-}z)_{y=0}$ (a), $(y\text{-}z)_{x=0}$ (b) and $(x\text{-}y)_{z=10\text{ cm}}$ (c) planes.

The discharge structure is illustrated by the distribution of the electron temperature T_e (figure 2) and density n_e (figure 3) in the $(x\text{-}z)_{y=0}$, $(y\text{-}z)_{x=0}$ and $(x\text{-}y)_{z=10\text{ cm}}$ planes obtained at $p = 8$ mTorr. A complicated structuring of the discharge due to the spatial interplay of particle- and electron-energy fluxes across the magnetic field replaces the smooth spatial variation of the plasma parameters (smooth decrease of n_e towards the walls and almost constant T_e, due to comparatively high electron density combined with comparatively long mean free path, resulting in high thermal conductivity), well known for plasmas without magnetic field and shown here by the changes along the magnetic field (the y-direction) in figures 2(b), 2(c), 3(b) and 3(c).

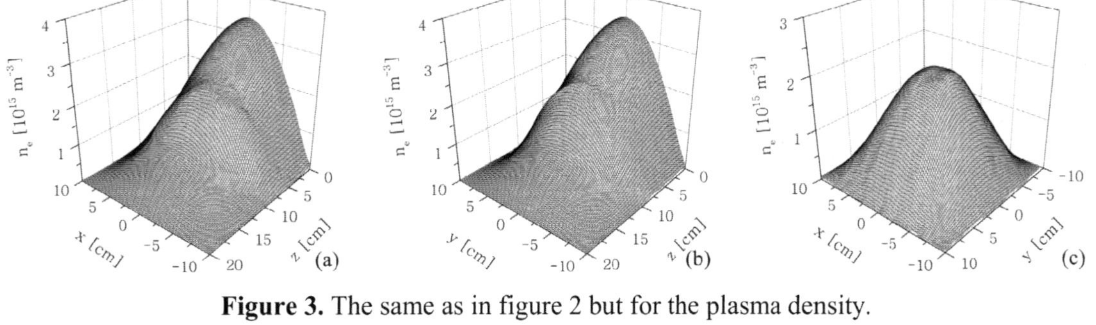

Figure 3. The same as in figure 2 but for the plasma density.

Starting with a slight variation of in the region of the rf power deposition at $z = 0$, T_e drops sharply in the z-direction (figures 2(a) and 2(b)) due to the reduction of the thermal conductivity coefficient across the magnetic field. The result is a reduced electron energy flux and, thus, reduced nonlocality of the electron heating. The axial drop of T_e drives the thermal flux in the x-direction which is related to the diamagnetic drift and leads to an increase of T_e in the $(+x)$-direction (figures 2(a) and 2(c)). The latter drives a thermal flux again related to the diamagnetic drift, but now in the z-direction which leads to electron heating behind the filter and formation of a groove in the filter region (figure 2). Thus, the reduced – by the filter field – electron-energy flux and thermal fluxes related to the diamagnetic drift are responsible for the electron cooling by the filter with formation of a minimum of T_e.

The changes of n_e caused by the magnetic field (figure 3) are not so drastical as those of T_e. A formation of a maximum of n_e in the filter region is the main effect of the magnetic field. It results from simultaneously acting thermal diffusion and diffusion: Thermal diffusion related to the axial gradient of T_e drives the electrons along the discharge (in the z-direction) and the reduced – by the filter field – diffusion permits the accumulation of electrons in the filter region (figures 3(a) and 3(b)). Therefore, the filter operation, involving strong reduction of T_e, is an example of the importance of the thermal diffusion that usually appears to be smaller than the diffusion.

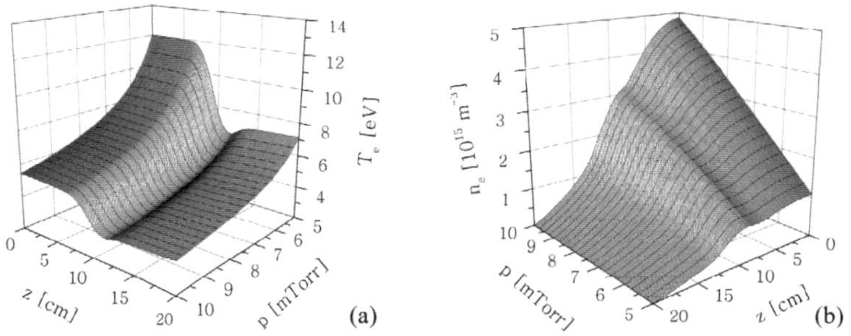

Figure 4. Changes in the axial profiles (at $x = 0$, $y = 0$) of the electron temperature and density as the gas pressure is varied.

Figure 4 shows the changes in the axial variations of T_e and n_e under varying gas pressure. With the increase of p, T_e and its axial gradient in the filter region decrease and the maximum of n_e there becomes less pronounced. This is in agreement with the results of probe diagnostics along the discharge axis [4].

4. Conclusions

The analysis of the results of the 3D model of the magnetic filter operation in the sources of negative hydrogen ions, developed for describing the complete plasma structure in the filter region in the source, emphasizes the formation of a minimum of the electron temperature and a maximum of the electron density in the filter region. The former is due to suppressed – by the magnetic field – electron energy flux causing a drop in the electron temperature and consequent action of thermal fluxes related to the diamagnetic drifts, and the latter is formed by thermal diffusion and diffusion fluxes. Accounting for the negative ions and, respectively, for the vibrationally excited molecules, including rovibrational excitation [6], would be the further extension of the work.

Acknowledgments

The work was performed within the Bulgarian Association EURATOM/INRNE program (task P2).

References

[1] Bacal M 2006 *Nucl. Fusion* **46** S250-9

[2] Holmes A J T, McAdams R, Proudfoot G, Cox S, Surrey E and King R 1994 *Rev. Sci. Instrum.* **65** 1153-8

[3] Kolev St, Lishev St, Shivarova A, Tarnev Kh and Wilhelm R 2007 *Plasma Phys. Control. Fusion* **49** 1349-69

[4] Lishev St, Shivarova A and Tsankov Ts 2009 *AIP Conf. Proc.* **1097** 127-36

[5] Golant V E, Zhilinskiy A P and Sakharov I E 1980 *Fundamentals of Plasma Physics* (New York: Wiley)

[6] Gabriel O, Schram D C and Engeln R 2008 *Phys. Rev.* E **78** 016407

Analytical calculation of the gas temperature and measurement of the electron temperature for gas discharges in Ne and He mixtures with copper, bromine, hydrogen and strontium

K A Temelkov[1], N K Vuchkov, I Freijo-Martin and R P Ekov

Metal Vapor Lasers Laboratory, Georgi Nadjakov Institute of Solid State Physics, Bulgarian Academy of Sciences, 72 Tzarigradsko Chaussee, 1784 Sofia, Bulgaria

E-mail: temelkov@issp.bas.bg

Abstract. Thermal conductivities of binary gas systems are calculated on the basis of 12-6 Lennard-Jones and rigid sphere inter-atomic interaction approximations for the case of gas discharges in He and Ne with small admixtures of copper, bromine, hydrogen and strontium. Assuming that the gas temperature varies only in the radial direction and using the calculated thermal conductivities, analytical solutions of the steady-state heat conduction equation are found for two cases of uniform and non-uniform power input, respectively. For both cases the average gas temperature is found by averaging the radial gas temperature distribution over the radius. Measurement of the relative intensities of some He and Ne spectral lines originating from different upper levels has enabled us to determine the average electron temperature.

1. Introduction

One of the main problems in plasma physics is the determination of characteristic constants for basic processes in plasma, such as asymmetric charge transfer, Penning ionization, electron-heavy particle collisions, diffusion and heat conduction, which are fundamentally important for gaseous discharges, laser physics, plasma technologies, gas-discharge mass spectroscopy, absorption and emission spectroscopy and plasma in general. It is well known that characteristic constants for the heavy particle interaction depend on the gas temperature. The experimental technique for gas temperature determination using measurement of Doppler broadening of spectral lines is definitely imprecise.

Electron impact excitation, de-excitation and ionization, as well as three-body recombination, depend on the electron temperature. In particular, in laser physics these processes directly influence the creation of population inversion, or indirectly through excitation transfer by the above-mentioned heavy particle collisions. Measurements of the electron temperature by a Langmuir probe are not suitable for pressures of helium or neon above 10 Torr, particularly in high-voltage and high-current nanosecond pulsed longitudinal discharges. In the literature there are several models of varying degrees of complexity, which predict, among other parameters, the electron temperature values with considerable variation without any overlap.

[1] To whom any correspondence should be addressed.

2. Theoretical and experimental results

In order to obtain the gas temperature distribution, the following steady-state heat conduction equation is solved:

$$div.\left(k\ gradT_g\right)+q_v=0, \tag{1}$$

where k is the thermal conductivity, T_g is the gas temperature and q_v is the power deposited into the discharge zone per unit volume. Considering cylindrical geometry and assuming that the gas temperature varies only in the radial direction, equation (1) assumes the following form:

$$\frac{1}{r}\frac{d}{dr}\left(rk\frac{dT_g}{dr}\right)+q_v=0\cdot \tag{2}$$

The dependence of the thermal conductivity k of gases and gas mixtures on the gas temperature has the form $k=B\cdot T_g^{\ a}$, where B and a are constants (within a certain temperature range) specific for each gas or gas mixture. The constants B and a could be obtained through fitting the existing experimental data taken from the Handbook-tables of Physical Constants. Unfortunately, except for rarified gases, data for the thermal conductivities of the chemical elements of our interest are either very scarce in the literature or are within a narrow temperature range. This is why it is necessary to calculate the thermal conductivities to obtain the gas temperature distribution via solving the steady-state heat conduction equation. There are two widely used theoretical approaches for thermal conductivity determination – rigid, i.e. hard, sphere (R Sph) and 12-6 Lennard–Jones (L–J) approximations, which consider different interactions between the particles. Following [1] the thermal conductivities k_1 and k_2 for R Sph and L-J approximations are expressed as follows:

$$k_1=0.083264\frac{T_g^{\frac{1}{2}}}{\mu^{\frac{1}{2}}.d^2},\qquad k_2=0.083264\frac{T_g^{\frac{1}{2}}}{\mu^{\frac{1}{2}}\cdot\sigma^2\cdot\Omega_V\left(\frac{k_bT_g}{\varepsilon_0}\right)}, \tag{3}$$

where μ is the particle mass in amu, T_g is the gas temperature in K, d is the R Sph diameter (sum of atomic radii of the interacting particles) in Å, σ is the inter-atomic distance in Å at which the potential energy is zero, ε_0 is the potential well depth, Ω_V is the collision integral, which depends on the interaction potential. Ω_V for the R Sph model is unity, while for the L–J potential the following expression is used: $\frac{1}{\Omega_V}=0.697\left[1+0.323\ln\left(\frac{k_bT_g}{\varepsilon_0}\right)\right]$. Thermal conductivities of binary gas systems are calculated on the basis of the empirical method of Brokaw [1] for the case of gas discharges in He (10 Torr) and Ne (16.7 Torr) with small admixtures of hydrogen (0.03 Torr), copper (0.3 Torr), bromine (0.3 Torr and 1.4 Torr) and strontium (0.7 Torr). The binary gas mixtures considered in this paper are optimal for laser operation in the deep ultraviolet - Cu^+ Ne-H_2-CuBr, and in the middle infrared - He-$SrBr_2$ lasers. The results for the thermal conductivities are presented in table 1.

For the case of uniform power input, i.e. q_V = *constant*, in a gas discharge with radius R and taking into account the following boundary conditions $T_g\big|_{r=R}=T_w$ experimentally measured and $\frac{dT_g}{dr}\big|_{r=0}=0$, the solution of equation (2) has the form:

$$T_g(r)=\left(T_w^{\ 1+a}+\frac{1+a}{4B}q_v\left(R^2-r^2\right)\right)^{\frac{1}{1+a}}. \tag{4}$$

For calculation of characteristic constants it is convenient to use the average gas temperature in the discharge zone, which is found by averaging (4) over the radius ($T_0 = T_g(r=0)$):

$$\langle T_g \rangle = \frac{4B}{(2+a)\cdot q_v R^2}\left(T_0^{\;2+a} - T_w^{\;2+a}\right) \quad . \tag{5}$$

Table 1. *B* and *a* – constants, which are for *k* in $W.m^{-1}.K^{-1}$, determining thermal conductivity, RSph – Rigid Sphere approximation, L–J – Lennard–Jones approximation.

gas or gas mixture	*B* (exp. fit)	*a* (exp. fit)	*B* (RSph)	*a* (RSph)	*B* (L–J)	*a* (L–J)
Ne	$9.7x10^{-4}$	0.685	$38.3x10^{-4}$	0.500	$12.3x10^{-4}$	0.648
He	$34.9x10^{-4}$	0.670	$90.9x10^{-4}$	0.500	$46.2x10^{-4}$	0.626
H [800K, 1600K]	$4.4x10^{-4}$	1				
H [1600K, 2600K]	$4.4x10^{-25}$	7.56	$130x10^{-4}$	0.500	$31x10^{-4}$	0.649
H [80K, 2000K]	$11.3x10^{-4}$	0.873				
Cu	-	-	$58.6x10^{-4}$	0.500	$0.9x10^{-4}$	0.982
Br	-	-	$3.3x10^{-4}$	0.500	$0.6x10^{-4}$	0.740
Ne-H [800K, 1600K]	$9.7x10^{-4}$	0.686	-	-	-	-
Ne-H [1600K, 2600K]	$0.7x10^{-4}$	1.042	-	-	-	-
Ne-Cu	-	-	$10.1x10^{-4}$	0.680	$9.3x10^{-4}$	0.690
Ne-Br	-	-	$9.7x10^{-4}$	0.679	$9.1x10^{-4}$	0.687
He-Cu	-	-	$35.3x10^{-4}$	0.666	$31.4x10^{-4}$	0.679
He-Br	-	-	$32.7x10^{-4}$	0.658	$29.0x10^{-4}$	0.675
He-Sr	-	-	$31.9x10^{-4}$	0.660	$26.8x10^{-4}$	0.680
$He-Br_2$	-	-	$32.6x10^{-4}$	0.658	$28.9x10^{-4}$	0.675

Following [2], from $q_V = j.E$ and $E(r) = E_0 .J_0((2.4/R)r)$, where $J_0((2.4/R)r)$ is the Bessel function of the first kind of zero order, one can obtain $q_V = Q_0 .[J_0((2.4/R)r)]^2$, where the constant Q_0 remaines to be obtained. In order to obtain an analytical solution, a polynomial fit of the third degree is used instead of $[J_0(x)]^2$, i.e. $[J_0(x)]^2 \approx b + c.x + d.x^2 + e.x^3$, where $x = (2.4.r)/R$. Through fitting, the following values of *b*, *c*, *d* and *e* are obtained: $b = 1.005$, $c = -0.016$, $d = -0.5702$ and $e = 0.1687$.

In this way, the following expression for q_V is obtained:

$$q_V(r) = Q_0 \cdot \left[b + c\frac{2.4}{R} r + d\left(\frac{2.4}{R}\right)^2 r^2 + e\left(\frac{2.4}{R}\right)^3 r^3 \right] \cdot \tag{6}$$

Comparing the areas bounded between the functions $q_0 = constant$ and $q_V = q_V(r)$ and the variable axis *r*, the following expression for Q_0 is obtained:

$$Q_0 = \frac{q_0}{R\left(\frac{b}{2} + \frac{c}{3}2.4 + \frac{d}{4}2.4^2 + \frac{e}{5}2.4^3\right)} \cdot \tag{7}$$

For this case of non-uniform power deposition and taking into account the above-mentioned boundary conditions, the solution of (2) has the following form:

$$T_g = \left\{ T_w^{\;a+1} + \frac{(a+1)Q_0}{B}\left[\frac{b}{4}\left(R^2 - r^2\right) + \frac{2.4c}{9R}\left(R^3 - r^3\right) + \frac{2.4^2 d}{16R^2}\left(R^4 - r^4\right) + \frac{2.4^3 e}{25R^3}\left(R^5 - r^5\right) \right] \right\}^{\frac{1}{a+1}} \cdot \tag{8}$$

In this case the average gas temperature is found by comparing the areas bounded between the gas temperature profiles for $q_0 = constant$ and $q_V = q_V(r)$ and the variable axis *r*. The gas temperature distributions (4) and (8) for uniform and non-uniform power input, respectively, could be easily determined, because they both have analytical solutions. Equations (4) and (8) are obviously different, but the attention must be drawn to the result that the average gas temperatures in both definitely different cases are almost equal. The discrepancy varies from 0.2 % to 1 % - the averaged value is about 0.6 %.

In the case of local thermodynamic equilibrium (LTE), a measurement of the relative intensity ratio of two lines originating from different upper levels *i* and *j*, usually, but not necessarily, terminating at the same lower levels, would be sufficient for the determination of the electron temperature T_e. The following expression is used [3]:

$$T_e = \frac{\left(E_i - E_j\right)}{k_b}.\ln\left(\frac{I_i \lambda_i g_j A_j}{I_j \lambda_j g_i A_i}\right), \tag{9}$$

where for the measured transitions from upper levels *i* and *j* λ is the wavelength, *I* is the measured intensity, *A* is the transition probability, i.e. spontaneous rate coefficient, *g* is the statistical weight, i.e. degeneracy of the corresponding upper level and *E* is the energy of the respective upper level.

In table 2 the data required for the experimental determination of T_e, as well as average T_e obtained experimentally and from averaging the time dependence of T_e presented in various kinetic models, are shown for some Ne and He transitions. It should be noted that average T_e means non-time-resolved T_e, i.e. averaged over the time. The experimental error decreases as the energy gap E_i - E_j increases. This is why the combinations are made between wavelengths given in bold and all the others. As one can easily observe the experimentally determined T_e is in fair agreement with the calculated values. The discrepancies are 0.01 eV and 0.02 eV, respectively, for gas discharges in Ne and He, which are in the range in the experimental error of about 0.1 eV. The results obtained satisfy the condition for LTE $n_e >> 1.6\times10^{18}.(\Delta E)^3.\sqrt{T_e}$, where n_e is the electron density in m^{-3}, while the units of T_e and ΔE are eV. Using the energy gaps of 2.401 eV and 1.325 eV, respectively for Ne and He, as well as the average T_e determined experimentally, one can easily obtain that $n_e >> 1.534\times10^{19} m^{-3}$ for Ne and $n_e >> 2.859\times10^{18} m^{-3}$ for He, which is confirmed by the electron density of about 3.5×10^{20} m^{-3} for Ne and 7.9×10^{20} m^{-3} for He calculated through the above-mentioned kinetic models.

Table 2. λ – wavelength at which emission intensity is measured, A_i – radiative transition probability, E_i – energy of transition upper level, g_i – statistical weight, I_i – relative intensity, T_e^{exp} – electron temperature experimentally determined, T_e^{th} – electron temperature from averaging the time dependences of T_e presented in various kinetic models.

atom	λ (nm)	A_i (s)	E_i (eV)	g_i	I_i (arb. units)	T_e^{exp} (eV)	T_e^{th} (eV)
Ne	352.047	9.3x10^6	20.369	1	4	0.49 / 0.63	
Ne	470.886	4.2x10^6	21.014	3	5	0.54	
Ne	534.109	4.2x10^6	21.014	3	6.3	0.48 / 0.52	0.485
Ne	**540.056**	9.0x10^5	18.966	1	6.4	-	
Ne	**702.405**	1.89x10^6	18.613	3	77.8	-	
Ne	748.887	3.49x10^7	20.037	5	18.8	0.32	
Ne	753.577	4.3x10^7	20.026	3	24.4	0.36	
		3.25x10^5		3		-	
He	**402.619**	2.93x10^6	24.043	5	4	-	
		1.17x10^7		7		-	
He	587.562	2.94x10^7	23.074	3	30.1	0.53 / 0.66	0.610
He	587.597	3.92x10^7	23.074	3	30.8	0.57 / 0.63	
He	706.519	1.54x10^7	22.718	3	47.9	0.53 / 0.55	
He	728.135	1.81x10^7	22.920	1	9.2	0.61 / 0.65	

3. Conclusions

Using the thermal conductivities of binary gas systems calculated on the basis of 12-6 Lennard-Jones and rigid sphere inter-atomic interaction, analytical solution of the steady-state heat conduction equation and the average gas temperature are found for uniform and non-uniform power input in gas discharges in He and Ne with small admixtures of copper, bromine, hydrogen and strontium. The average electron temperature is also determined through measurements of the relative intensities of some He and Ne spectral lines.

Acknowledgements

This work was supported by project BK-03-08 of Georgi Nadjakov Institute of Solid State Physics, Bulgarian Academy of Sciences and ARGO Project within the Spanish Leonardo Da Vinci Program.

References

[1] Reid R C and Sherwood T K 1966 *The properties of gases and liquids* (New York: McGraw-Hill Book Company) pp 492-525
[2] Blau P 1996 *Pulsed Metal Vapour Lasers*, ed C E Little and N V Sabotinov (Dordrecht: Kluwer Academic Publishers) pp 215-20
[3] Loveland D G, Zerrouk A F and Webb C E 1992 *Plasma Sources Sci. Technol.* **1** 141-6

The "cometary" discharge, a possible new type of DC electric discharge in air at atmospheric pressure, and its bactericidal properties

V Scholtz[1,3] and J Julák[2]

[1]Department of Physics and Measurements, Faculty of Chemical Engineering, Institute of Chemical Technology, Technická 5, 166 28 Praha 6, Czech Republic
[2]Institute of Immunology and Microbiology, 1st Faculty of Medicine, Charles University, Studničkova 7, 128 00 Praha 2, Czech Republic

E-mail: scholtzv@vscht.cz

Abstract. A new type of electric discharge was observed during studies of a DC discharge of the point-to-point type in air at atmospheric pressure. Some of its properties were studied with respect to its bactericidal properties.

1. Introduction

In previous papers we described the design of a simple setup generating a negative corona discharge in the point-to-plane arrangement [1, 2] and the microbicidal properties of the low-temperature plasma generated by this discharge in air [3, 4]. In so far unpublished works, we also attempted to employ other discharge types and arrangements, namely the positive corona of the point-to-plane type and the discharge of point-to-point type. The overall physical properties of various known types of DC discharge, namely the negative and positive corona, were previously described in details in several works, e.g. Sigmond et al. [5]. However, studying the point-to-point discharge, we observed a new phenomenon, which we provisionally called the "cometary" discharge in view of its shape; some of its properties are described below.

2. Experimental arrangement

We used the simple electric circuit drawn schematically in figure 1(a). An HT 2103 apparatus (Utes Brno, Czech Republic) was used as the source of variable high voltage, with maximal voltage 10 kV and current up to 0.5 mA. The electrodes were arranged in parallel; the positive one was tilted approximately 30° from the vertical. Ordinary medical injection needles served as electrodes, but very similar results were obtained with electrodes made from tailor's pins. The setup was used to study the microbicidal effects of low-temperature plasma generated the between electrodes in samples placed below electrodes. The detailed results of this research will be published elsewhere as soon as they will be compiled.

[3] Towhom any correspondence should be addressed.

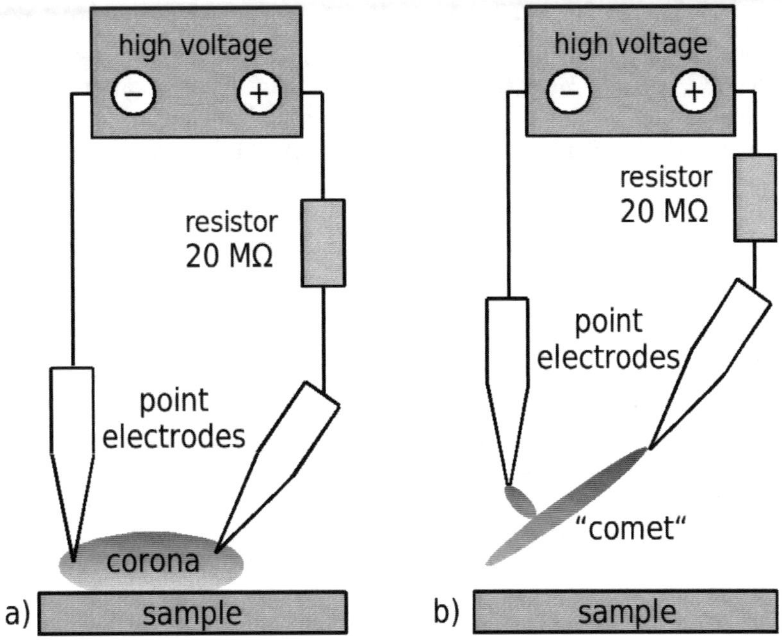

Figure 1. Schematic diagram of the setup for generation of low-temperature plasma by the common point-to-point (a) and "cometary" (b) discharges.

3. Generation of the "cometary" discharge

The ordinary point-to-point discharge is shown in figure 2, where the ellipsoid of low-temperature plasma between the electrodes can be seen. A new type of discharge is formed if the positive electrode tip is lifted 1 to 2 mm above the tip of negative one, figure. 1(b), and the voltage and the inter-electrode distance (see below) are maintained within the values specified in table 1. Under these conditions, an additional cloud of plasma arises, streaming not from the positive to the negative electrode, but rather into the space below the electrodes. This cloud, resembling the tail of a comet, is shown in figure 3.

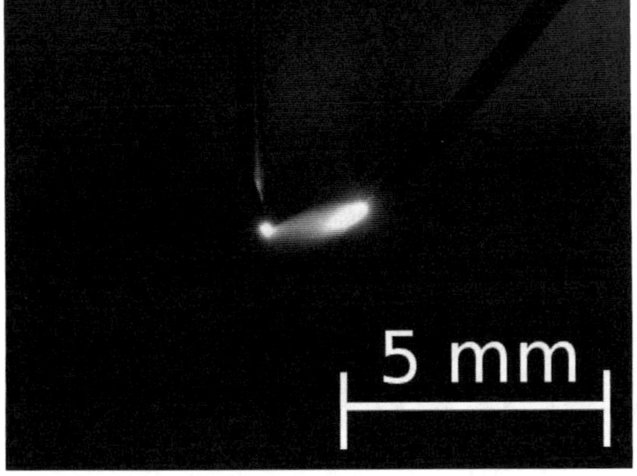

Figure 2. An ordinary point-to-point discharge between the positive (at right) and the negative electrode (at left).

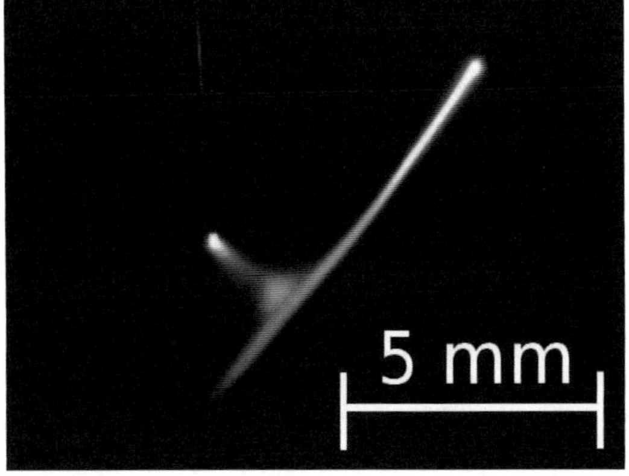

Figure 3. A "cometary" discharge streaming from the positive electrode (at right) below the negative electrode (at left).

Journal of Physics: Conference Series **223** (2010) 012005 doi:10.1088/1742-6596/223/1/012005

This "cometary" discharge is formed if the inter-electrode distance is greater than 4 mm and is sustainable up to a distance of 12 mm. The voltage applied *U* must be kept within the minimal and maximal limiting values given in table 1 for the various inter-electrode distances *d*. The common point-to-point discharge appears at the lower *U* value and is extinguished at the higher one. The values of the corresponding DC current *I* are also given in table1 and plotted in figure 4. Under these conditions, the burning of discharge is temporally steady. The current has a direct component without a superposed alternating one with a frequency lower that 20 kHz. Nevertheless, the presence of higher frequencies cannot be excluded. The plateau and the consecutive decrease of the current values at higher *d* are caused by the limits of the source used, which cannot deliver more than 10 kV.

Table 1. Distance – voltage – current conditions necessary to sustain the "cometary" discharge. The "cometary" discharge only burns between the minimal and maximal voltage and current values. The maximal voltage and current values for distances over 9 mm were not measured due to the 10 kV voltage limit of the source used.

	Minimal		**Maximal**	
***d* /mm**	***U* /kV**	***I* /μA**	***U* /kV**	***I* /μA**
4	4.4	50	6.0	200
5	6.2	50	7.0	270
6	6.2	50	8.1	330
7	6.3	30	8.2	360
8	6.5	30	9.8	400
9	7.7	30	> 10	400
10	8.8	30	> 10	300
11	9.0	30	> 10	250
12	9.2	30	> 10	250

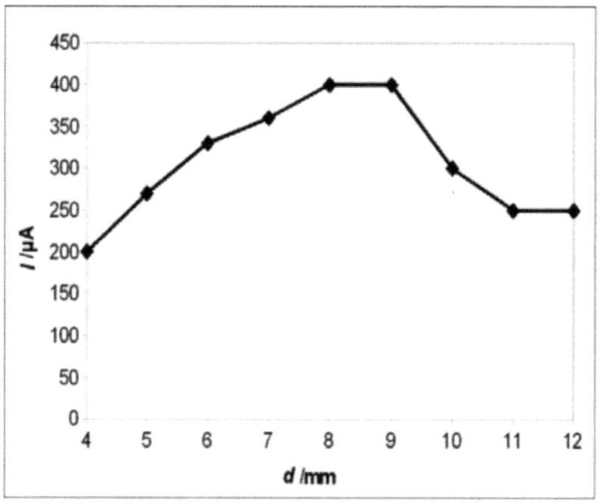

Figure 4. Maximal current (*I*) of the „cometary" discharge as a function of the inter-electrode distances (*d*). The plateau and consecutive decrease of the current are caused by the 10 kV limit of the source used.

4. Comparison of bactericidal effects

The bactericidal effects of the common point-to-point and the "cometary" discharge were assessed on cultures of *Staphylococcus epidermidis* and *Escherichia coli* bacteria. The agar plates were inoculated with bacterial suspensions and exposed to both discharges generated at 10 kV and acting for 15 minutes. After incubation of the plates, inhibition zones appeared in the bacterial cultures. The shape of the zones differs markedly: after action of the common point-to-point discharge, a divaricating zone appeared resembling a butterfly. On the contrary, the „cometary" discharge yielded elliptical zones occupying a larger area and no divarication was apparent. Images of the inhibition zones of both discharges with the position of electrodes indicated are shown in the figure 5.

In the next experiments, we placed a drop of bacterial suspension onto an inert surface and exposed it to the common point-to-point and to the tip of the „cometary" discharge. The complete sterilization of the liquid occurred within about 2 minutes in both cases. These effects were of comparable magnitude with those observed after action of other DC discharges described in our previous works [1, 3 and 4].

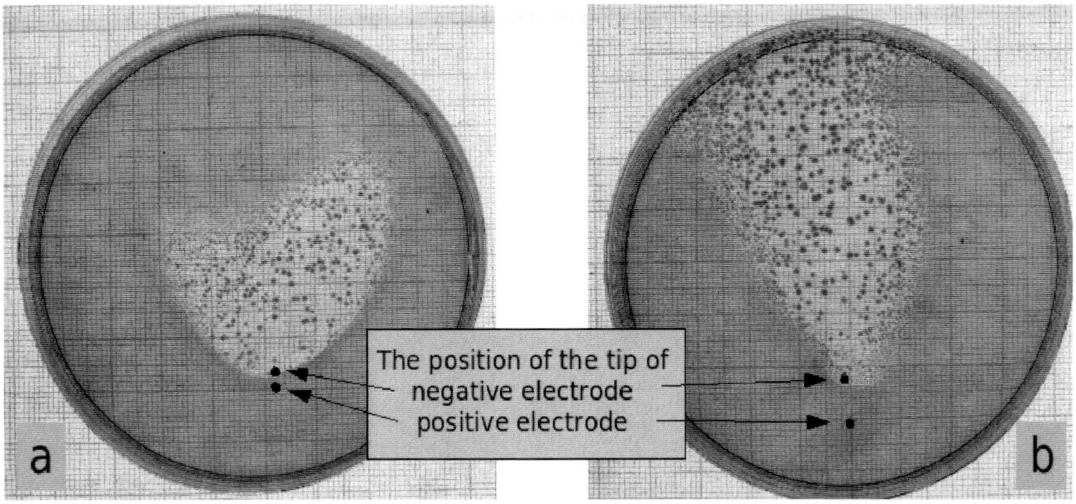

Figure 5. Images of inhibition zones of *Escherichia coli* bacteria of ordinary point-to-point (a) and "cometary" (b) discharges with the position of electrodes indicated.

5. Conclusions

The "cometary" discharge represents a new type of electric discharge producing low-temperature plasma and displaying bactericidal effect. We still have no theoretical explanation of its origin and we continue to investigate its properties and compare them with those of other discharge types.

Acknowledgments

This study has been supported by grants MSM ČR 0021620806, MSM ČR 6046137306.

References

[1] Julák J, Kříha V and Scholtz V 2005 Germicidal effects of atmospheric corona discharge studied on *Escherichia coli* bacteria *Proc. XXVII ICPIG* (Eindohoven, The Netherlands) pp 4-177

[2] Julák J, Kříha V and Scholtz V 2006 Corona discharge: A simple method of its generation and study of its bactericidal properties *Cz. J. Phys.* **56** 1332-8

[3] Scholtz V, Julák J, Kříha V and Mosinger J 2007 Decontamination effects of low-temperature plasma generated by corona discharge. Part I: An overview *Prague Med. Report* **108** p 115-27

[4] Scholtz V, Julák J, Kříha V, Mosinger J and Kopecká S 2007 Decontamination effects of low-temperature plasma generated by corona discharge. Part II: New insights *Prague Med. Rep.* **108** 128-46

[5] Sigmond R S, Kurdelova B and Kurdel M 1999 Action of corona discharge on bacteria and spores *Cz. J. Phys.* **49** 405-20

Journal of Physics: Conference Series **223** (2010) 012006 doi:10.1088/1742-6596/223/1/012006

Impedance-stabilized positive corona discharge and its decontamination properties

P Horák[1] and J Khun

Department of Physics and Measurements, Faculty of Chemical Engineering, Institute of Chemical Technology, Technická 5, 166 28 Praha 6, Czech Republic

E-mail: pavel.horak@vscht.cz

Abstract. The point-to-plane DC corona discharge in air at atmospheric pressure was stabilized by a serially connected ballast impedance. The ballast impedance was implemented by a resistor–capacitor group connected in parallel. In the case of connecting the serial impedance into the electric circuit of a negative corona, the transition into a spark takes place at parameters similar to those of a non-stabilized discharge. In contrast, in the case of a positive corona, the discharge does not undergo a transition into a spark, but rather into a mode of periodic streamers. We measured the bactericidal effect of the stabilized discharge. The experiments showed that after a 2-minute exposure the quantity of surviving bacteria decreased from 95% for a non-stabilized discharge down to 5% for a stabilized one.

1. Introduction

Non-thermal plasma decontamination is among the intensively studied techniques described in a large number of publications, e.g. [1, 2]. The advantage of using non-thermal plasma for decontamination purposes is that most of the energy introduced into the system is used mainly for ionization, dissociation, excitation, creation of charged or reactive species, etc. Therefore, non-thermal plasma decontamination is useful on heat sensitive surfaces since it does not degrade the material and no energy is wasted for heating. The operation expenses are thus much lower in comparison with the classic means of thermal decontamination, such as autoclaving. Due to its easy implementation, the corona discharge is a very suitable source of non-thermal plasma. Even though the mechanisms of positive and negative corona discharge formation are different, increasing the discharge voltage leads in both cases inevitably to a transition into a spark, which is the basic limitation of the corona discharge [3]. This breakdown occurs at a very low discharge currents (up to hundreds μA depending on the gap width and discharge polarity). Stabilization is, therefore, required in order to achieve higher current and power values thereby increasing the efficiency of future applications.

This article deals with the stabilization by means of an in-series impedance implemented by the parallel connection of a resistor and a capacitor.

We achieved a streamer mode of operation by choosing a proper impedance value. This mode is called "flashing corona" in the literature [5] (even though it has not been thoroughly described) and

[1] To whom any correspondence should be addressed.

consists of periodical current pulses – streamer channels. These thin weakly-ionized plasma channels [5, 6] grow in the gap between the electrodes thus making a conductive connection. This connection may increase the electric current significantly and cause a streamer-to-spark transition. In comparison with the spark, the streamer channel provides a relatively low current of about 10 mA [5], while a current of $10^4 - 10^5$A with current density $j = 10^4$ A·cm^{-2} flows through the spark channel [5,6]. This is sufficient to heat up significantly the surrounding gas so that the spark discharge tends to generate thermal plasma while the streamer remains almost at room temperature [4].

In order to find a suitable combination, several capacitance-resistance values were tested and the current, power and discharge stability achievable were measured. In the second part of this work, the decontamination properties of the discharges investigated were compared; finally, the optimal parameters of the discharge and of the ballast impedance are presented with respect to the decontamination energy yield and efficiency.

2. Experimental arrangement

The experimental setup is shown in figure 1. The corona discharge in ambient air was formed by using a DC high-voltage power source (DC HVPS) (current range 0 – 350 μA, voltage range 0 – 25 kV). The point electrode consisted of a hypodermic needle (N). It was fixed into a micrometric screw and placed over the plane electrode - water surface (W). The stabilizing impedance consisting of a parallel R–C combination was connected serially into the circuit. The resistance values were varied between 4 and 10 MΩ, those of the capacitance, between 0 and 100 pF. The voltage drop on the stabilizing impedance U_z and its time dependence were measured using a voltmeter and an oscilloscope connected through a high-voltage probe (HVP) (HVP–40, 1:1000, max. 40 kV DC). The DC HVPS was equipped by built-in volt- and ammeters. To minimize the parasitic capacitance and inductance in the circuit, the cables were shortened to the smallest possible lengths. The discharge stabilization by the in-series impedance was tested for positive as well as negative point electrode polarity. During all experiments, the distance between the tip of the needle and the water surface was set at d = 4 mm. The decontamination properties of stabilized and of non-stabilized discharges were tested on cultures of the *Escherichia coli* bacteria. One ml of water suspension of these bacteria was placed into a dish with diameter of 2 cm and depth of 1 cm thus forming a plane electrode as in the previous experiments. The suspension samples were exposed to a positive or negative corona discharge and, after appropriate dilution, were cultivated overnight on the surface of a Mueller-Hilton cultivation medium. Finally, the number of colonies grown was counted.

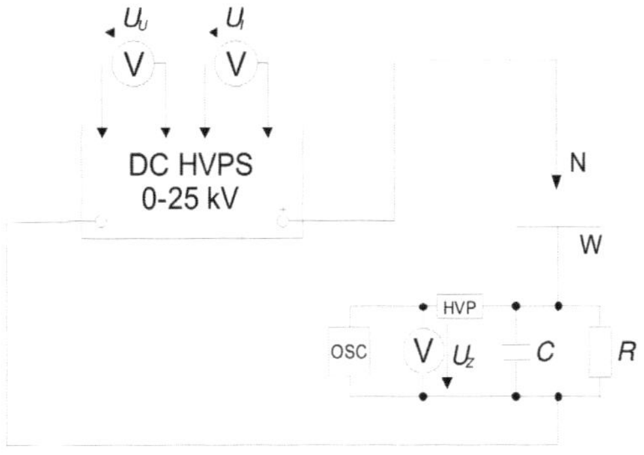

Figure 1. Schematic view of the experimental setup.

3. Experimental results and discussion

3.1. Discharge stabilization

In the case of a positive corona, the inclusion of a suitable impedance increased the discharge current by an order of magnitude without a transition into a spark taking place. Furthermore, this discharge did not undergo a transition into a spark, but rather into a periodic streamer mode. In contrast, in the case of a negative corona, the in-series impedance did not affect significantly the discharge current or the discharge mode. Examples of V-A characteristics of non-stabilized and stabilized positive and negative corona discharges for one typical stabilizing impedance are shown in figure 2.

It can be seen that for a non-stabilized positive discharge the current slowly increases with the discharge voltage. A transition into spark occurs at values of approximately U = 6.2 kV and I = 25 μA.

In contrast, for a stabilized discharge, the current increases with the discharge voltage rapidly up to values of $U = 5$ kV and $I = 80$ μA. At this point, a qualitative change occurs of the discharge mode. The discharge transits into a pulsed streamer mode (values of around $U = 4$ kV, slowly decreasing, and I between 210 and 340 μA) and does not change into a spark. The streamer frequency as measured by the oscilloscope was in the range from 800 to 2500 Hz and increased with the current. Current values over $I = 350$ μA could not be measured due to the voltage and current limits of the high-voltage supply used.

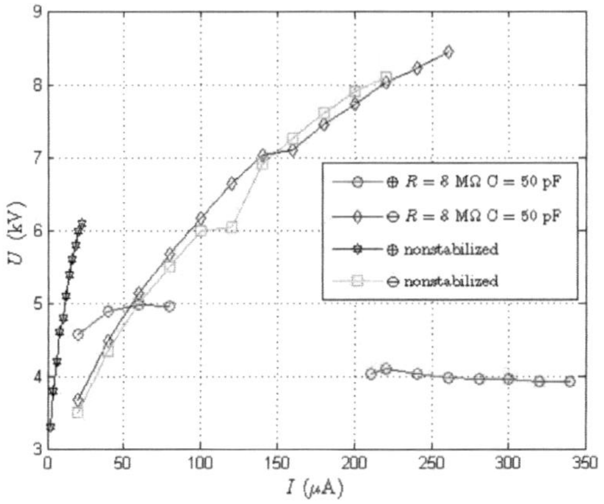

Figure 2. V-A characteristics of impedance-stabilized and non-stabilized positive and negative corona discharges (red circles - positive with impedance, blue diamonds - negative with impedance, black inverted triangles – positive without impedance, green squares – negative without impedance).

In the case of negative corona discharges, the V-A characteristics were similar. The stabilized corona underwent a transition into a spark at slightly higher values. It is interesting to note that the plateau measured in the case of a non-stabilized discharge at current values near $I = 110$ μA was shifted to the higher values. This is probably related to a change in the discharge mechanism. However, studying the origin of this phenomenon was not among the aims of this work.

3.2. Decontamination

Typical time dependencies of the decontamination efficiency η (amount of inactivated bacteria divided by amount of bacteria before exposition) are shown in figure 3. Due to the very similar electrical properties of the stabilized and non-stabilized negative discharges, the efficiency is almost the same and after 3 minutes of exposure increases up to around 60%. In contrast, the decontamination efficiencies of the stabilized and non-stabilized positive discharges differ as markedly as their electrical properties. The stabilized corona efficiency increases up to almost 100 % after 3 minutes of

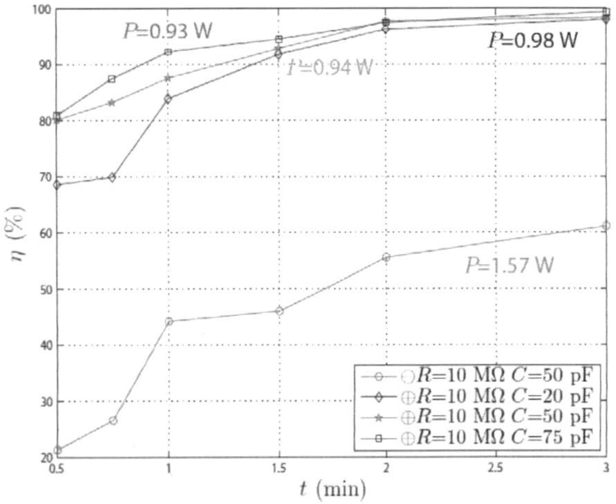

Figure 3. Decontamination efficiencies for stabilized positive and negative corona. $I = 200$ μA.

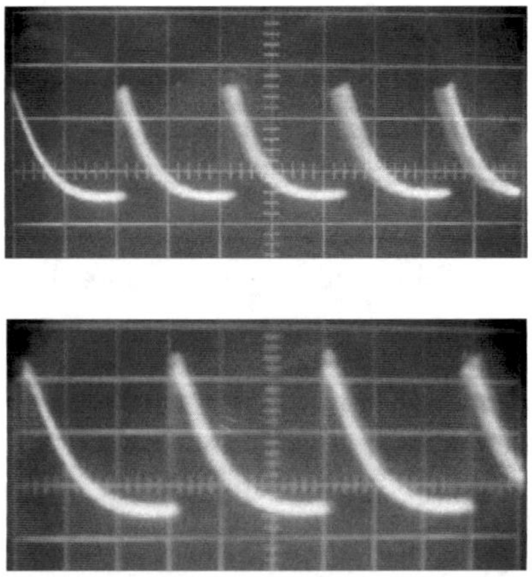

Figure 4. Typical waveforms of the impedance voltage. $I = 250$ μA, $R = 10$ MΩ, $C = 50$ pF (top), $C = 75$ pF (bottom).

exposition, while in the case of non-stabilized one the efficiency after 3 minutes is immeasurable or is lower than 10 %. The decontamination efficiency slightly increases with the ballast impedance capacitance and unexpectedly decreases with the discharge power (figure 3). This may be explained by the voltage waveforms as seen on the oscilloscope (figure 4). Assuming that the waveform of the voltage across the impedance correlates with the waveform of the current in the discharge, one could conclude that the frequency decreases as the capacitance of the ballast impedance is raised; thus, the current pulse width increases. The decontamination efficiency as a function of the capacitance is shown in figure 5. The decontamination efficiency increases with the capacitance; it is higher, almost 100%, for the value of $C = 75$ pF and is and does not depend on the resistance used.

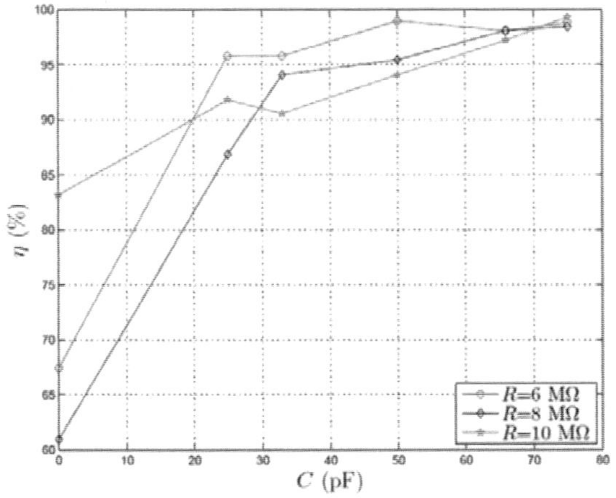

Figure 5. Decontamination efficiencies for stabilized positive corona as a function of the capacitance included in the ballast impedance. $I = 300$ μA, $t = 45$ s.

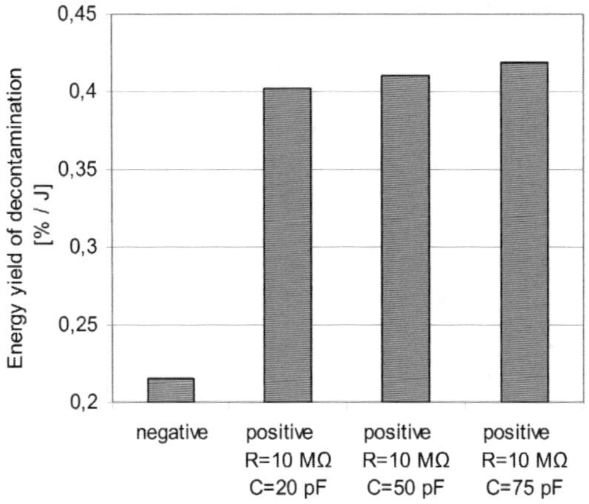

Figure 6. Energy yield of decontamination for $t = 3$ min.

It is clear that the positive corona discharge is more efficient than the negative one. This proposition is supported by the following calculations. Figure 6 presents the energy yields of decontamination for several discharges. These are calculated as the ratio of the percentage of inactivated bacteria to the total energy delivered by the power supply for 3 minutes (necessary to inactivate almost all bacteria). The discharge current was set at 0.2 mA; in the case of negative corona there was no need to use ballast impedance, while the ballast impedance parameters in the case of positive polarity are shown in the figure 6. Although in the case of positive polarity some part of energy is lost in the ballast impedance, it is clear that the energy yield is almost twice as high as that of the negative corona.

4. Conclusions

By inserting a suitable in-series impedance, implemented by an R-C group connected in parallel, into the electrical circuit of a positive corona discharge we were able to prevent the discharge from undergoing a transition into a spark and obtained a periodic streamer mode. The discharge current was thus increased by a factor of nearly 13, with the further increase of the current being limited by the high-voltage power supply used only. This change in the discharge mode is illustrated by the V-A characteristics presented. For negative corona, the effect of the impedance on the current and power increase was not significant. However, a shift was observed of the plateau in the V-A characteristics in comparison with the non-stabilized case, which will be investigated in future studies.

By using suitable ballast impedance, the decontamination efficiency of the positive corona discharge reached almost 100 % after 3 minutes of exposure; in contrast, the negative corona discharge efficiency for the same exposure time was 60 % only. We found that the suitable ballast impedance from the viewpoint of decontamination efficiency was capacitance $C = 75$ pF and resistance $R = 6–10$ MΩ. Moreover, the energy yield of the positive corona discharge was almost twice as high compared with the negative one. For the negative corona, the decontamination efficiency at the particular current was the same as that in the non-stabilized case.

References

[1] Julák J, Kříha V and Scholtz V 2006 *Czechoslovak J. Phys.* B **56** 1333
[2] Laroussi M and Leopold F 2004 *Int. J. Mass Spectr.* **233** 81
[3] Pekárek S and Khun J 2004 *Czechoslovak J. Phys.* D **54** 784
[4] van Veldhuizen E M and Rutgers W R 2002 *J. Phys. D: Appl. Phys.* **35** 2169
[5] Fridman A, Chirokov A and Gutsol A 2005 *J. Phys. D: Appl. Phys.* **38** R1
[6] Bazeylan E M and Raizer Yu P 1998 *Spark Discharge* (Boca Raton: CRC Press LLC)

Evaluation of the mean argon plasma conductivity and temperature at the exit of an atmospheric DC arc plasma torch

V A Riaby[1,4], V P Savinov[2], A V Patziora[1] and D P Tkachenko[3]

[1]Research Institute of Applied Mechanics and Electrodynamics, Moscow Aviation Institute, State Technical University, 5 Leningrad Road, 125080 Moscow, Russia
[2]Physics Department, Moscow State University, Vorobiovy Gory, 119899 Moscow, Russia
[3]Moscow Aviation Institute, State Technical University, 4 Volokolamskoe Road, 125871 Moscow, Russia

E-mail: riaby2001@yahoo.com

Abstract. A new long-life DC arc T-plasma torch [1] was studied. The well-known method of electric field strength measurements in a stabilized arc channel [2] was applied in a modified form as a consequence of the specific shape of the presumably diffuse anode spot attached to a gas vortex on the external surface of the anode unit. The electric field strength was determined assuming that the potential drop across the diffuse anode spot in the new plasma torch was small. This yielded a mean argon plasma conductivity $\sigma \leq 118$ Ohm^{-1}cm^{-1} for arc current $I \leq 180$ A which agrees with the independent experiment [2] thus affirming the correctness of the above assumption. The analysis of the known experimental and theoretical data on atmospheric argon plasma conductivity resulted in the selection of R.S. Devoto's theoretical dependence $\sigma(T)$ [3] as the most reliable one for T = 800 - 20000 K at P = 1 atm, which allowed us to evaluate the mean argon plasma temperature at the exit of the plasma torch, namely, $T \leq 19\,500$ K.

1. Introduction

The new DC arc plasma torch with a tungsten rod cathode and a copper nozzle anode was described in [1]. Its main feature distinguishing it from others is the special design of the anode unit – see the schematic diagram in figure 1.

Here l_{AC} is the inter-electrode distance and l_{AN} is the anode channel length. The exit from the anode channel is implemented as a step of abrupt expansion that generates a gaseous vortex. In addition, it is deflected by the transverse flow of technologic gas that tends to deform the vortex and presumably adds to its twist. In such a situation and if l_{AN} is not too long, this vortex attracts an anode arc spot that expands to become diffuse thus lowering the anode material erosion by several orders of magnitude. Due to the shape of the working gas path, this device was named a T-plasma torch.

In contrast with conventional plasma torches with constricted anode spots having an erosion rate of $m_A \sim 10^{-5}$ - 10^{-6} g/C [4], in the T-plasma torch this parameter turned out to be some 4 - 5 orders of magnitude lower: $m_A \sim 10^{-10}$ g/C [1] which meant a high purity of the plasma and a substantial increase in the device's design durability (about 10 000 hours at moderate arc currents I ~ 50 - 100 A).

---

[4] To whom any correspondence should be addressed.

Journal of Physics: Conference Series **223** (2010) 012007 doi:10.1088/1742-6596/223/1/012007

2. Determination of the mean argon plasma conductivity

The mean electrical conductivity of arc plasmas is usually measured employing a stabilizing channel in a metallic inter-electrode insert operating under a floating electrical potential where the electrical field E is measured using small or large electrostatic probes. In the present version of the T-plasma torch such a technique appears problematic due to the small size of the discharge space while a similar outcome can be provided by the anode channel that can act as an arc column stabilizer when the anode spot of the discharge is attached to the external anode surface. This special feature of the present experiment allowed measurements of E in the anode channel.

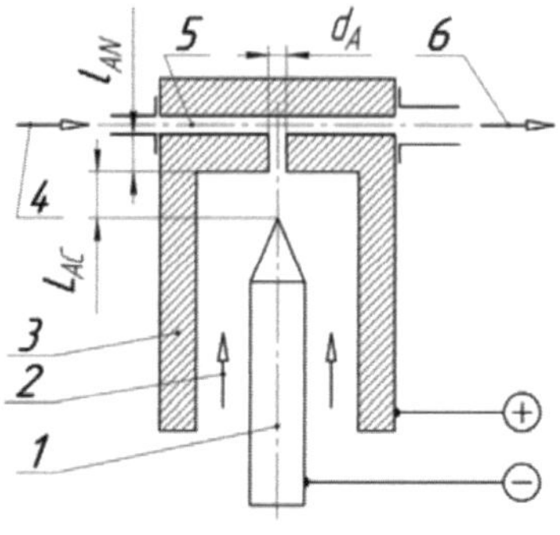

Figure 1. Schematic diagram of the T-plasma torch: 1- cathode, 2- argon supply, 3- anode, 4- technologic gas supply, 5- flat technologic channel, 6- plasma jet exit.

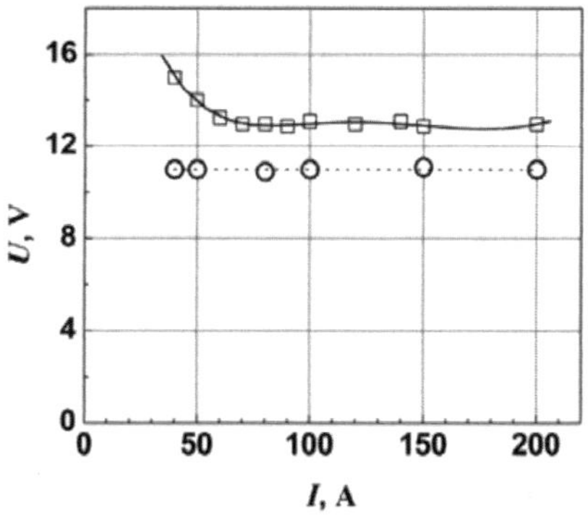

Figure 2. VA-characteristics of the atmospheric argon arc for two different anode channel lengths: l_{AN}=4 mm – squares; l_{AN}=8 mm – circles.

In the T-plasma torch with an external diffuse anode spot, the arc positive column is divided into two parts: a free arc between cathode and anode and a stabilized arc inside the anode channel. The task is to measure the potential difference ΔU_{AN} along the stabilized part of the arc column. It was done here using arc VA-characteristics for the same d_A = 2.5 mm and l_{AC} = 2 mm and for two different anode channel lengths l_{AN} = 4 and 8 mm. The location of the anode spot was determined by checking the state of the anode unit after each run of the plasma torch. The plasma pressure was kept atmospheric using rather small values of the argon flow rate Q_{Ar} = 2 l/min: according to [2] this condition corresponds to "ventilative" argon flow rates $Q_{Ar} \leq 6$ l/min at which the arc VA-characteristics remain unchanged. The present experiment was arranged so that argon was supplied into the cathode unit and no gas was fed into the technologic channel. The VA-characteristics for two lengths of the anode channel were registered at currents I = 40 - 200 A. The results are presented in figure 2.

These curves were obtained by the point-to-point method with a pause of some 10 min. at the top arc current I = 200 A to allow the anode spot to leave its trace on the anode surface. After-run inspections showed that for the long anode channel l_{AN} = 8 mm arc traces could be seen at the anode channel inlet as a melted spot on the channel's edge surrounded with a dark stain. Thus, in this case we had a free arc with the conventional constricted anode arc root. In the case of the short anode channel l_{AN} = 4 mm, an arc trace appearing as a dark ring around the anode exit was located on the external anode surface while the anode inlet remained clean. Moreover, in this latter case, the arc burned inside the anode channel and the arc positive column was stabilized by the anode's cold wall.

The situation described is illustrated in figure 3 where a schematic diagram of the discharge space along with the axial voltage distributions is shown. Here line OG represents the arc voltage for the experiment with the long anode channel l_{AN} = 8 mm. Its voltage distribution appears schematically as graph ABCE. Line OF corresponds to arc voltage in the experiment with l_{AN} = 4 mm and its voltage distribution is shown as graph ABCD. It is well known that for a conventional anode spot its voltage drop should be considerably larger than the one for a diffuse anode spot, which is reflected in figure 3; here the line CE is much longer than the short line beside point D representing ΔU_{ADif}.

The mean electrical conductivity of the argon plasma was determined through the voltage difference U_D-U_C that was found as $U_F - (U_G - U_{CE})$, where $U_{CE} \approx 6$ V is the voltage drop across the conventional anode spot that had been measured previously for the same discharge conditions [5]. In this operation the difference $U_D - U_C$ was considered to be equal to ΔU_{AN} assuming that the voltage drop across a diffuse anode spot is small: $\Delta U_{ADif} \approx 0$. The resulting dependence σ(I) marked with circles is shown in figure 4.

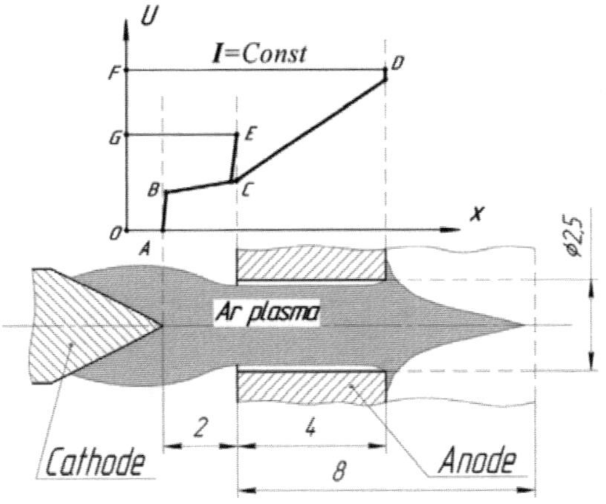

Figure 3. Schematic diagram of the discharge space, arc plasma, and arc voltage axial distributions for two anode channel lengths and a constant arc current.

Note that the real σ values should lie lower due to the effect of discharge current diversion into the copper anode body [6]. This effect could be more or less seriously evaluated for the inter-electrode inserts of [2] by using the physical concept in [6]. But here we have a conductive body collecting electrons by its entire internal surface which makes this parasitic current indefinitely large. In this situation and in view of our assumption $\Delta U_{ADif} \approx 0$ we needed some independent check. It was found in the experimental data in [2] that were extrapolated from arc stabilizing channel diameters d_A= 8, 6, and 4 mm to d_A= 2.5 mm of the present work and yielded the data marked with triangles in figure 4. A general agreement in the conductivity magnitude that confirms the above assumption of a small potential drop across a diffuse anode spot seems evident. In more detail, the results in [2] show lower σ values for the high current range I ≥ 150 A where the discharge current diversion should be noticeable. Considering this possible effect in [2] to be less severe than that with big parasitic currents to the anode in our case, we corrected our results to the thick solid line in figure 4 that goes lower by ~40% at I = 180 A (accounting for the data [2]) and by ~10% at I = 40 A (just applying common sense). Now the mean argon plasma conductivity at the exit of the T-plasma torch can be evaluated as $\sigma \leq 118$ $Ohm^{-1}cm^{-1}$ for $I \leq 180$ A.

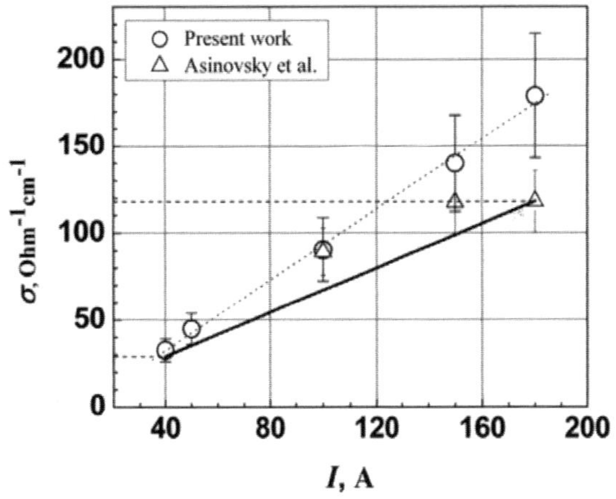

Figure 4. Mean electrical conductivity of argon plasma against arc current in the stabilizing channel d_A = 2.5 mm: (dotted line – initial results, thick solid line – correction in view of extrapolated data of [2]).

3. Evaluation of the mean temperature at the exit of the T-plasma torch

For this purpose, a reliable dependence of the argon plasma electrical conductivity at atmospheric pressure is necessary. It would be better if such dependence were experimental. The necessary information was found in [7-11]. Note that in [7] the final analytical function σ(T) has been obtained as a result of strict numerical processing of argon plasma conductivity/temperature measurements similar to the ones made in [2]. As for the theoretical data, three works were analyzed published by R.S.Devoto [3], L.I. Grekov et al. [12] and P.P Kulik [13]. All these results are compared in figure 5.

It can be seen that R.S. Devoto's curve [3] fits the experimental data better than the results of [12, 13].

Note that Baturin's experimental function [7] nearly coincides with Devoto's predictions, although in the paper [7] such a remarkable comparison is absent. It thus can be concluded that R.S. Devoto's calculations [3] can be used as the sought after reliable temperature dependence $\sigma(T)$ for argon at atmospheric pressure and temperatures in the range T = 800 - 20 000 K.

Figure 5 shows the boundaries of the present work's data on argon conductivity as horizontal dotted lines. They indicate that the mean argon plasma temperature at the exit of the T-plasma torch lies in the range T ~ 1 000 – 19 500 K. Note that these temperatures characterize electrons (T_e) as well as heavy particles ($T_{a,i}$), as it has been determined previously that the argon plasma state in the atmospheric arc is very close to a local thermodynamic equilibrium [7].

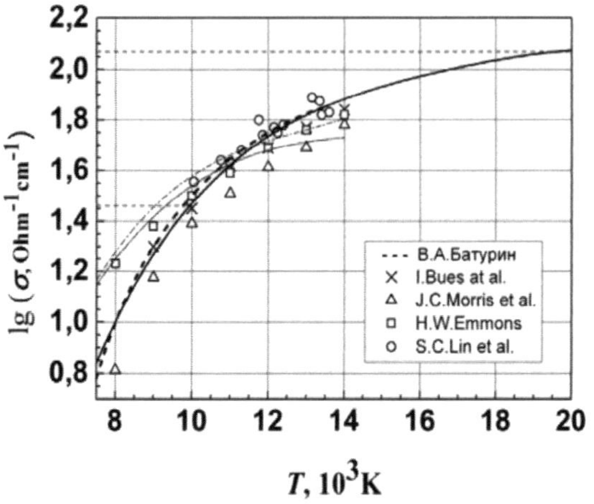

Figure 5. Comparison of the experimental and theoretical data on argon plasma conductivity at atmospheric pressure: thick solid line – R.S. Devoto curve [3], dashed line – L.I. Grekov et al. [12], thin solid line – P.P. Kulik [13].

4. Conclusions

The VA-characteristic method of electrical field strength measurements in the arc stabilizing channel was modified using an anode channel instead of a hole in the insulated inter-electrode insert.

Comparison of the present experiment with an independent one showed that the voltage drop across a presumably diffuse anode spot in the T-plasma torch should be small.

The analysis of the known literature data on $\sigma_{Ar}(T)$ at atmospheric pressure showed that R.S. Devoto's calculations [2] gave the most realistic function for $T = (8 - 20)\cdot 10^3$ K.

The present experiment showed that the mean conductivity of argon plasma at the exit of the T-plasma torch reaches about 118 $Ohm^{-1}cm^{-1}$ while its mean temperature does not exceed 19 500 K for arc currents $I \leq 180$ A.

Acknowledgments

The authors wish to express their gratitude to Academician Professor G.A. Popov, Professor D.V. Rutovsky and Senior Scientist Dr. I.V. Kravchenko for the friendly support and fruitful discussions during the present investigation.

References

[1] Lee H J, Plaksin V Yu and Riaby V A 2007 The volt-ampere characteristics of a DC arc plasmatron with a distributed anode spot *Thin Solid Films* **515** 5197-5201
[2] Asinovskiy E I et al. 1992 *Stabilized electrical arcs and their application in thermophysical experiments* (Moscow: Nauka) pp 105-15 (in Russian)
[3] Devoto R S 1967 Transport coefficients of partially ionized argon *Phys. Fluids* **10** 354-64
[4] Raizer Yu P 1987 *Physics of the gas discharge: A manual* (Moscow: Nauka) (in Russian)
[5] Finkel'nburg V and Mekker G 1961 *Electrical arcs and thermal plasma* (Moscow: IIL) p 332 (in Russian)
[6] Iermolova N A et al. 1991 Physical aspects of dry plasma-chemical etching of photoresists in barrel type reactors *Proc. 20th ICPIG* (Il Ciocco, Italy) **1** pp 325-6
[7] Baturin V A 1970 Experimental determination of argon plasma electrical conductance in a stabilized arc *J. Appl. Mech. & Techn. Phys.* **11**/2 153-61 (in Russian)
[8] Bues I et al. 1967 *Zeitschrift fur angewandte Physik* **22/**4 345

[9] Morris J C et al. 1970 Measurements of electrical and thermal conductivity of hydrogen, nitrogen and argon at high temperatures *Phys. Fluids* **13**/3 608-17

[10] Emmons H W 1967 *Phys. Fluids* **10/**6 1125

[11] Lin S C et al. 1955 Electrical conductivity of highly ionized argon produced by shock waves *J. Appl. Phys.* **26/**1 95

[12] Grekov L I et al. 1964 *Basic properties of some gases at high temperatures* (Moscow: Mashinostroenie) pp 37-8 (in Russian)

[13] Kulik P P 1971 *Studies on the physics and chemistry of low-temperature plasma* (Moscow: Nauka) p 5 (in Russian)

Journal of Physics: Conference Series **223** (2010) 012008 doi:10.1088/1742-6596/223/1/012008

Multi-viewpoint imaging based simulations of sensors for APS jet monitoring

B E Djakov[1,3], D H Oliver[1], R Enikov[1], E Vasileva[1], O Chumak[2] and M Hrabovsky[2]

[1]Emil Djakov Institute of Electronics,
Bulgarian Academy of Sciences, 72 Tzarigradsko Chaussee, 1784 Sofia, Bulgaria
[2]Institute of Plasma Physics, Academy of Sciences of the Czech Republic,
Za Slovankou 3, P.O. Box 17, 182 00 Prague 8, Czech Republic

E-mail: boyan_d@yahoo.co.uk

Abstract. Test results are presented for different configurations of newly developed optical sensors consisting of several narrow-collimated photo detectors pointed at a plasma jet along different directions in its cross section. The sensors are capable of detecting time-variations of the position, width and asymmetry of a plasma jet. Two plasma torches with different plasma forming gas, air and argon/water vapor mixture, were used in our experiments under typical atmospheric plasma spraying (APS) conditions. The tests are based on fast, side-on imaging of the plasma jet from two or three angles (viewpoints). Reference values for the jet center coordinates, width and maximum brightness were deduced by processing the full information in the images. Next, data from small areas were processed to obtain values of the same parameters that simulate the sensor actual performance. It is commented that the accuracy of the new sensors seems reasonable for the purpose of on-line control of APS.

1. Introduction

In a number of thermal plasma technological processes, geometrical properties of the plasma jet such as position, width and asymmetry can strongly influence the quality of the end product. In the atmospheric plasma spraying (APS) [1], powder particles with 10 to 100 μm diameters are injected into a thermal plasma jet, subsequently melted and then deposited to a surface to form coatings of 100 μm to a few millimeters thickness. Such coatings have specific functional properties that are of interest for a wide range of applications, from wear and corrosion protection to biocompatible materials and solid oxide fuel cells.

Our publications [2 – 5] describe simple optical sensors of geometrical properties: three to six narrow-collimated photo detectors (PD) are pointed at the radiating plasma object along different directions within a cross section of the object (plus one wide-angle PD in some cases). It is essential that different sensor configurations undergo extensive testing under conditions as close as possible to real situations in APS, as all geometrical and physical properties of the jet are strongly influenced by (i) plasma torch characteristics such as nozzle outlet diameter, (ii) torch regime – electric power and gas flow rate, (iii) operational drift and fluctuations inherent to the plasma torch, (iv) injection of powder particles.

[3] To whom any correspondence should be addressed.

An efficient way to do such tests is to use fast, side-on imaging of the plasma jet, then select small areas of the images that correspond to portions of the jet which will eventually be "seen" by the individual PDs. The photographic density within such areas simulates the PD signals. After appropriate calculations that simulate signal processing in the real case, the jet center coordinates and width are found within the chosen cross section of the jet. To simulate different orientation of the individual detectors, the plasma was imaged via a system of mirrors thus taking photographs at two or three different angles (aspects).

Two plasma torches with different plasma forming gas, air and argon/water vapor mixture, are used in the experiments under typical APS conditions. Reference values for the geometrical parameters are obtained by processing the full information in the images. Next, data from small areas are processed to obtained simulated values for the same parameters.

2. Principles of the new sensors

The measurements are based upon optical observation by several photo detectors (PD) pointed at the radiating plasma object (RPO) along three different directions in a plane cross section of the RPO. All PDs have identical spectral response. For our purposes, a certain type of light emission spatial distribution within the RPO is assumed. The characteristics of such a distribution are directly related to the RPO geometrical properties. Further, we describe how the PD output signals can be utilized to determine the geometrical properties by an appropriate signal fusion algorithm.

The PDs receive light each via a narrow pipe which restricts the field of view practically to a line of sight. A number of such lines are arranged in the jet cross-sectional plane with co-ordinate system Oxy, figure 1a. (Ideally, the cross section would be a circle with a centre at the origin O.) Let θ_i be the angle of the i-th line (i=1,2,…,N) with the x-axis and r_i' the distance of the line to the jet cross-section center, C. It is assumed that

i. the jet cross-section is an ellipse, figure 1b, with semi-axes ξ, η at an angle χ with the x-axis;
ii. the radiation leaving the jet, I_k, is described by the Gaussian

$$I_k = I_m \exp(-r_k'^2/R_k^2), \quad k = 1, 2, ..., N, \tag{1}$$

where I_m is the signal when the PD is pointed at the jet cross-section center, R_k is the jet radius as "seen" by the k-th PD. For a special orientation of PD $R_k=\xi$ or $R_k=\eta$ it was found that experimental data on radial brightness distribution correlate with the predictions of equation (1), with correlation coefficient of about or greater than 0.95 [2, 3].

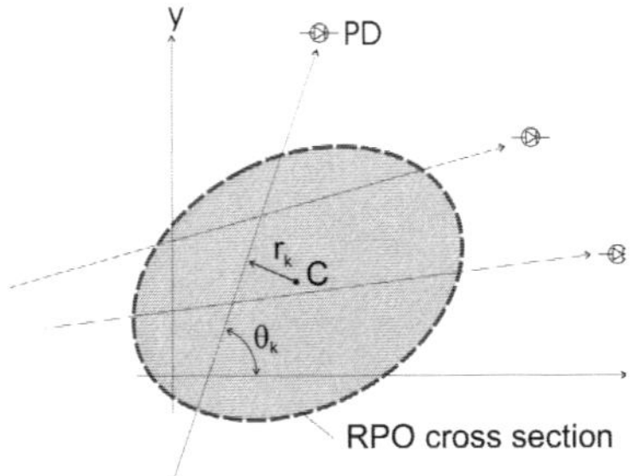

Figure 1a. Radiating plasma object (RPO) and photo detectors (PD).

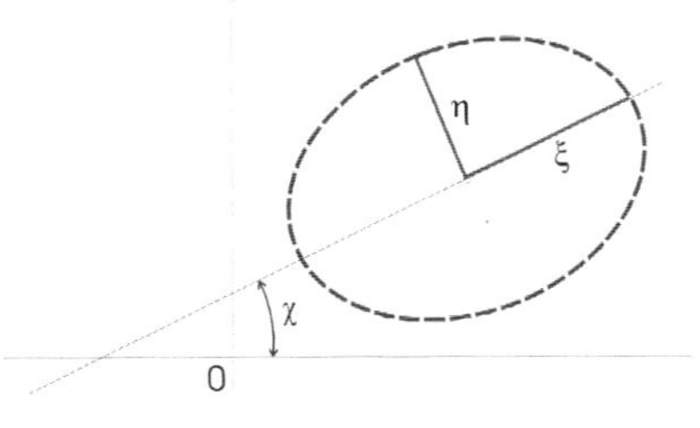

Figure 1b. Elliptical cross section of RPO.

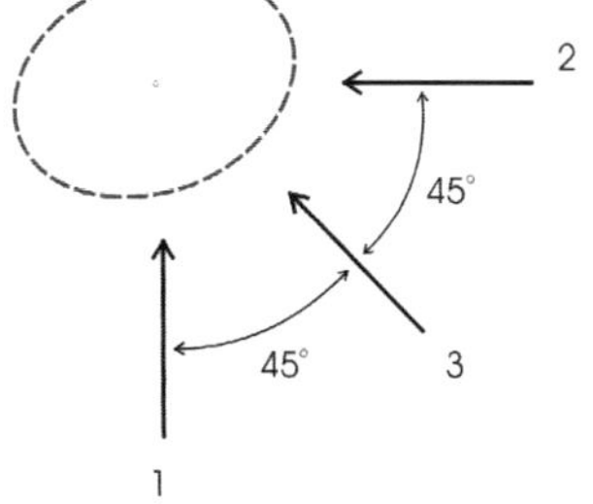

Figure 2. Torch A (aspects 1, 2), torch B (aspects 1,2,3).

Let the origin of xy-coordinates coincide with the time-averaged position of the center, C, its instantaneous position being at coordinates x_c, y_c.

Thus we have defined five geometrical properties of the jet cross section: x_c, y_c (characterizing position), ξ, η, χ (characterizing size and shape). These quantities are related to the PD signals in a way that depends on the overall sensor configuration. For convenience, all distances and signals are normalized by division by the time-averaged values of R_k and I_m.

3. Sensor configurations and sensor fusion algorithms

Several types of sensors with 3 to 6 narrow-angle PDs are arranged so as to "see" the object from 2 or 3 angles (aspects). Table 1 shows a schematic outline of the sensors to be studied here and the formulae used to compute the RPO geometrical parameters:

$$x_c,\ y_c,\ \chi,\ \rho = (\xi + \eta)/2 - 1,\ \varepsilon = (\xi - \eta)/2 \qquad (2)$$

I_k, I_k' and I_k'' are PD output signals. One wide-angle PD with output I_4 is included in Sensor IV.

Table1. Sensors.

	Configuration	Formulae for calculations of geometrical parameters	
Sensor I: 6 PD 3 aspects;	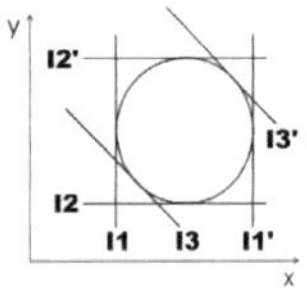	$x_c=\frac{1}{4}(I_1'-I_1),\ y_c=\frac{1}{4}(I_2'-I_2)$, $\omega+1=\frac{1}{4}(I_1+I_1'+I_2+I_2'),\ \rho=\omega/3$ $\mu=\frac{1}{4}[(I_1+I_1')-(I_2+I_2')],\ \nu=\frac{1}{2}(I_3+I_3')-(\omega+1)$ $\chi=\frac{1}{2}\tan^{-1}(\nu/\mu),\ \varepsilon=\mu/\cos(2\chi)$ (see [3])	(3) (4) (5) (6)
Sensor II: 4 PD 2 2 aspects	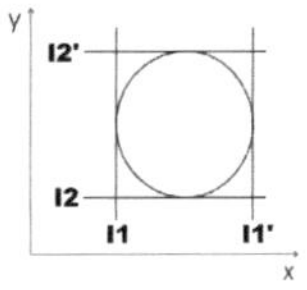	$x_c,\ y_c,\ \rho$ $\chi=0,\ \varepsilon=\mu$	(3), (4), (5) (7)
Sensor III: 6 PD 2 aspects	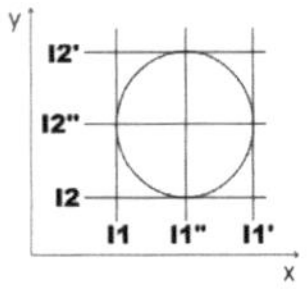	$\psi_1=[\ln(I_1/I_1'')/\ln(I_1'/I_1'')],\ \psi_2=[\ln(I_2/I_2'')/\ln(I_2'/I_2'')]$ $x_c=(\psi_1-1)/(\psi_1+1),\ y_c=(\psi_2-1)/(\psi_2+1)$ $\eta=(1+x_c)/[-\ln(I_1/I_1'')]^{1/2}\equiv(1-x_c)/[-\ln(I_1'/I_1'')]^{1/2}$ $\xi=(1+y_c)/[-\ln(I_2/I_2'')]^{1/2}\equiv(1-y_c)/[-\ln(I_2'/I_2'')]^{1/2}$ $\chi=0,\ \varepsilon=\eta\ (1+\rho)\equiv(1+\rho)-\xi$ (see appendix)	(8) (9) (10) (11)
Sensor IV: 3 (+1 wide angle) PD 3 aspects;	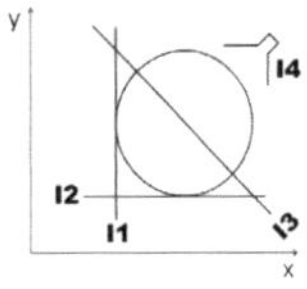	$W=I_4/[(\pi/2)^{1/2}I_3],\ \rho=W/2^{3/2}-1$ $X=W(1-\frac{1}{2}e^{1/2}I_1/I_3),\ Y=W(1-\frac{1}{2}e^{1/2}I_2/I_3)$ $x_c=X-1,\ y_c=Y-1$ $\chi=0,\ \varepsilon=0$ (see [5])	(12) (13) (14)

4. Results and discussion

Experiments were carried out using free jets produced by two APS torches, a water-stabilized arc torch [6] (torch A) operated at power 100 kW and a wall-stabilized arc torch [7] (torch B) at 40 kW, with steam-argon mixture and air as a plasma-forming gas, correspondingly. Numerical simulations of sensors I to IV, table 1, were carried out using sequences of images filmed by fast CCD cameras from two (torch A) and three (torch B) aspects as shown in figure 2. Experimental and data processing details for some torch/sensor combinations have been reported elsewhere [2, 3, 5].

Tables 2 to 4 present typical results of simulation. Reference data are provided directly from the gaussian fits of photographic density obtained by scanning cross sections of the images, see multimedia enhancement *Sequence of image M*. Selected pixels from the scans simulate PD signals. The equations of table 1 are used to find the simulated values of the geometrical parameters.

Table2. Comparison of sensor performance for displacement measurements.

Torch, image#		Simulated parameter	Reference value	Values from simulation of sensor operation			
				Sensor I	Sensor II	Sensor III	Sensor IV
A	19	x_c	0.161	*	0.237	0.153	0.236
		y_c	-0.083	*	-0.115	-0.100	-0.066
B	91	x_c	-0.075	-0.100	-0.100	-0.109	-0.077
		y_c	-0.044	**0.200**	**0.200**	**0.204**	-0.077
B	94	x_c	0.005	-0.027	-0.027	-0.046	0.013
		y_c	-0.168	-0.245	-0.245	-0.179	-0.135
B	112	x_c	0.279	0.191	0.191	0.207	0.210
		y_c	0.017	0.112	0.112	0.076	-0.010

Table3. Comparison of sensor performance for width measurements.

Torch, image#		Simulated parameter	Reference value	Values from simulation of sensor operation			
				Sensor I	Sensor II	Sensor III	Sensor IV
A	19	ρ	0.009	*	0.015	0.046	-0.013
	91	ρ	-0.139	-0.092	-0.083	-0.142	-0.184
B	94	ρ	0.102	-0.046	-0.026	0.228	**-0.160**
	112	ρ	-0.052	-0.058	-0.026	0.058	-0.116

Table4. Comparison of sensor performance for ellipticity measurements

Torch, image#		Simulated parameter	Reference value	Values from simulation of sensor operation			
				Sensor I	Sensor II	Sensor III	Sensor IV
A	19	ε	-0.144	*	-0.140	-0.161	*
	91	ε	0.090	0.090	0.050	0.064	*
		χ	-36.6°	-28.2°	*	*	*
B	94	ε	0.387	**0.181**	**0.135**	0.350	*
		χ	-22.2°	-21.1°	*	*	*
	112	ε	0.169	0.194	0.039	0.142	*
		χ	-36.7°	-39.2°	*	*	*

As x_c, y_c, ρ, ε are normalized to the time averaged jet radius, accuracy of about 0.1 seems reasonable for providing feedback in on-line automatic control. Except for a few cases, the simulated values for the linear parameters differ from the reference by less than 0.1, and for the angle χ by less than 10°. In tables 2 to 4, cases of greater difference are shown by bold lettering. Preliminary analysis of such cases shows that

i. Excessive inaccuracy is possible, if the signal is so low that error in digitizing takes place (illustrated by #91: I_1=6 with full range 0 to 255 for 8 bit processing);
ii. Exaggerated value for radius ρ (#94) is probably due to concentration of directions of observation (figure 2) along the longer semi-axis of the ellipse;
iii. Excessive errors in ellipticity ε are possible, if $\varepsilon \sim 0.5$ (illustrated by #94).

5. Conclusions

Some general trends are obvious. Complex sensor configurations such as I (table 1) are not always more accurate than simpler ones, unless information on the angle χ is needed. Whenever certain type of asymmetry is expected e.g. due to specific arc electrode behavior or powder injection, simple configurations such as IV may serve the purpose of automatic control, and strategic orientation of the lines of observation may also help. Studies of such trends are under way.

The accuracy of the new sensors seems reasonable for the purpose of on-line control of APS.

Appendix: the 6 PD, 2 aspect sensor

The derivation of formulae for sensor III, table 1, follows.

As detectors PD1″ and PD2″ are pointed at the RPO central zone, their outputs are, approximately, $I_1''=I_{1m}$ and $I_2''=I_{2m}$. Equation (1), for detectors PD1, PD1′, PD2 and PD2′, then becomes

$$\ln (I_1/I_1'') = -(1+x_c)^2/\xi^2 \quad \text{(A1)} \qquad \ln (I_2/I_2'') = -(1+y_c)^2/\eta^2 \quad \text{(A3)}$$
$$\ln(I_1'/I_1'') = -(1-x_c)^2/\xi^2 \quad \text{(A2)} \qquad \mathrm{Ln}(I_2'/I_2'') = -(1-y_c)^2/\eta^2 \quad \text{(A4)}$$

where ξ and η are defined in figure 1b by assuming $\chi = 0$, equation (11). Equation (A1) is divided by equation (A2), and (A3) by (A4). This yields

$$[\ln (I_1/I'')/\ln(I_1'/I_1'')]^{1/2} = (1+x_c)/(1-x_c) \quad \text{(A5)}$$

and, similarly, for PD2 and PD2′. Left hand sides are then denoted ψ_1 and ψ_2, see equation (8). It now remains to solve the resulting equations for x_c and y_c thus obtaining formulae (9). Back to equations (A1) to (A4), ξ and η can be expressed in terms of PD signals as well as x_c and y_c, equation (10). Equation (11) relates ξ and η to ellipticity ε, the radius ρ being defined by equation (2).

References

[1] Montavon G 2004 *High Temperature Material Processes* **8** 45
[2] Djakov B E, Oliver D H, Enikov R and Vasileva E 2004 *Proc. 12th Int. Workshop Plasma Technology* (Ilmenau, Germany) pp 111-8
[3] Djakov B E, Oliver D.H., Enikov R and Vasileva E 2006 *Plasma Proc. & Polymers* **3** 160–4
[4] Djakov B E, Jones G R and Yokomizu Y 2008 *Chromatic Monitoring of Electrical Plasmas: Chromatic Monitoring of Complex Conditions (Series in Sensors)* ed G R Jones, A G Deakin et al (Routledge: CRC Press) Chapter 4 ISBN 978-1-58488-988-5
[5] Djakov B E, Oliver D H and Vasileva E 2009 Operational Principles and Preliminary Tests of a Sensor for Monitoring the Position and Size of Dynamic Plasma Objects *16th Int. Summer School VEIT* (28 Sept–2 Oct 2009 Sunny Beach, Bulgaria) Abstracts book p 54
[6] Hrabovsky M, Konrad M, Kopecky V and Sember V 1997 *IEEE Trans. on Plasma Sci.* **25** 833–9
[7] Enikov R, Nedkov I and Oliver D H **1997** *J. Phys. IV* **C1** (France) 603

Three dimensional optical imaging of irregularly shaped plasma jets

B E Djakov[1], D H Oliver, E Vasileva and R Enikov

Emil Djakov Institute of Electronics,
Bulgarian Academy of Sciences, 72 Tzarigradsko Chaussee, 1784 Sofia, Bulgaria

E-mail: boyan_d@yahoo.co.uk

Abstract. This communication presents an algorithm for reconstructing the 3D distribution of optical radiation from a plasma object of irregular shape. The object is seen by a camera at three different angles. The photo density profiles in each single line of the three images are presented as sums of Gaussians; the reconstruction is made by means of a set of equations that relate radiation intensity in a 2D mesh to the Gaussian amplitudes. An example based on photographs of a typical plasma jet used for atmospheric plasma spraying is worked out.

1. Introduction

The shape and structure of complex plasma objects has successfully been studied by tomography [1]. This methodology requires acquisition and processing of a large amount of data. Here, we report the use of three optical images obtained simultaneously from different angles to reconstruct the spatial distribution of the radiation power density (RPD) from a plasma jet produced by an arc torch. If the jet is nearly axi-symmetrical, the problem can be solved by approximating the emission distribution within each jet cross section by a Gaussian [2]. Here, jets of more irregular structure are studied the RPD being represented as a sum of Gaussians. (There have been reports on multi-aspect imaging techniques used in other fields such as medicine, e.g. [3].)

2. Geometrical model of the jet

The plasma jet issued from the nozzle of an arc torch propagates (figure 1) along the line AA'. Here A is the center of the nozzle outlet. Light is intensively radiated from a zone (jet core) bounded by the contour line BB'A'D'D (jet boundary). Every cross section of this zone is of elliptical [2] or more complex shape described by parameters Z_i, i=1,2,…,N, that are functions of the axial distance z = AO and time t. Let P be the radiation power density (RPD). The quantity that is experimentally measured is the radiant exitance which for optically thin radiator is given by

$$I(x,z,t) = \int P(x,y,z,t)\,\mathrm{d}y\,, \tag{1}$$

for an observer with line of sight parallel to y-axis (figure 1).

[1] To whom any correspondence should be addressed.

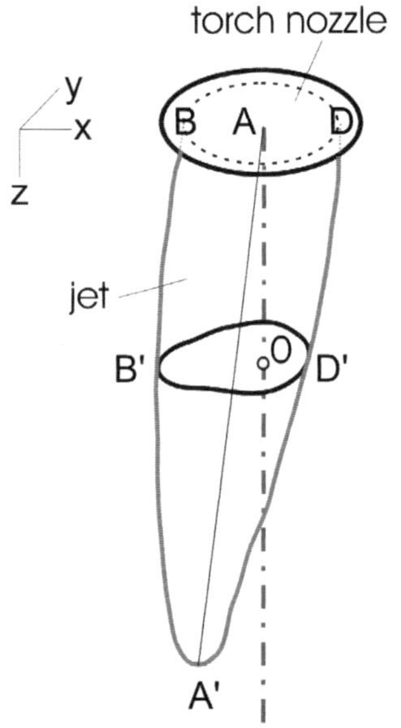

Figure 1. Geometrical model of a free jet.

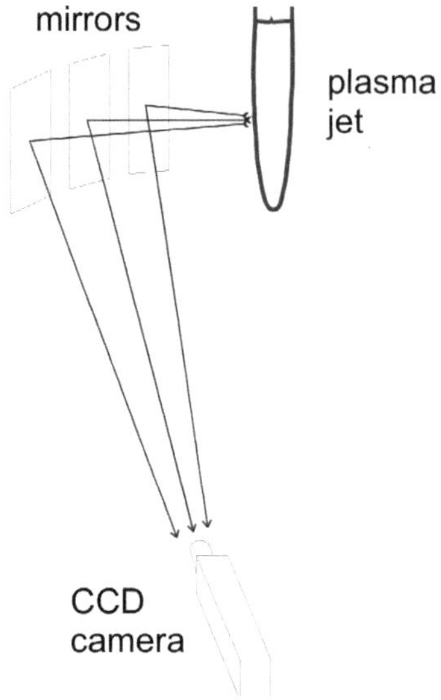

Figure 2. Imaging technique.

3. Light emission distribution and reconstruction algorithm

For any fixed value of z and t, the radiant exitance from a thin layer enclosing the corresponding cross-section (figure 1) is distributed along x as $I(x)$, which is approximated by a sum of $2n$+1 Gaussians (Gabor type transform [4]),

$$I(x) = \sum_{i=-n}^{n} A_i \exp\left[-\frac{1}{2}\left(\frac{x - iB_0}{B_1}\right)^2\right]. \tag{2}$$

Here A_i are amplitudes, B_1 semi-width and iB_0 ($-n \leq i \geq n$) displacements of Gaussians relative to the point $x = 0$.

In this paper we propose a technique to find RPD from measured values of the radiant exitance, see equation (1). Thereupon the aim of our reconstruction is to identify the jet boundary by setting this at a fixed value (e.g. $1/e$ of maximum) in the experimentally measured radiant exitance profile. Figure 3 shows such a profile fitted by (2) with n=3.

Our experimental set up is shown in figure 2. A single imaging device – a fast CCD video camera – views the jet via three mirrors positioned so that three aspects, at 45° from one another, are obtained.

Firstly, a sequence of triple images of the jet is taken (e.g. see multimedia enhancement *Sequence of Images M*). Each photograph is then used to make a three-dimensional reconstruction of the jet shape (e.g. figure 5) as follows (see also the multimedia enhancement *figure 4M*):

(I) For several (e.g. about 10 to 20) axial positions along the jet, z = const, the cross-sectional plane is divided into a number of cells (e.g. in figure 4: 6x6 cells); the "reconstructed" RPD values at cell centers are denoted $Z_1, Z_2, \ldots, Z_m$;

(II) For each axial position the photographic density, known also as pixel value, is measured at a number of points along the x-axis, within each of the three images;

(III) The pixel values obtained, for each image, are approximated by a sum of Gaussians (2);

(IV) From the amplitudes A_i (2) the values of Z_i are found (ill-posed problem) by solving a set of equations which in the case of figure 4 takes the form

$$A_1 = Z_1, A_2 = Z_2 + Z_4 + Z_7, \ldots, A_6 = Z_3 + Z_6 + Z_{10}, \ldots,$$
$$A_{10} = Z_1 + Z_2 + Z_3, \ldots, A_{13} = Z_{11} + Z_{12} + Z_{13} \quad ; \tag{3}$$

(V) If the number of equations (as in other cases) does not match the number of unknowns, a suitable minimization procedure is used;

(VI) Knowing Z_i, the lines of constant RPD, including the jet boundary defined above, are drawn.

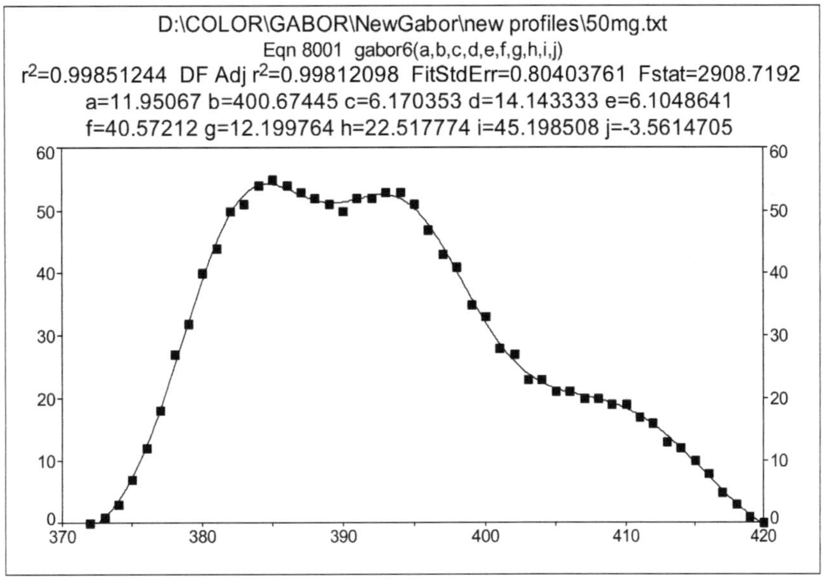

Figure 3. Traverse scanning of a jet image and its Gabor transform (n=3). Horizontal axis: position in pixel, vertical axis: radiant exitance (pixel value) 8 bit digitized.

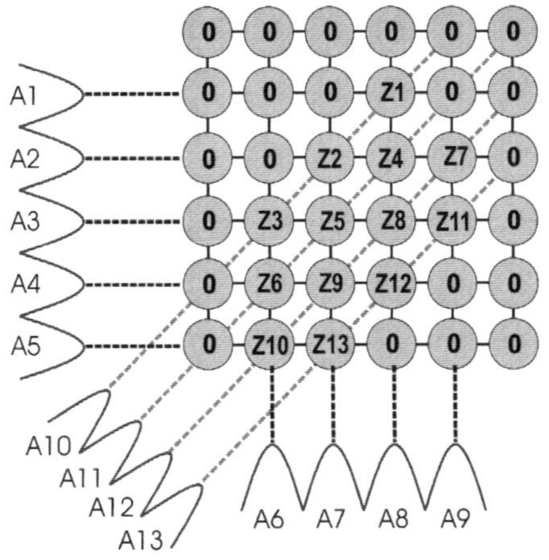

Figure 4. Example of image processing.

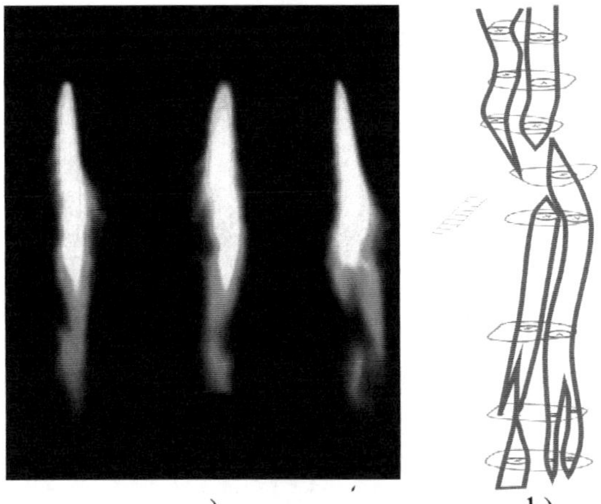

a) b)

Figure 5. 3D reconstruction of a plasma jet; a) original image; b) reconstructed image.

4. Example of reconstructed plasma jet shapes

Our DC plasma torch was operated at a power of about 40 kW with air as plasma forming gas, and powder injection that caused severe asymmetry of the plasma flow. The jet was filmed with framing rate of about 1 frame/s and exposure time of about 1 ms. Significant differences in the jet position, dimensions and luminosity were observed from frame to frame (e.g. see multimedia enhancement *Sequence of Images M*). An example of 3D reconstruction for frame # 5 (t=const) is shown in figure 5. The multimedia enhancement *figure 5M* presents different aspects (angles of view) of the 3D reconstruction.

5. Conclusions

Such 3D reconstructions can be used to estimate, on-line, features of plasma geometry. In the case of plasma spraying, asymmetric propagation of the jet may influence the process of powder particles entrainment and heating. Jet asymmetry may also serve as an indicator of excessive nozzle erosion that precedes torch failure.

References

[1] Ingesson L C and Pickalov V V 1996 *J. Phys. D: Appl. Phys.* **29** 3009
[2] Djakov B E, Oliver D H, Vasileva E and Enikov R 2007 Three dimensional optical imaging of plasma jets *Proc. 14th Workshop on Plasma Technology* (Ilmenau, Germany) pp 21–27
[3] Borodina E L, Gorskaya N V, Gorskii S M, Zverev V A, Nikolaev I N and Khil'ko G N 1995 Multiaspect imaging systems in a plane layered medium *Acoustical Physics* **41**/2 157-64
[4] Gabor D 1946 *J. IEE* **93** 429

Plasma nitridation of silicon by N_2 and NH_3 in PECVD reactor

V S Bakardjieva [1,4], Z I Alexieva[1], G D Beshkov[2] and E S Mateev[3]

[1]Central Laboratory of Solar Energy and New Energy Sources,
Bulgarian Academy of Sciences, 72 Tzarigradsko Chaussee, 1784 Sofia, Bulgaria
[2]Georgi Nadjakov Institute of Solid State Physics,
Bulgarian Academy of Sciences, 72 Tzarigradsko Chaussee, Sofia 1784, Bulgaria
[3]Emil Djakov Institute of Electronics,
Bulgarian Academy of Sciences, 72 Tzarigradsko Chaussee, Sofia 1784, Bulgaria

E-mail: vlabakar@phys.bas.bg

Abstract. The effect was investigated of nitrogen and ammonia plasma treatment of monocrystalline Si wafers. The experiments were carried out in a plasma-enhanced chemical vapor deposition reactor. The wafers were subjected to N_2 and NH_3 plasma treatment for varying times at temperature of 380 °C. The plasma treated surfaces were studied by transmission electron microscopy with C-Pt replicas, reflection high-energy electron diffraction and Auger electron spectroscopy. The results point to the growth of an amorphous layer on the surface. The Auger electron spectroscopy depth profiles obtained by sputtering show the presence of an oxynitride layer with varying composition depending on the time of plasma treatment. The Auger electron spectroscopy analysis shows that after 60 s of treatment in N_2 plasma, the nitrogen content is 8 at.%, while after 300 s it is 22 at.%, the thickness of the oxynitride nanolayer being 2.5-7.2 nm. In the case of NH_3 plasma the thickness calculated from the sputtering time (from 50 s to 15 min) varies between 2 and 12 nm, and the nitrogen content, between 5 and 35 at.%.

1. Introduction

In order to reduce the thermal budget in the fabrication of current and future microelectronic devices and integrated circuits, high-temperature process steps must be minimized in number and duration, and low-temperature alternatives, such as plasma-enhanced chemical-vapor deposition (PECVD), applied [1]. The basic advantages of this method consist in the possibility of activating the operating gas and the substrate and depositing device-quality film in a single run process at low temperature [2]. Exposing the substrate to a controlled N_2 plasma atmosphere is effective in reducing the hydrogen content in silicon nitride films and improving its dielectric properties [2]. Compared to thermal processes at same substrate temperatures, film growth rates are accelerated by the plasma-enhanced generation of reactive chemical species or by the presence of electric fields to assist the charged-particle transport during plasma processes [1]. In this study, nitrogen (N_2) or ammonia (NH_3) plasma nitridation of monocrystalline Si was performed in a PECVD reactor using radio-frequency (RF) plasma at low temperature.

[4] To whom any correspondence should be addressed.

2. Experimental

All plasma treatments were carried out in a capacitively-coupled parallel-plate ASM RF plasma reactor operating at a frequency of 13.56 MHz, and RF power of 60 W. Polished p-type (100) oriented silicon wafers, with resistivity of 10 - 15 Ohm.cm were used. They were subjected to a standard chemical surface cleaning and then placed in the reactor and subjected to N_2 (99.999%) plasma treatment for 15, 30, 60 and 300 s at a temperature of 380°C, nitrogen flow-rate of 1.2 standard liter per minute (slm) and working pressure of 0.2 Torr. The influence of the NH_3 plasma on the silicon surface was also investigated. The conditions were the same but for the working pressure, which was kept at 1.52 Torr. The plasma treated surfaces were studied by transmission electron microscopy (TEM) with C/Pt replicas, reflection high energy electron diffraction (RHEED) and Auger electron spectroscopy (AES). The AES depth profiles were obtained by grazing-incidence Ar-ion sputtering at 15° to the surface. In order to minimize the depth of the damaged surface layer, the Ar ion energy was chosen to be 1 KeV. The AES spectra were taken in dN(E)/dE mode with a modulation amplitude of 4 V. The calculations were made using the element sensitivity coefficients obtained at the same experimental conditions for Si LVV line – 92 eV from monocrystalline Si, taken as a reference sample, Si LVV - 79 eV and O KLL from SiO_2 and Si LVV - 86 eV and N KLL – from Si_3N_4 stoichiometric layers. The inaccuracy of the calculations is estimated to be ±3 at.% for N and O content and as ±5 at.% for Si.

3. Results and discussion

TEM of C/Pt replicas is superior at resolving nanoscale surface structures, including the development of new faces and the different microtopography of individual particles [3]. The TEM images did not show any changes induced by the plasma treatment when the duration of treatment was less than 30 s. Cone-like defects were detected on the surface when the N_2 plasma treatment was prolonged. Figure 1 shows the surface morphology prior and after plasma treatment for 300 s. Defects as high as 3 nm with density of 2.5×10^8 cm^{-2} are observed in figure 1 b). RHEED studies were carried out on the same surfaces. The electron pattern of the clean silicon surface prior to plasma treatment showed Kikuchi lines characteristic for a monocrystalline surface. After the treatment, the sharpness of the lines decreased and a hallo pattern was observed indicating a process of amorphization. The mechanism of initial defect formation is not quite clear but probably active nitrogen atoms, which are the dominant active species, can react with the silicon [3-6]. The AES analysis showed that after 60 s treatment in N_2 plasma, the nitrogen content was 8 at.%, while after 300 s, 22 at.%, the thickness of the $Si_xO_yN_z$ nanolayer being 2.5 - 7.2 nm.

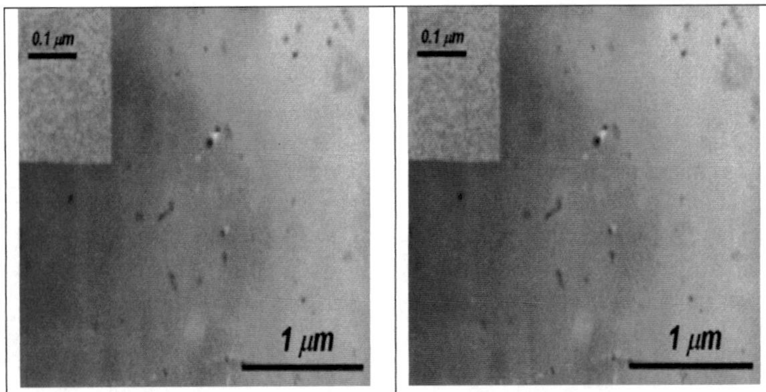

a) b)

Figure 1. Surface morphology of a clean Si monocrystalline wafer a) prior to plasma treatment and b) after N_2 plasma treatment for 300 s.

The influence of NH_3 plasma on the silicon surface morphology was also investigated by TEM with C/Pt replicas. Figure 2 shows the silicon surface after 1 min. and 15 min. The roughness is due to an increased layer growth in defect sites, as well as to erosion in the vicinity of the defects caused by the ion bombardment.

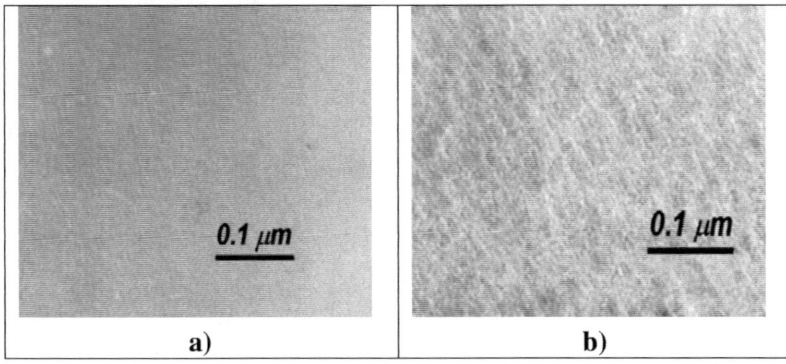

Figure 2. Surface micrograph of a NH_3 plasma treated silicon wafer after a) – 1 min and b) – 15 min.

Figure 3 shows the AES profiles of the silicon surface after different times of NH_3 plasma treatment (the sputtering rate is 1 nm/min). The concentration is shown of O, Si and N measured in at.% as a function of the sputtering time.

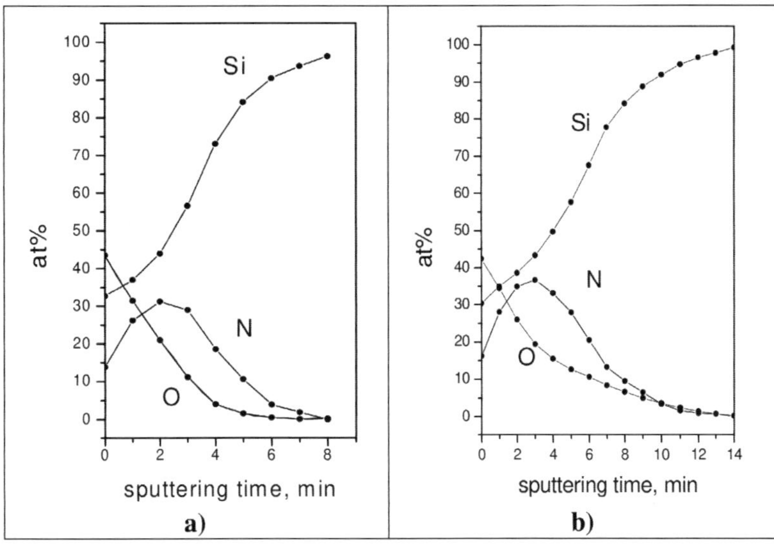

Figure 3. AES depth profile of a Si wafer after NH_3 plasma treatment for a) – 5 min and b) – 15 min.

The comparison of the figures shows that as the plasma treatment time is increased (5 min. and 15 min) the maximum nitrogen concentration rises and moves away from the outer surface. The thickness of the nano-layers obtained can be calculated from the sputtering time – they are respectively about 7 nm and 12 nm. The AES results indicate that N_2 and NH_3 plasma treatment induces a growth of oxynitride layers on the silicon surface. Direct growth of pure nitride layers onto silicon is difficult for two reasons. In spite of the preliminary etching for native oxide removal, the latter grows rapidly in air during the wafer loading in the reactor. In addition, there are trace amounts of oxygen and water vapors in the N_2 or NH_3 gas, which lead to the presence of oxygen in the nitride bonding structures during plasma growth processes [4-8].

The layers grown by N_2 plasma treatment are thinner and with lower N content than the layers grown by NH_3 gas. This is due to the fact that N_2 has higher dissociation energy than the NH_3 molecule. As a result, a lower density of active particles is achieved at the same plasma conditions. The higher pressure of NH_3 in the discharge acts in the same direction.

4. Conclusion

The effect of nitrogen and ammonia plasma treatment of silicon surfaces for varying times was investigated. The results indicate an amorphous layer growth on the surface. The AES depth profiles obtained by sputtering show the presence of a SiO_xN_y layer with varying composition according to the time of N_2 and NH_3 plasma treatment. The layers grown by N_2 plasma treatment are thinner and with less N content than the layers grown by NH_3 plasma treatment, because N_2 has higher dissociation energy than the NH_3 molecule, resulting in lower density of active particles.

Acknowledgments

This work was supported by the National Science Fund at the Ministry of Education and Science of Bulgaria under Grant X 1505.

References

[1] Heiss D W 1999 *IBM J. Res. Development* **43** 127
[2] Franco C, Cioffi N, Ditaranto N, Vitiello M, Sibilano M, Torsib L and Scamarcio G 2008 *J. Mater. Process. Tech.* **206** 462
[3] Chien Y, Mucci A, Paquette J, Sears S and Vali H 2000 *Microsc. Microanal.* **12/**4 302
[4] Nakae M; Hayakawa R; Yoshimura T; Fujimura N; Kunugi S and Uehara T 2007 *J. Appl. Phys.* **101** 023513
[5] Wong S and Oldham W 1984 *IEEE Electron Device Lett.* **EDL-5** 175
[6] Hirayama M, Matsukawa T, Arima H, Ohno Y, Tsubouchi N and Nakata H 1984 *J. Electrochem. Soc.* **131** 663
[7] Hirayama M, Matsukawa T, Arima H, Ohno Y and Nakata H 1985 *J. Electrochem. Soc.* **132** 2494
[8] Reisman A, Berkenblit M, Ray A and Merz C 1984 *J. Electron. Mater.* **132** 505

Dry etching of thin chalcogenide films

Kiril Petkov[1,3], Gergo Vassilev[2] and Venceslav Vassilev[2]

[1]Acad. J. Malinowski Central Laboratory of Photoprocesses,
Bulgarian Academy of Sciences, Acad. G. Bonchev Str., Bl. 109, 1113 Sofia, Bulgaria
[2]Department of Semiconductors, University of Chemical Technology and Metallurgy,
8 Kl. Ohridsky Blvd., 1756 Sofia, Bulgaria

E-mail address: kpetkov@clf.bas.bg

Abstract. Fluorocarbon plasmas (pure and mixtures with Ar) were used to investigate the changes in the etching rate depending on the chalcogenide glasses composition and light exposure. The experiments were performed on modified commercial HZM-4 vacuum equipment in a diode electrode configuration. The surface microstructure of thin chalcogenide layers and its change after etching in CCl_2F_2 and CF_4 plasmas were studied by SEM. The dependence of the composition of As-S-Ge, As-Se and multicomponent Ge-Se-Sb-Ag-I layers on the etching rate was discussed. The selective etching of some glasses observed after light exposure opens opportunities for deep structure processing applications.

1. Introduction

Chalcogenide glasses are among the most widely known families of amorphous materials and have been extensively studied for several decades because of their many potential applications in optical imaging, optical memories and sensing elements. Most of these applications are based on the wide variety of light-induced effects exhibited by these materials [1-4]. In our previous investigations [5, 6] of the changes in the dissolution rate of thin As-S-I(Bi) we drew some conclusions connected with their practical application.

There exist experimental indications that amorphous chalcogenides have rather high non-linear optical properties and, hence, these materials are promising candidates as non-linear optical elements [7, 8].

The interest in chemical dry etching has grown considerably in recent years due to its advantages over the wet etching process. Chemical dry etching, unlike the wet stripping methods, reduces the handling and disposal of toxic chemicals and can easily be integrated into modern multi-chamber systems [9-11].

The aim of this work is to present the results of a study of the dry etching process of chalcogenide thin films depending on their composition and on the etching conditions.

2. Experimental

The bulk glasses were synthesized by direct monotemperature synthesis in vacuumed (residual pressure of 10^{-3} Pa) and sealed quartz ampoules. The As-Se glassy samples were prepared from elements of purity 5N. As initial components for the synthesis of samples of the Ge-Se-Sb-Ag-I

[3] To whom any correspondence shoul be addressed.

system we used AgI (99.99 %, Merck), $GeSe_2$ and Sb_2Se_3 (synthesized by us from Ge, Sb and Se with purity of 5N, 4N and 5N, respectively).

Thin films of both systems were deposited by thermal evaporation in vacuum (6 - 8×10^{-6} Torr) from a Mo crucible on soda-lime and graphite substrates at room temperature. The deposition rate of 5 nm/s was monitored by a quartz thickness and deposition rate monitor (MIKI, Hungary). The thin films thickness was about 800 nm, as measured by a Talystep profilometer (Taylor-Hopson, England). The thin films were illuminated by a halogen lamp (20 $mW.cm^{-2}$) for two hours.

The dry etching was carried out in a modified commercial vacuum installation HZM-4 (Germany). The diode electrode configuration was connected to a 2.5 MHz rf power supply. The samples were placed on the water-cooled rf electrode with a diameter of 200 mm. The rf power density was varied in the range 50 - 350 $mW.cm^{-2}$. Most of the experiments were performed at a pressure of 8 - 10 Pa and a power density of 130 $mW.cm^{-2}$. The SEM studies were carried out on a Philips scanning electron microscope.

3. Results

The composition of the bulk glasses, as well as of the thin films deposited on graphite substrates, was determined by scanning electron microscopy by an X-ray microanalyser (Joel Superprobe 733, Japan) – table 1. The results show that the elemental composition of the As-containing films is very close (within ±1 at.%) to that of the bulk samples, while the Ge-containing samples show a difference between the thin films and bulk samples content.

Table 1. Data for the elemental composition of bulk samples and thin films. Dry etching of unexposed and exposed (800 nm) thin chalcogenide films in CCl_2F_2 and CF_4 plasmas: pressure 8 - 10 Pa; rf power density 130 mW/cm^2; etching time 120 s; gas flow 15 ccm/min.

Composition of bulk samples (at.%)	Composition of thin films (at.%)	CCl_2F_2		CF_4	
		Unexp	Exp	Unexp	Exp
$As_{28}Se_{72}$	$As_{29}Se_{73}$	31	17	20	10
$As_{40}Se_{60}$	$As_{41}Se_{59}$	35	27	25	28
$As_{45}Se_{55}$	$As_{46}Se_{54}$	61	32	23	32
$As_{40}S_{60}$	$As_{41}S_{59}$	33	8	33	22
$As_{30}Ge_{10}S_{60}$	$As_{32}Ge_{10}S_{58}$	65	18	20	5
$As_{20}Ge_{20}S_{60}$	$As_{21}Ge_{21}S_{58}$	52	18	16	8
$As_{10}Ge_{30}S_{60}$	$As_{11}Ge_{33}S_{56}$	25	16	13	18
$Ge_{40}Se_{60}$	$Ge_{42}S_{58}$	9	18	21	5
$Ge_{33}Se_{67}$	$Ge_{31}Se_{69}$	3.3	-	7.2	-
$Sb_{40}Se_{60}$	$Sb_{43}Se_{57}$	1.6	-	0	-
$Ge_{19}Se_{64}Sb_{17}$	$Ge_{14}Se_{60}Sb_{26}$	5	-	55	-
$Ge_{29}Se_{57}Ag_7I_7$	$Ge_{28}Se_{62}Ag_3I_7$	4.1	-	5	-
$Ge_{17}Se_{56}Sb_{15}Ag_6I_6$	$Ge_{17}Se_{57}Sb_{21}I_5$	6.7	-	45.8	-

In our previous paper [12] we showed that the optimal conditions for dry etching of chalcogenide thin films are pressure of 8-10 Pa; rf power density of 130 mW/cm^2; gas flow of 15 ccm/min. The changes in the etching rate depending on the film composition and exposure to light for a wide range of film compositions were studied. We did not find any difference in the etching rate of As-containing chalcogenide films using a mixture of Ar and Freon gases. We showed experimentally that the etching rate of all chalcogenide films decreases when the etching time increases. This is why in this study we compared the etching rate after 120 s etching time. The results are shown in table 1.

It is seen that the etching rate of unexposed thin As-S-Ge films decreases as the Ge content in the films is increased. After exposure, a negative etching effect was observed after applying CCl_2F_2 or CF_4

plasmas due to densification in their structure. The etching rate of As-Se films is higher when a CCl_2F_2 plasma is applied for etching than when CF_4 gas is used, which could be explained by the higher reactivity of the former gas. We observed that illumination causes photodarkening of As-containing films (associated with structural changes). This is the reason for the difference between the etching rate of as-deposited and exposed thin films. We also observed that the selective etching (the ratio between the etching rate of exposed to light and unexposed films) in CCl_2F_2 plasma is the highest for thin $As_{46}Se_{54}$ layers (negative etching process), while in CF_4 plasma for films with the same composition the etching process is positive (higher etching rate for exposed layers). The etching rate depends on the chemistry of the reactions taking place between CCl_2F_2 (CF_4) and the thin films from the As-Se (As-Ge-S) systems, as a result of which areas with different structural units ($AsSe_{3/2}$, $GeSe_{4/2}$, $SeSe_{2/2}$ and $SS_{2/2}$) are attacked and volatile (AsF_6, SeF_6, GeF_4, SF_6, SF_4 and S_2F_2) and condensed (pasivating: AsF_3, $AsCl_3$, $AsCl_5$, SeF_4, $GeCl_4$ and S_2F_{10}) components are produced.

For the Ge-containing samples the dry etching rate in CF_4 is much higher than that in CCl_2F_2 with the exception of composition $Sb_{43}Se_{57}$, which is related to the high chemical activity of the F-atoms. The "anomaly" of the films with composition $Sb_{43}Se_{57}$ is due to the specific features of the chemical reaction's products formed on the etched surface – $SbCl_3$ and SbF_3. $SbCl_3$ is characterized by a lower melting temperature and a higher tendency for sublimation compared to SbF_3. The etching rate for the film with composition $Ge_{28}Se_{62}Ag_3I_7$ is the same in CF_4 and CCl_2F_2 plasmas since the Ag in the form of AgF and AgCl blocks the film surface.

We used scanning electron microscopy to observe how the surface microstructure changes during the etching process. We plan to continue this study by using AFM to quantify the surface topology. SEM micrographs of the surface of as-deposited, etched and etched after illumination samples are shown in figures 1-3. It is seen that the surface of unexposed thin As-Se films is smooth. It remains smooth after etching in CF_4, while when the layers are etched in CCl_2F_4 grains with size of 100-200 nm appear due to the chemical reaction on the chalcogenide layer surface. No difference between the surface microstructure of unexposed and exposed films after etching is seen.

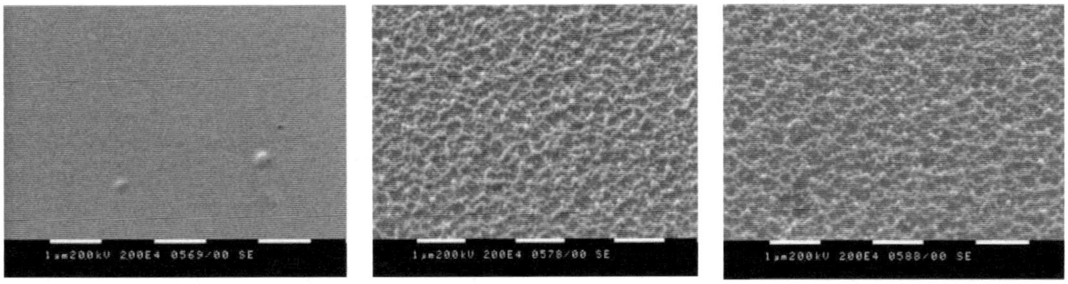

(a) (b) (c)

Figure 1. SEM micrographs of the surface of thin $As_{40}Se_{60}$: (a) as-deposited, unetched; (b) etched and (c) exposed and etched in CCl_2F_2 plasma.

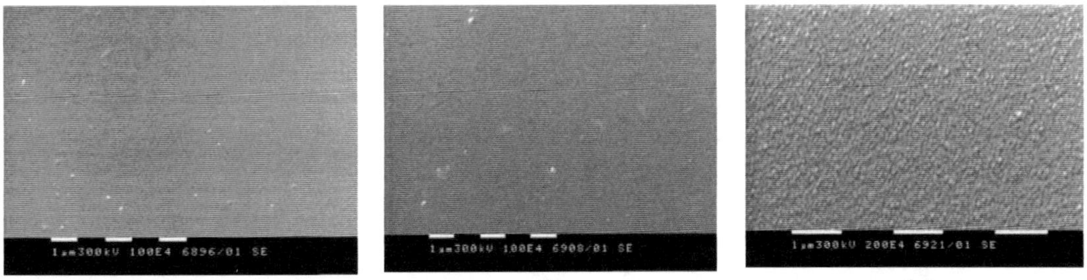

(a) (b) (c)

Figure 2. SEM micrographs of the surface of thin $As_{45}Se_{55}$films: (a) as-deposited, unetched; (b) etched in CF_4 and (c) etched in CCl_2F_2 plasma.

(a) (b) (c)

Figure 3. SEM micrographs of thin as-deposited Ge-containing films etched in CCl_2F_2 plasma: (a) $GeSe_2$, (b) Sb_2Se_3, (c) $Ge_{28}Se_{62}Ag_3I_7$.

4. Conclusions

We studied the dry etching process of a wide range of film compositions (binary thin As-Se, ternary As-S-Ge and multicomponent Ge-Se-Sb-Ag-I chalcogenide films) and we showed that the etching rate depends on their composition, etching time and light exposure. For the first time we deposited thin layers from such complex systems and showed that we can control the film composition by changing the evaporation conditions. The chalcogenide layers surface is very smooth; after etching, and depending on the film composition and the etching gas, grains with different sizes appear. The light exposure causes changes in the thin layers structure and leads to different changes in the etching rate depending on the plasma chemical composition. Good selectivity is demonstrated for some composition, which gives a possibility for applying such chalcogenide layers in planar optical technologies.

Acknowledgements

The study was supported by the Bulgarian National Science Fund at the Ministry of Education, Youth and Science under contract DO02-123/15.12.2008.

References

[1] Ewen P J S and Owen A E 1992 Photoinduced Changes in Chalcogenide Glasses *High Performance Glasses* ed M Cable and J M Parker (Glasgow: Blackie) pp 287-309
[2] Pfeiffer G, Paesler M A and Aggarwal S C 1991 *J. Non-Cryst. Solids* **130** 111
[3] Kolobov A and Elliot S R 1991 *Advances in Physics* **40** 625
[4] Kolobov A V 2003 *Photo-Indiced Metastability in Amorphous Semiconductors* (Weinheim: Wiley-CH)
[5] Petkov K, Ewen P J S and Vlcek M 1997 *Future Directions in Thin Film Science and Technology* ed J Marshall, N Kirov and A Vavrek (World Sci Ltd, Singapure) pp 397-400
[6] Petkov K, Dinev B 1994 *J. Mater. Sci.* **29** 468
[7] Nasu H, Ibara Y and Kubodera K 1989 *J. Non-Cryst. Solids* **110** 353
[8] Cardinal T, Richardson K A, Shin H, Schutte A, Beatty R, Foulgoe K I, Meneghini C, Viens J F and Villeneuve A 1999 *J. Non-Cryst. Solids* **256/257** 353
[9] Kastenmeier B E E, Matsuo P J, Beulens J J and Oehrein G S 1996 *J. Vac. Sci. Technol* **A14** 2802
[10] Hajito E, Belford R E, Ewen P J S E and Owen A E 1989 *J. Non-Cryst. Solids* **115** 129
[11] Duk-Yong C, Maden S, Rode A, Wang R and Luther-Davis B 2008 *J. Non-Cryst. Solids* **354** 3179
[12] Petkov K 1999 *Le Vide: Science, Technique et Applications Suppl.* **291** 308

16th International Summer School on Vacuum, Electron, and Ion Technologies (VEIT 2009) IOP Publishing
Journal of Physics: Conference Series **223** (2010) 012012 doi:10.1088/1742-6596/223/1/012012

Intermolecular interactions and thermophysical properties of O_2

M Damyanova[1,3], U Hohm[2], E Balabanova[1] and L Zarkova[1]

[1]Emil Djakov Institute of Electronics,
Bulgarian Academy of Sciences, 72 Tzarigradsko Chaussee, 1784 Sofia, Bulgaria
[2] Institut für Physikalische und Theoretische Chemie der TU Braunschweig,
Hans-Sommer-Str. 10, D-38106 Braunschweig, Germany

E-mail: milenadam@gmail.com

Abstract. We studied the applicability of the spherically-symmetric (*n*-6) Lennard-Jones temperature dependent potential (LJTDP) for describing the binary interactions between O_2 molecules in the temperature range between 200 K and 1300 K for pressures below 0.1 MPa. We observed a reasonable agreement between the thermophysical properties calculated using the (*n*-6) LJTDP (second virial, viscosity, and diffusion coefficients) and the corresponding measurements. A similar good agreement is also observed between our results and other correlations given in the literature.

1. Introduction

Even in the low pressure regime (p<0.1 MPa), the thermophysical gas properties depend on intermolecular interactions. The relationship between equilibrium (e.g the second *pVT* virial coefficient *B*) and non-equilibrium (e.g. the viscosity η and the self-diffusion ρD) properties and the underlying intermolecular forces might become complicated even for small molecules. Especially for open-shell systems like O_2 in its electronic ground state ${}^3\Sigma_g^-$, it is difficult to find the required ab initio potential-energy functions which are able to reproduce the thermophysical gas properties. For an accurate description of the properties of the O_2-O_2 dimer, three different intermolecular potentials with singlet, triplet and quadruplet multiplicities are required. Due to this difficulty, potential energy models have been used in the past to describe the thermophysical properties of O_2. However, in most cases either equilibrium or transport properties of O_2 have been considered. Since the transport properties depend more on the repulsive part whereas the equilibrium properties reflect strongly the attractive part, the resulting potential models look quite different [1, 2].

In our previous works we have observed that the intermolecular interactions between non-spherical molecules (halogens [3], *n*-alkanes [4], alkenes [5] and perflourinated *n*-alkanes [6]) can be described with acceptable accuracy by using a spherically-symmetric potential with temperature-dependent parameters. Therefore, in this work our (*n*-6) Lennard-Jones temperature dependent potential (LJTDP) model is extended to the case of O_2. It is used to describe some thermophysical properties in the temperature range between 200 K and 1300 K.

[3] To whom any correspondence should be addressed.

2. Background

Oxygen is a small open-shell molecule with an electronic ground state $^3\Sigma_g^-$. As has been done previously, we use a rotationally averaged spherically symmetric intermolecular interaction energy function $\varphi(R, T)$ (Lennard-Jones temperature dependent potential, LJTDP)

$$\varphi(R,T) = \frac{\varepsilon(T)}{n-6}\left[6\left(\frac{R_m(T)}{R}\right)^n - n\left(\frac{R_m(T)}{R}\right)^6\right]. \tag{1}$$

Here n is the repulsive parameter (assumed to be independent from the temperature) and R is the distance between the centers of mass. The temperature dependence of the potential well depth $\varepsilon(T)$ and equilibrium distance $R_m(T)$ is described by:

$$R_m(T) = R_m(\mathrm{T}=0\,\mathrm{K}) + \delta\cdot \mathrm{f}(T), \tag{2}$$

$$\varepsilon(T) = \varepsilon(\mathrm{T}=0\,\mathrm{K})\left(\frac{R_m(\mathrm{T}=0\,\mathrm{K})}{R_m(T)}\right)^6. \tag{3}$$

In equation (2), δ is a temperature independent fit parameter, whereas the function $f(T)$ takes into account the influence of the vibrational excitation of the molecule on its geometrical radius. In the framework of the harmonic oscillator approximation, $f(T)$ can be calculated at any T via the vibrational partition function $Z(T)$. More details on the LJTDP model can be found in [4].

3. Procedure

First, we collected all measured thermophysical properties B (second pVT virial coefficient), η (viscosity) and ρD (diffusion coefficient) from the literature. Only low pressure (p≤0.1 MPa) data were taken into account, in order to avoid influence of higher than binary interactions. The number of experimental data points for B [7] is N_B = 121, for η [8-16] it is N_η = 158 and for ρD [8, 14] it is $N_{\rho D}$ = 7. Altogether we have $N = N_B+N_\eta+N_{\rho D}$ = 286 data points which were subsequently used in our minimization procedure. The data for B have been obtained in the narrow temperature range between 77 K and 500 K in contrast to the viscosities given in the range between 90K and 1300 K. The lack of measured self-diffusion data is due to the difficulties in their experimental determination. This is also the reason for their comparatively large experimental uncertainties [8].

The second step is to minimize the sum F of the squared deviations between the measured values ("*exp*") and those calculated with the (n-6) LJTDP ("*calc*") normalized to their experimental uncertainties a_{exp}

$$F = \sum_1^{N_B}\left[\frac{\ln\left(\frac{B_{\exp}}{B_{calc}}\right)}{a_{\exp B}}\right]^2 + \sum_1^{N\eta}\left[\frac{\ln\left(\frac{\eta_{\exp}}{\eta_{calc}}\right)}{a_{\exp\eta}}\right]^2 + \sum_1^{N\rho D}\left[\frac{\ln\left(\frac{\rho D_{\exp}}{\rho D_{calc}}\right)}{a_{\exp\rho D}}\right]^2. \tag{4}$$

The goal is to find the potential parameters $\varepsilon(T = 0$ K$)$, $R_m(T = 0$ K$)$, n, and δ which correspond to the minimal value of F.

In the final step our calculated $B(T)$, $\eta(T)$, and $\rho D(T)$ are compared to experimentally measured data in order to check for systematic deviations. We also compare our findings to correlations given by other authors.

4. Results and Discussion

The results from the minimization are: $\varepsilon(T{=}0\text{K})/k_B$= (185.53±2.01) K, R_m (T=0 K) = (3.460±0.009).10^{-10} m, n =23.20±0.54 and δ= (0.075±0.017).10^{-10} m with a root-mean-square deviation $RMS=\sqrt{F/N}=0.82\ a_{exp}$.

The relative deviations (normalized to a_{exp}) between the measured and calculated thermophysical properties of O_2, are plotted in figure 1. It can be seen that the deviations of the experimental points are situated symmetrically with respect to the zero line (our solution) and most of them are within ±1 a_{exp}.

In order to provide a simple way for obtaining $R_m(T)$ it is fitted to the function:

$$R_m(T) = R_m(0) + A_1 \cdot \exp(-B_1/T) + A_2 \cdot \exp(-B_2/T). \quad (5)$$

The fitting constants are: A_1=(0.15373±0.00014)·10^{-10}m, B_1 =(2283.935±0.49) K, A_2=(0.16961±0.00073)·10^{-10}m, and B_2 =(5156±11) K. These coefficients then allow for an easy calculation of $\varepsilon(T)$ via equation (3).

In figure 2, absolute deviations between our results ("LJTDP") and other reference values ("Correlations") for B are shown, whereas in figure 3 relative deviations for η are displayed.

A good agreement can be observed for both properties. However, the deviations shown in figure 3 (viscosities) show a slight systematic behavior.

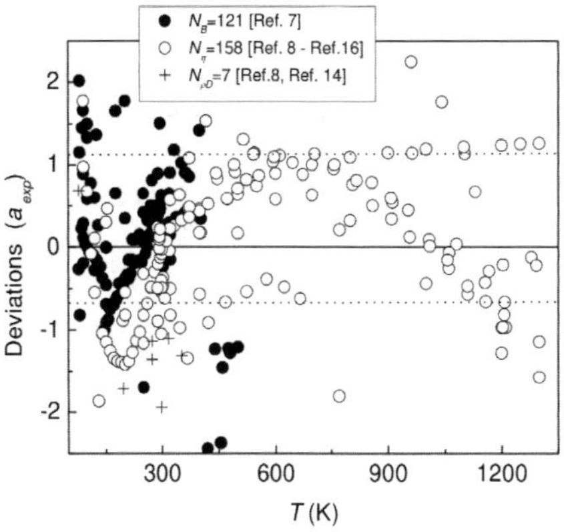

Figure 1. Relative deviations between the experimentally measured (symbols) and the thermophysical properties calculated (zero line) in this work.

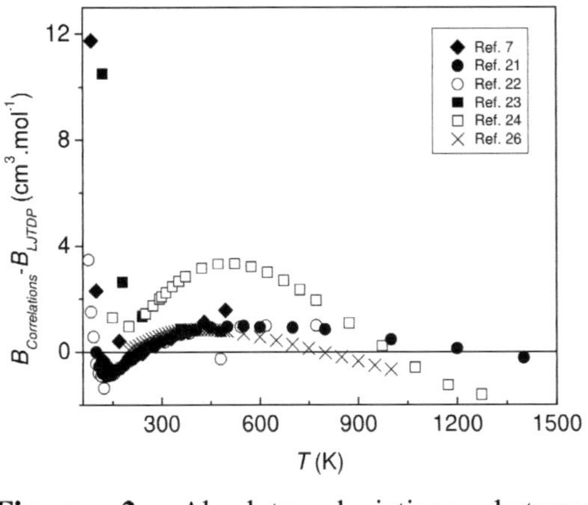

Figure 2. Absolute deviations between correlations of other authors (symbols) and results for B obtained in this work (zero line).

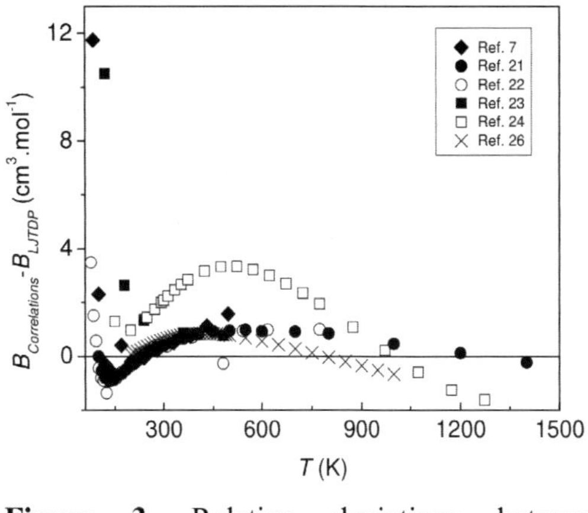

Figure 3. Relative deviations between correlations of other authors (symbols) and results for η obtained in this work (zero line).

5. Conclusions

The (n-6) Lennard-Jones temperature-dependent potential proposed for O_2 adequately reproduces the existing measured data for the second virial coefficient and the viscosity within their experimental error bounds, which is also reflected by the value obtained of RMS= 0.82 a_{exp}.

The algorithms presented for $R_m(T)$ and $\varepsilon(T)$ might be useful for the calculation of other thermophysical properties of O_2 and its mixtures with other gases.

References

[1] del Rio F, Ramos J E and McLure I A 1999 *Phys. Chem. Chem. Phys.* **1** 4937
[2] Maghari A and Jalili A H 2004 *Bull. Chem. Soc. Jpn.* **77** 1297
[3] Damyanova M, Zarkova L and Hohm U 2009 *Int. J. Thermophys.* (in press)
[4] Zarkova L, Hohm U and Damyanova M 2006 *J. Phys. Chem. Ref. Data* **35** 1331
[5] Zarkova L and Hohm U 2009 *J. Chem. Eng. Data* **54** 1648
[6] Hohm U, Zarkova L and Stefanov B B 2007 *J. Chem. Eng. Data* **52** 1539
[7] Dymond J H, Marsh K N, Wilhoit R C, and Wong K C 2002 *Landolt–Börnstein: Virial Coefficients of Pure Gases and Mixtures – Virial Coefficients of Pure Gases* New Series IV/21A (Springer Berlin)
[8] Landolt-Börnstein 1969 *Transportphänomene. Kinetik. Homogene Gasgleichgewichte* **5a** (New York Springer)
[9] Bonilla C F, Brooks R D and Walker P L 1951 *Proc. of General Discussion of Heat Transfer,* (Inst. Mech. Eng. London) p 167
[10] Kestin J and Yata J 1968 *J. Chem. Phys.* **49** 4780
[11] Kestin J, Ro S T and Wakeham W A 1971 *Trans. Faraday Soc.* **67** 2308
[12] Clifford A A, Gray P and Scott A C 1975 *J. Chem. Soc. Faraday Trans.* **71** 875
[13] Matthews G P, Thomas C M S R, Dufty A N and Smith E B 1976 *J. Chem. Soc. Faraday Trans.* **72** 238
[14] Winter E R S 1951 *Trans. Faraday Soc.* **47** 342
[15] Wobster R and Muller F 1940 *Kolloid-Beihefte* **52** 165
[16] Lavushev A A and Lyusternik V E 1976 *Russ. J. Phys. Chem.* **50** 1795
[17] Trautz M and Heberling R 1931 *Ann. Phys.* **10** 155
[18] Yen K L 1919 *Phil. Mag.* **38** 582
[19] Van Itterbeek R and Keesom W H 1935 *Physica* **2** 97
[20] Hirschfelder J O, Curtiss C F and Bird R B 1954 *Molecular Theory of Gases and Liquids* (Wiley London)
[21] *American Institute of Physics Handbook* 3rd ed 1972 (McGrew-Hill New York)
[22] *Data Project "Selected Values of Properties of Chemical Compounds"* 1980 (Thermodynamics Research Center, Texas A&M University, College Station, Texas, 1980-extant) Loose leaf data sheets
[23] Bohn M, Lustig R and Fischer J 1986 *Fluid Phase Equilibria* **25** 251
[24] Boushehri A, Bzowski J, Kestin J and Mason E A 1987 *J. Phys. Chem. Ref. Data* **16** 445
[25] *NIST Thermophysical Properties Database for Semiconductor Process Gases* 2004 Release: http://properties.nist.gov/fluidsci/semiprop/gases/O2.html
[26] *Thermophysical Properties of Matter* 1970 **13** (IFI/Plenum New York)
[27] Vasserman A and Rabinovich V 1970 *Thermophysical Properties of Liquid Air and its Components* (National Science Foundation, Special Foreign Currency Science Information Program, Washington DC)
[28] Maitland G C and Smith E B 1972 *J. Chem. Eng. Data* **17** 153
[29] Hanley H J M and Ely J F 1973 *J. Phys. Chem. Ref. Data* **2** 735
[30] Hanley H J M, McCarty R D and Haynes W M, *J. Phys. Chem. Ref. Data* **3** 979
[31] Stephan K and Lucas K 1979 *Viscosity of dense Fluids* (New York: Plenum Press)
[32] Younglove B A 1982 *J. Phys. Chem. Ref. Data* **11** Suppl.1
[33] Cole W A and Wakeham W A 1985 *J. Phys. Chem. Ref. Data* **14** 209
[34] Laesecke A, Krauss R, Stephan K and Wagner W 1990 *J. Phys. Chem. Ref. Data* **19** 1089
[35] *Spravochnik chimika* (Gosudarstvennoe Nauchno-technicheskoe Izdatelstvo Chimicheskoj Literaturi, Moskva, Leningrad) book I p 1002 (in Russian)

Structural and optical properties of poly-Si thin films obtained by aluminium induced crystallization

D Dimova-Malinovska

Central Laboratory of Solar Energy and New Energy Sources,
Bulgarian Academy of Sciences, 72 Tzarigradsko Chaussee, 1784 Sofia, Bulgaria

E-mail: doriana@phys.bas.bg

Abstract. The structural properties were studied of poly-Si films prepared by aluminium induced crystallization (AIC) of amorphous Si films (a-Si:H) deposited on glass substrates covered with Al layer. Raman and XRD spectroscopy were used for characterization of their short and long range order, respectively. The UV-reflectance spectra of poly-Si films were measured as well, and their surface morphology was observed by optical microscopy. The dependence was revealed of the structural and optical properties of the poly-Si films obtained on the annealing gas atmosphere and the hydrogen pressure during the deposition of the a-Si:H precursor, together with the correlation between them. The poly-Si films were applied in p-poly-Si/n-ZnO heterostructures; the I-V characteristics of the latter are presented and discussed.

1. Introduction

Recently aluminium induced crystallization (AIC) has attracted attention as a promising technology for low-temperature fabrication of large-area polycrystalline silicon (poly-Si) with grain sizes larger than the thickness of the Si layers [1-6]. It is an economically attractive process for producing poly-Si at a low temperature (<550^{0}C) in a short time starting from small-grained or amorphous Si deposited on a low-cost substrate such as glass or plastic. The preparation of poly-Si films by AIC is based on an annealing of amorphous Si films deposited by different methods. When Si is in contact with certain metals, the a-Si crystallization temperature is drastically reduced. This is due to a change in the Si-Si covalent bonds at the interface Si/metal due to the free electrons from the metallic phase. During the isothermal heat treatment, the semiconductor film is dissolved into the metal film, where it diffuses and precipitates. In spite of the large variety of metals used, Al has found the largest applications in the MIC of a-Si films. The main parameters involved in the AIC, i.e. the annealing temperature, duration and atmosphere, the Al/Si thickness ratio, the interface oxide layer, the layer sequence, the method of Al and Si deposition, etc. have been intensively studied [3-6].

In this review the influence is presented of the annealing atmosphere and the hydrogen concentration in the a-Si:H precursor films on the structural and optical properties of poly-Si films prepared by AIC.

2. Experimental

Poly-Si films were prepared by AIC of the glass/Al/a-Si:H(or a-Si) structures. The precursor Al layers were deposited at a substrate temperature T_s^{Al} = 300 ^{0}C and were kept for 24 hours in air before the deposition of the a-Si:H (or a-Si) films. Unhydrogenated (a-Si) and hydrogenated (a-Si:H) films were deposited on top of the Al by magnetron sputtering at 130 W rf power, different substrate temperatures, $T_s^{a\text{-}Si}$, and different H_2 partial pressures in the gas mixture $Ar+H_2$. The precursor layers, Al and a-Si (or a-S:H) had equal thickness of about 100-150 nm. The structures prepared were annealed isothermally at different temperatures in the range 480-530°C for different times and in different annealing atmospheres: N_2, forming gas (N_2+H_2) and H_2. During the annealing, crystalline silicon grains were formed in the bottom layer, while Al atoms moved to the top surface, which resulted in a layer exchange. After annealing, the aluminium was removed from the surface of the poly-Si films formed by etching in a chemical solution based on phosphoric acid. The structures of the samples were studied by Raman or microprobe Raman scattering, optical light microscopy (OM) and XRD as described in [2, 6]. The peak positions and the full-width at half-maximum (FWHM) of the Raman bands were measured with a mean error of 0.5 cm^{-1}. An estimate of the grain size can be deduced from both the red shift and the FWHM of the Raman peak [7]. It should be noted that accurate values for the grain size cannot be determined from the relationships described previously in the literature [7, 8], as they depend on the structures of the grains and their boundaries. Despite this, comparisons with spectra obtained from similar materials are valid. The UV hemispherical reflectance spectra in the range 250 - 400 nm were measured by a Perkin-Elmer UV/VIS/NIR Lambda 900 spectrometer.

3. Results and discussion

3.1. Influence of the annealing atmosphere

Optical micrographs of poly-Si films annealed in different ambiences using a-Si precursor are shown in figure 1. The poly-Si obtained by annealing in N_2 (figure 1a) have large grains (dendrites), with 20-30 μm size, separated by inter-crystalline spaces of about 10-20 μm. When annealing is performed in an atmosphere containing H_2 (figure 1b and c), the surface of the resulting poly-Si is very smooth. The grains are very closely packed and continuous poly-Si films are obtained. The size of the inter-grain space decreases to about 2-3 μm. The poly-Si surface in these cases is more uniform. In order to analyze the grain orientations shown in the OM-images, X-ray Diffraction Spectroscopy (XRD) measurements were performed. Figure 2 shows the XRD-pattern of a poly-silicon film. It is evident that the crystallographic peak of the silicon is (111).

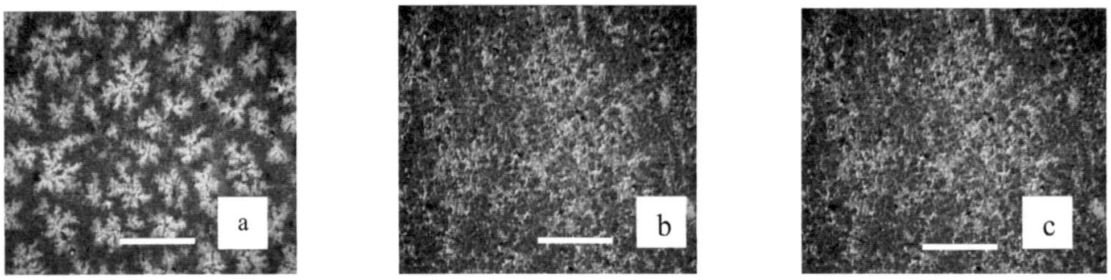

Figure 1. Optical images of poly-Si films after annealing the glass/Al/a-Si structure at 500°C in: N_2 for 6 h (a); N_2+H_2 for 5 h (b); H_2 for 3 h (c). The markers correspond to 100 μm.

Microprobe Raman spectra of poly-Si layers prepared by isothermal annealing in different atmospheres at 500°C are shown in figure 3. Two spectra were measured for each sample – on the grain surface and between the grains. All of the spectra of the annealed samples have a Si-Si TO-like phonon band centered in the range 520-521.3 cm^{-1}, showing the presence of the Si crystalline phase only. The microprobe Raman spectrum taken from the grain (figure 3, curve 1a) of the poly-Si film prepared by annealing of the structure Al/a-Si in N_2 for 6 hours has a band centered at 520 cm^{-1}, with a FWHM of 10 cm^{-1}. When the annealing is performed in an atmosphere of forming gas (N_2+H_2) for 5h, the positions of both Raman peaks, from the grain surface and from the inter grain material,

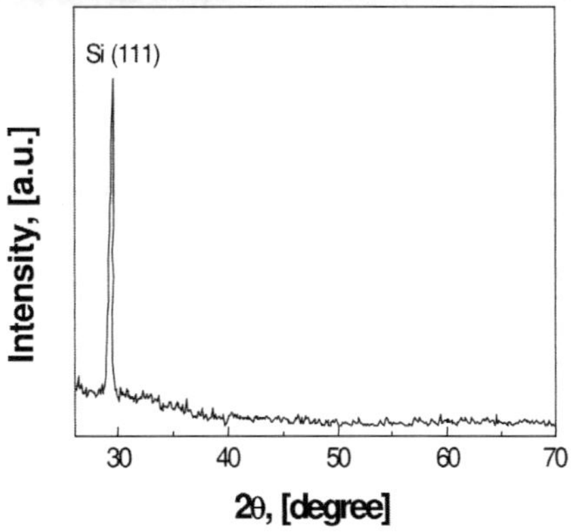

Figure 2. X-ray diffraction spectrum of a glass/Al/a-Si sample annealed at 500^{0}C in N_2+H_2 for 5 hours.

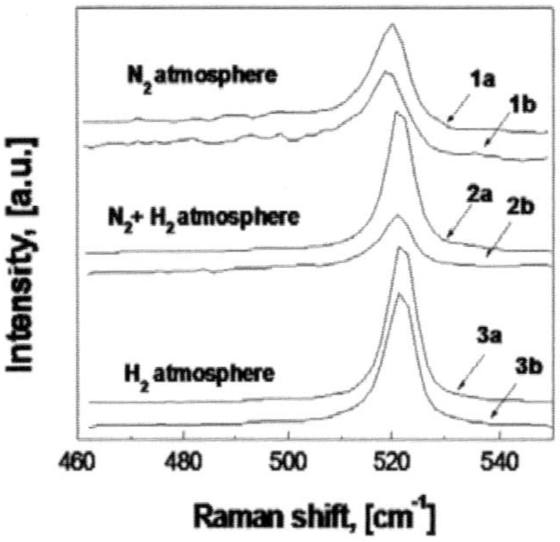

Figure 3. Microprobe Raman spectra of glass/Al/a-Si samples annealed in different atmospheres measured at intra-grain (a) and inter-grain (b) positions.

(figure 3, curves 2a and 2b) are centered at 521 cm^{-1}, with decreased FWHMs of 7 cm^{-1} and 7.5 cm^{-1} respectively. The narrowest Si-Si TO-like band centered at 521.3 cm^{-1} with a FWHM of 7 cm^{-1} (figure 3 curve 3a, 3b) is exhibited by the sample annealed in H_2 for both grain and inter-grain material. The results demonstrate that annealing in the presence of H_2 results in an improvement of the structure of both the grain and the inter-grain materials. The grain size is inversely proportional to the FWHM of the peak [9], so that one can assume that the grain size increases when the annealing has been performed in the presence of H_2, as confirmed by the OM images. The results obtained can be explained by supposing that H_2 stimulates the crystalline grain growth, probably by increasing the diffusion rates of Al and Si, which exchange their places during the AIC. It is possibly due to the reducing properties of H_2 - it reduces the Al_2O_3 film formed during the storage of the Al films in air for 24 hours before the a-Si films deposition. As a result, aluminium hydroxide is formed, leading to an enhancement of material transport. This results in an acceleration of the AIC process and an increase of the grain size in poly-Si films when annealing is performed at the same conditions, as was observed and deduced from the OM and microprobe Raman studies.

3.2. Influence of the hydrogen content in the a-Si:H precursor

To study the influence of the hydrogen content in the a-Si:H precursor, unhydrogenated (a-Si) and hydrogenated (a-Si:H) films were deposited at different substrate temperatures: RT (without heating of the substrates), 250, 300, 350 and 400 ^{0}C. Three different sets of samples were prepared with a-Si:H precursors deposited without hydrogen (a-Si) and at H_2 partial pressure of 0.05, and 0.1 Pa in the Ar+H_2 gas mixture. These conditions resulted in a hydrogen concentration in the amorphous Si precursor layers from 4 to 25 at.% (measured by ERDA [10]) and from 0 to 19 at % (calculated from IR spectra). The hydrogen concentration increased as the H_2 partial pressure was increased and the substrate deposition temperature, decreased (table 1) [11].

The Raman spectra of films obtained from a-Si or a-Si:H precursors deposited at different $T_s^{a\text{-}Si}$ and three different hydrogen partial pressures are shown in figure 4. All of the samples display Raman spectra typical of the crystalline Si structure - a Si-Si TO band, centered between 518.5 and 520.5 cm^{-1}. The Si-Si TO peak for crystalline silicon, measured under the same conditions, is at 521 cm^{-1} and has a FWHM of 4.5 cm^{-1}. The dependence of the Si-Si TO peak position and its FWHM (estimated from a Lorentzian fit) on the a-Si (a-Si:H) substrate temperature and the H_2 partial pressure are presented in figure 5a and figure 5b, respectively. The accuracy of the values is indicated by the error bars. As is mentioned before, the grain size is inversely proportional to the FWHM of the peak [8, 9]. On the other

Table 1. Deposition conditions of the precursor layers: φ and ξ - H concentration calculated from FTIR absorption spectra and from ERDA, respectively; 2θ - (111) XRD peak position; Δ2θ - its FWHM; D - average grain size; σ - the stress in the poly-Si films from the XRD data; ω_{TO} - the position of the Si-Si TO-like Raman band; and its FWHM are given as well.

T_s^{a-Si}, °C	P_{H2}, Pa	φ, at. %	ξ, at. %	2θ, deg.	Δ2θ, deg.	D, Å	σ, MPa	ω_{TO}, cm^{-1}	FWHM, cm^{-1}
No heat.	0.10	19	25	28.44	0.13	630	- 680	519	8.0
250	0.10	12	18	28.49	0.13	630	- 680	518.5	7.0
300	0.10	15	19	28.49	0.13	630	- 700	519.5	7.0
400	0.10	11	16	28.40	0.15	546	- 393	520	7.0
250	0.00	0	4	28.50	0.21	431	- 629	520	7.7
250	0.05	9	14	28.48	0.23	372	- 661	519.5	7.5
300	0.20	18	21	28.50	0.24	341	- 710	518	7.0

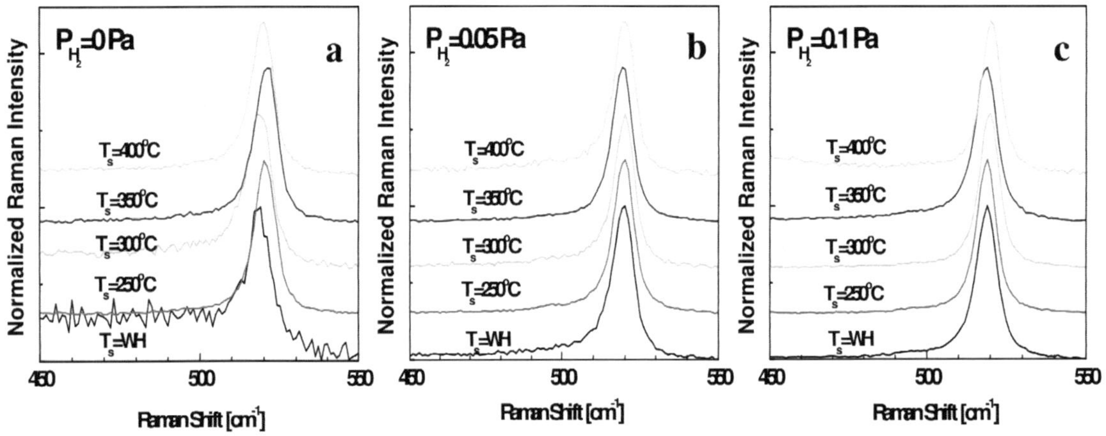

Figure 4. Raman spectra of poly-Si films obtained from a-Si and a-Si:H precursors deposited at different T_s^{a-Si} and different H_2 pressure: 0 Pa (a), 0.05 Pa (b) and 0.1 Pa (c).

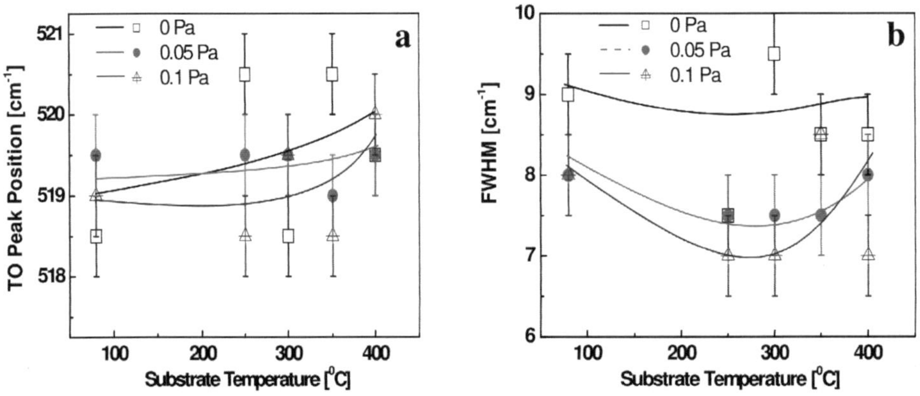

Figure 5. Si-Si TO-like peak position (a) and FWHM (b) taken from Raman spectra of the samples presented in figure 4. The lines are just guides to the eye.

hand, a shift of the Si-Si TO peak position, ω_{TO}, to a lower wave number could be related to an increase in the value of the tensile stress [12]. The following tendencies can be seen. The Si-Si TO peak position shifts slightly to a higher wave number as $T_s^{a\text{-}Si}$ is raised. This is an indication of a reduction in the tensile stress in the poly-Si samples as $T_s^{a\text{-}Si}$ is increasing [13]. A weak tendency for increased stress in the poly-Si films, when a-Si:H precursors were used, can be noticed and could be explained by the effusion of H during the AIC. This could leave pinholes and microvoids in the poly-Si films resulting in a higher tensile stress.

The FWHM of the Si-Si TO peak passes through a minimum for poly-Si films obtained from hydrogenated precursor layers deposited at $T_s^{a\text{-}Si}$ between 250 and 350° C - in this case larger grains were obtained in the poly-Si films. It should be noted that we observed similar results for the AIC of glass/a-Si(a-Si:H)/Al structures [14]. During annealing, effusion of H from the a-Si:H films takes a place. It is possible to suppose that this will enhance the disorder in the a-Si:H precursor during the process of annealing. This stimulates the dissolution of Si atoms into the Al and the consecutive rearrangement into a Si crystalline structure, resulting in a higher growth rate and larger grains in the poly-Si films. This suggestion is based on the fact that the higher degree of disorder creates energetically more favourable conditions for the transformation of the a-Si:H film into a poly-Si one.

It is known that the H concentration in a-Si:H films decreases with increasing the substrate temperature, so that the precursors deposited at RT (without heating) should have a higher H content [11]. In this case, the higher quantity of effusing hydrogen probably creates conditions for an increased diffusion rate of Al and Si. Thus, intermixing of Al and Si would take place within the bulk of the resulting poly-Si film, which would inhibit the exchange between Al and Si and would lead to a lower rate of crystallite growth and a smaller grain size. Deposition of the precursor layers at temperatures > 350° C would result in a reduced H content and in a better short range order of the a-Si:H films [12], which would slow down the re-arrangement into the crystalline structure – the poly-Si films would again have grains of a smaller size.

XRD and Raman spectra of the set of samples deposited at different P_{H2} and T_s = 250 °C are presented in figures 6 and 7. One reflection peak in XRD spectra at about $2\theta = 28.39^0$ - 28.48^0 is only observed for the poly-Si films, which corresponds to the (111) preferential orientation. The position of this peak in the c-Si is at $2\theta = 28.48^0$. The intensity of the peak is higher and is at $2\theta = 28.48^0$ in the poly-Si films obtained from a precursor deposited at P_{H2} = 0.1 Pa. This observation is an indication of a better crystalline quality. The Si-Si TO-like Raman peak for c-Si is at 521 cm^{-1} with FWHM = 4.5 cm^{-1} measured under the same conditions. The shift of this peak in the poly-Si films to lower wave numbers could be due to the tensile stress in the poly-Si films, [12, 13]. The Si-Si TO-like peak position is slightly shifted to the position of the c-Si as P_{H2} is raised to 0.1 Pa. Thus, the incorporation of H at

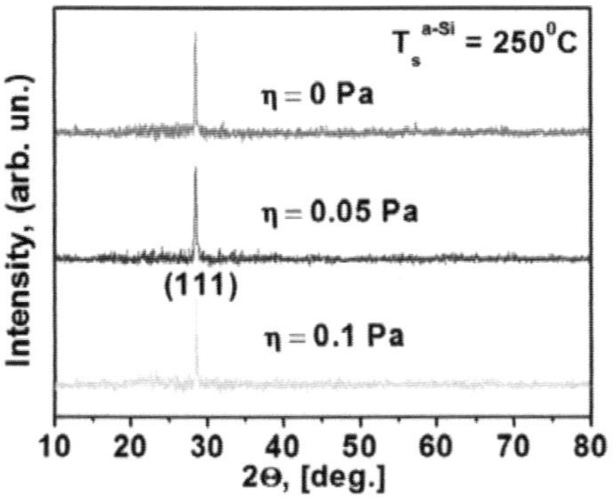

Figure 6. XRD spectra of poly-Si films obtained from a-Si:H precursors deposited at T_S = 250°C and different P_{H2}.

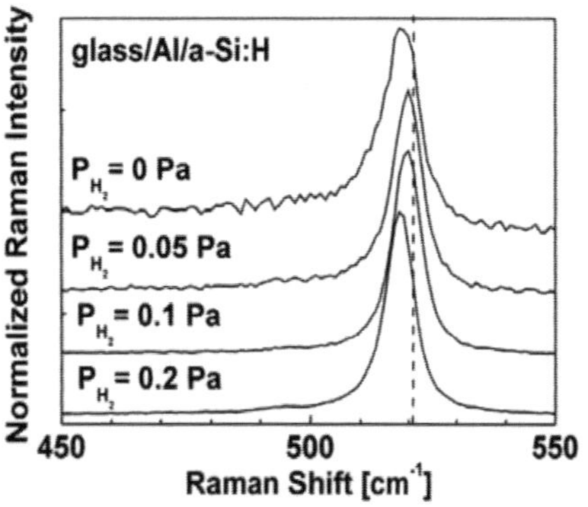

Figure 7. Raman spectra of poly-Si films obtained by AIC form a-Si:H precursors deposited at T_S = 250°C and different P_{H2}.

a moderate concentration in the a-Si:H precursor results in a lower tensile stress. However, a further P_{H2} increase to 0.2 Pa results in a higher stress value, probably as due to the creation of pinholes as a higher amount of H effuses. The FWHM is the lowest in the films obtained from a-Si precursors deposited at P_{H2} = 0.1 Pa and 0.2 Pa (table 1). The grain size tends to increase as P_{H2} is increased to 0.2 Pa, which could be explained by the influence of the hydrogen on the layer exchange process. The higher hydrogen concentration might accelerate the diffusion of Si in the Al layer, so that grains of a larger size are obtained under identical conditions of AIC [6, 13]. It is possible to assume that the effusion of H in the precursor layer enhances the disorder in the a-Si:H precursor during the process of AIC. This stimulates the diffusion of Si atoms into the Al and the subsequent re-arrangement into a Si crystalline structure, resulting in a higher growth rate and, thus, larger grains in the poly-Si films. This suggestion is based on the fact that the higher degree of disorder is energetically more favourable for the transformation of the a-Si:H film into a poly-Si one.

Images of the surface of the poly-Si films obtained from a-Si and a-Si:H precursors deposited at different $T_s^{a\text{-}Si}$, are shown in figures 8, 9 and 10. Optical micrographs of poly-Si films prepared from a- Si precursors deposited at $T_s^{a\text{-}Si} < 250^{\circ}$C show that they have smooth surfaces. Films deposited at $T_s^{a\text{-}Si}$ 250, 300 and 400 °C exhibit a high density of Si precipitates (hillocks or islands) on the poly-Si surface. However, in the case of poly-Si prepared from a hydrogenated amorphous silicon precursor which the precipitates appear on the poly-Si films surface increases with the increase of P_{H2}. The presence of numerous Si islands on the poly-Si surface after AIC has been reported when the Al precursor is thicker than the amorphous silicon one, in the case of precursors of a-Si:H with high short range disorder, and of microcrystalline Si films deposited by PE CVD methods [5]. It is reasonable to suppose that the growth of the silicon hillocks in our case is a result of the better structural order in the a-Si:H precursors deposited at higher temperatures. The influence of the hydrogen concentration in a-Si:H should be taken in account as well, because it depends on the $T_s^{a\text{-}Si}$ and the hydrogen pressure during the deposition. The deposition conditions of the precursor layer define the balance between the rate of the crystallite growth and the conditions for preparation of poly-Si with smooth surface.

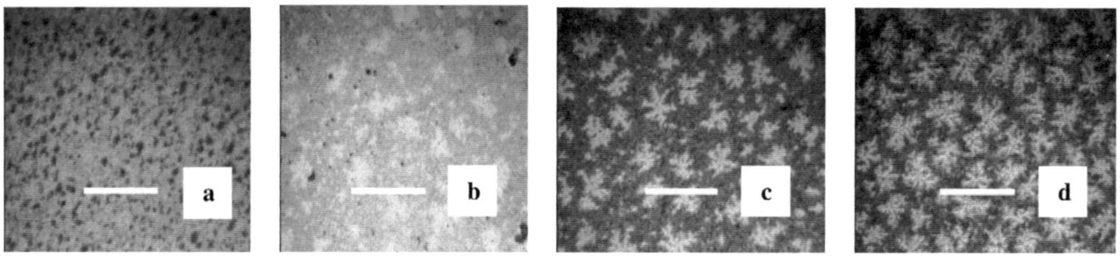

Figure 8. Optical micrograph images of poly-Si films using precursor layers of a-Si:H deposited with P_{H2}=0 Pa at different $T_s^{a\text{-}Si}$: RT (a), 250 °C (b), 300 °C (c) and 400 °C (d). The markers correspond to 100 μm.

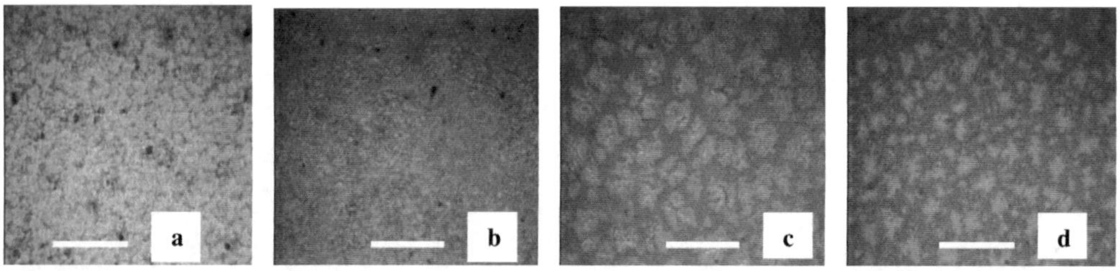

Figure 9. Optical micrograph images of poly-Si films using precursor layers of a-Si:H deposited with P_{H2}=0.1 Pa at different $T_s^{a\text{-}Si}$: RT (a), 300°C (b), 350°C (c) and 400°C (d). The markers correspond to 100 μm.

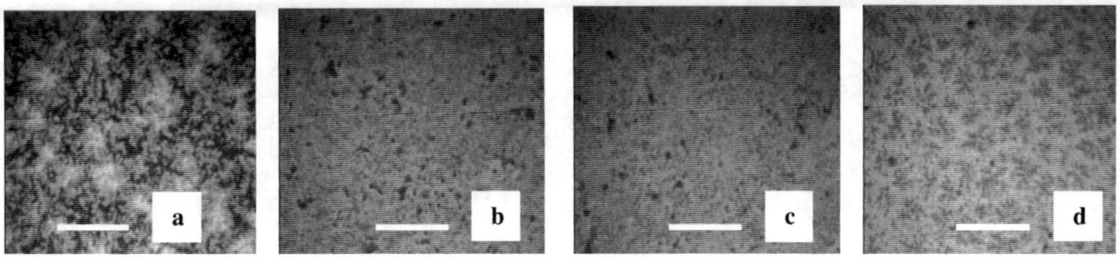

Figure 10. Optical micrograph images of poly-Si films using precursor layers of a-Si:H deposited with P_{H2}=0.2 Pa at different $T_s^{a\text{-}Si}$: RT (a), 300°C (b), 350° C (c) and 400°C (d). The markers correspond to 100 μm.

The UV hemispherical reflectance spectra of the poly-Si films prepared are presented in figure 11. For comparison, the corresponding spectra of a c-Si polished wafer measured under the same conditions are also shown. The two maxima in the spectrum of single-crystalline Si at 280 nm and 365 nm are caused by optical interband transitions at the X-point (band E_2) and along the Γ-L axis (band E_1) of the Brillouin zone, respectively [15, 16]. The deviation from the UV-R spectrum of bulk crystalline Si is related to the long range order deterioration or amorphization of the material [15]. The intensity of the E_2 band (280 nm) decreases and that of the E_1 (375 nm) increases with long range disorder according to the theory of long range order relaxation effects. Additionally, at short wave-lengths, in particular at 280 nm, the reflectance is largely determined by the high value of the absorption coefficient ($\alpha > 10^6$ cm^{-1}) corresponding to a penetration depth of less than 10 nm. Imperfect crystallinity in the near- surface region will cause a broadening and height reduction of this maximum [15, 16]. It is seen in figure 8a that for the poly-Si films obtained from a-Si deposited at $T_s > 250°$ C without hydrogen (P_{H2} = 0 Pa) the UV-R is very low in the whole photon range and smeared out maxima are observed. The maximal value of the UV-reflection and the best expressed maxima of E_1 and E_2 bands are observed for the case of using an a-Si:H precursor deposited at $T_s^{a\text{-}Si}$ = 300° C and P_{H2}=0.1 Pa; these values are the highest and closer to the reflection of the c-Si polished wafer – this sample has a better long range order and a better structure in the near surface region. This is in agreement with the XRD and Raman spectra and with the surface morphology observed in the optical light microscopy images.

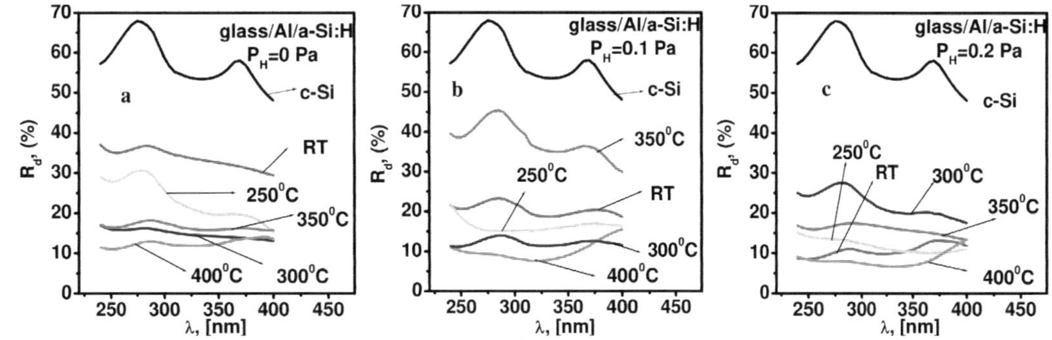

Figure 11. UV reflectance spectra of poly-Si films obtained by AIC from a-Si:H precursors deposited at different T_S and different H_2 partial pressure: 0 (a), 0.1 (b) and 0.2Pa (c).

3.3. Heterostuctures with participation of poly-Si films

The poly-Si films were used in the Al/poly-Si/n-ZnO/n-ZnO:Al thin film heterostructure. The a-Si:H film precursor contained 9 at.% hydrogen. The isotermal annealing of the glass/Al/a-Si:H structure was performed in forming gas (N_2+5%H_2) at 550°C for 21 h. The ZnO (300 nm thick) and ZnO:Al

(500 nm thick) polycrystalline films were deposited by rf sputtering in Ar atmosphere [17, 18] on the surface of a p-type poly-Si layer (100 nm thick). Al and ZnO:Al were used as electrodes. The scheme of the diode heterostructure is presented in figure 12. The I-V characteristics of the heterostructure measured in dark and under illumination by a halogen lamp (20 W) are shown in figure 13. The I-V curve of the p-poly-Si/n-ZnO thin film heterostructure exhibits rectifying behavior with ratio of about 5 and photosensitivity under illumination from the side of the ZnO:Al electrode.

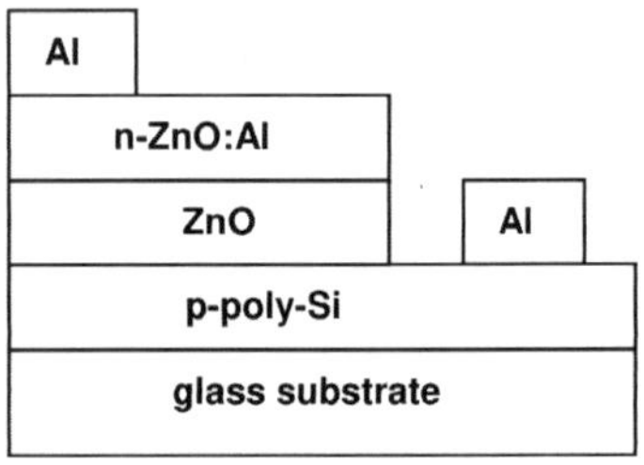

Figure 12. Structural scheme of the p-poly-Si/n-ZnO:Al heterostructure.

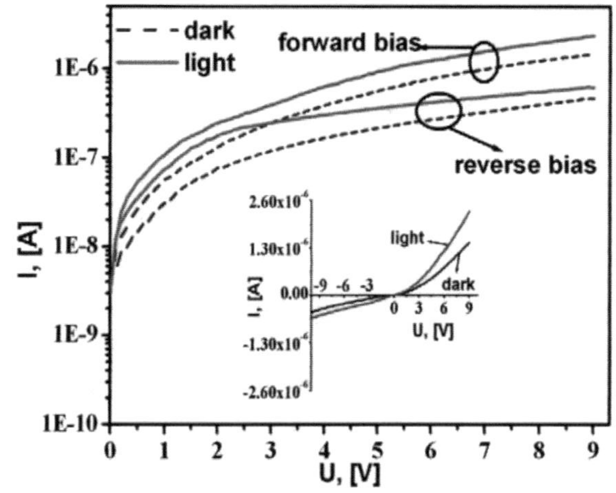

Figure 13. Room temperature I-V curves of p-poly-Si/n-ZnO heterostructure measured in dark and under illumination.

The diode turn-on voltage is about 0.2 V. The calculated ideality factor, n, was estimated to be about 2.95. The higher value of the ideality factor could be related to the presence of interface defect states between the ZnO and the poly-S films [19, 20]. The lattice mismatch between the poly-Si and the ZnO:Al thin films leads to formation of dislocation defects at the interface. The low electrical mobility in the polycrystalline ZnO film, and the lower donor density near the surface of the ZnO grains and near the poly-Si/n-ZnO interface may be other reasons. Additionally, the partial oxidation of poly-Si during the annealing and during the deposition of ZnO result in the creation of interface states with high density whose charge states dominate the potential distribution around the interface and influence the current transport across the junction. The photosensitivity is likely to be due to the ZnO.

4. Conclusion

The results demonstrate that the structure of poly-Si films improves when AIC is performed in an atmosphere containing H_2. It is assumed that H_2 stimulates the crystalline grain growth during the annealing by increasing the diffusion rate of Al and Si. The study shows that poly-Si films with larger grain sizes are obtained using a precursor of a-Si:H deposited at moderate $T_s^{a\text{-}Si}$ = 250-300°C and 0.05-0.2 Pa H_2 pressures in the sputtering chamber. The tensile stress in poly-Si films decreases with the increase of temperature of deposition of a-Si:H precursors. The differences in the structural properties of poly-Si films could be explained by different structural order and different content of H in the precursor a-Si (a-Si:H) layers deposited under different conditions. The presented results indicate that hydrogen has an important influence on the process of AIC. The poly-Si film obtained by AIC is applied to the diode poly-Si/ZnO heterostructure with rectifying behavior.

Acknowledgements
This work is funded by the Bulgarian National Scientific Fund under project X-1505.

References

[1] Gall S, Muske M, Sieber I, Nast O and Fuhs W 2002 *J. Non-Cryst. Solids* **299-302** 741-4
[2] Dimova-Malinovska D, Angelov O, Kamenova M, Sendova-Vassileva M and Vaseashta A 2003 *J. Mat. Sci: Mat. in Electronics* **14** 747-8
[3] Nast O and Hartmann A J 2000 *J. Appl. Phys.* **88** 716
[4] Schneider J, Klein J, Muske M, Gall S and Fuhs W 2004 *J. Non-Cryst. Solids* **338-340** 127
[5] Pihan E, Slaoui A, Roca I Cabarrocas P and Fosca A 2004 *Thin Solid Films* **451-452** 328
[6] Dimova-Malinovska D, Angelov O, Sendova-Vassileva M, Grigorov V and Kamenova M 2004 *Proc. 19-th EPVSEC* (Paris France) p 371
[7] Fauchet F M and Campbell I H 1988 *Crit. Rev. Solid State Mater. Sci.* **14** S79-S96
[8] Igbal Z and Vaprek S 1982 *J. Phys.* C **15**, 377-81
[9] Boultadakis S, Logothetidis S, Ves S and Kircher J 1993 *J. Appl. Phys.* **93** 914
[10] Cottereau P, Camplan J, Chaumonts J, Meunier R and Bernas H 1990 *NIM* B **45** 293
[11] Tzenov N, Tzolov M and Dimova-Malinovska D 1994 *Renewable Energy* **5** part III (UK: Elsevier Science Ltd Pergamon) 1685-7
[12] Langsfeld P and Nickel N 2002 *J. Non-Cryst. Solids* **299-302** 778
[13] Dimova-Malinovska D 2005 *J. Optoelectr. Adv. Mater.* **7** 99
[14] Grigorov V, Angelov O, Sendova-Vassileva M and Dimova-Malinovska D 2006 *Thin Solid Films* **511-512** 381
[15] Harbeke G and Jastrebski L 1990 *J. Electrochem. Soc.* **137** 696
[16] Greenaway D L and G. Harbeke G 1968 *Optical Properties and Band Structure of Semiconductors* (Oxford: Pergamon Press)
[17] Dimova-Malinovska D, Nichev H, Angelov O, Grigorov V and Kamenova K 2007 *J. Superlat. and Microstr.* **42** 123-8
[18] Dimova-Malinovska D, Nichev H, Georgieva V, Angelov O, Pivin J-C and Mikli V 2008 *Physica Status Solidi* (a) **205** 1993-7
[19] Dimova-Malinovska D and Nikolaeva M 2003 *Vacuum* **69** 227-31
[20] Ci-Hui L, Bing-Ce L and Zhu-Xi F 2008 *Chinese Physics* B **17**, 2292-6

Journal of Physics: Conference Series **223** (2010) 012014 doi:10.1088/1742-6596/223/1/012014

Sublimation epitaxy of cubic silicon carbide in vacuum

R Vasiliauskas[1,3], M Marinova[2], M Syväjärvi[1], A Mantzari[2], A Andreadou[2], E K Polychroniadis[2] and R Yakimova[1]

[1]Department of Physics, Chemistry and Biology, Linköping University, SE-581 83, Linköping, Sweden
[2]Department of Physics, Aristotle University of Thessalonica, GR54124, Thessaloniki, Greece

E-mail: remis@ifm.liu.se

Abstract. The epitaxial growth was compared of cubic silicon carbide on 6H-SiC substrates prepared in four different ways: (i) as received, (ii) re-polished, (iii) annealed and subsequently covered by a Si layer, (iv) with a (111) 3C-SiC buffer layer. The morphological details of the grown layers were studied by optical microscopy and their structure, by transmission electron microscopy. The influence of the substrate on the nucleation of 3C-SiC, on the homoepitaxial 6H-SiC nucleation before 3C-SiC and on the defect formation, primarily twinning, is discussed.

1. Introduction

Silicon carbide has long been considered a promissing semiconductor for high-temperature, high-power and high-frequency radiation-resistant device applications. The most common SiC polytypes (one-dimensional polymorphic modifications) are the two hexagonal ones, 4H-SiC and 6H-SiC, as well as the 3C-SiC, which is the only one exhibiting isotropic properties due to its cubic structure. Additionally, 3C-SiC is regarded advantageous over the other more hexagonal polytypes because it shows the highest electron mobility and lower density of the near-interface-traps in the SiO_2/3C-SiC system due to the smaller band gap. A crucial problem hindering the future development of a 3C-SiC technology is the unavailability of 3C-SiC substrates. Frequently, in order to overcome problems related to mismatch, as in the case of Si, the growth is performed on substrates from a hexagonal polytype (4H- or 6H-SiC). In this case, the 3C-SiC nucleates spontaneously on (0001) surfaces if the temperature is below 2000°C [1].

Efficient transport of the Si- and C-bearing species resulting in fast epitaxy is also obtained by implementing growth under vacuum at a short distance between the source and the substrate. Under such conditions, the growth rate is high (up to 1 mm/h for 3C-SiC) and the rate limiting mechanism in the growth configuration for sublimation epitaxy is the sublimation of the solid SiC source material. The use of a monolithic SiC plate gives more uniform sublimation than the traditional SiC powder. Earlier studies comparing epitaxy in vacuum and in argon atmosphere [2] found that the growth rate decreases by an order of magnitude at around 1–10 mbar argon pressure and the molecular vapors diffusion becomes the rate-determining step at higher vapor pressures.

[3] To whom any correspondence should be addressed.

Supersaturation is also dependant upon the medium (vacuum, argon, hydrogen or helium). Supersaturation certainly occurs since the Si/C ratio changes, as judged from the observation of step-bunching in sublimation epitaxy of 6H- and 4H-SiC [3].

The 3C-SiC is best formed under high Si/C ratio and high supersaturation conditions [4]; the best choice is, therefore, growth under vacuum conditions. The present paper presents investigation on the formation of 3C-SiC grown in vacuum on differently treated 6H-SiC substrates.

2. Experimental

The samples were grown in a vertical quartz tube reactor [2]. The cylindrical graphite growth cell consisted of a bottom piece and a lid mounted inside thermally insulating graphite foam heated by an inductive coil using an RF generator. Growth was performed in vacuum (pressure ~10^{-5} mbar). The source and the substrate were separated by a graphite spacer at a distance of 1 mm. A polycrystalline 3C-SiC plate was used as a source. When a temperature gradient is applied, with the temperature of the source being higher than that of the substrate, Si- and C- containing species are sublimed from the source and transported to the substrate where SiC nucleates.

The substrates used were subjected to different treatment prior to the growth experiment. Results were compared for (i) as received 6H-SiC substrates, (ii) 6H-SiC substrates re-polished by NovaSiC to minimize surface roughness; (iii) substrates annealed in a CVD reactor at 1450 °C for 10 min to reveal the growth steps and subsequently covered by a Si layer to protect the steps from the environment [5]; (iv) substrates with a 1.5 μm 3C-SiC buffer layer grown by means of the vapor-liquid-solid (VLS) method using a Si–Ge melt [6]. The entire set of growth experiments was performed on the Si-face of on-axis (0001) 6H-SiC substrates at identical conditions: growth temperature 1775°C, growth duration 30 min, and a temperature ramp of 5 K/min. The 3C- or 6H-SiC polytypes were identified by means of optical microscopy with Nomarski interference contrast, photoluminescence imaging at 77 K, and Fourier transform infrared spectroscopy. The structural quality of the layers was studied by transmission electron microscopy (TEM) in cross-section and plane-view configurations. The observations were performed on a JEOL 100X conventional TEM at acceleration voltage of 100 kV.

3. Results and discussion

In general, one can expect that the different surfaces would affect growth and nucleation in different ways. Rougher surfaces would act as nucleation sites for 6H-SiC instead of favoring formation of 3C-SiC, and smoother surfaces would favor formation of more 3C-SiC centers as these are usually created by two-dimensional island formation on smooth surfaces [7], i.e. would result in a larger percentage of 3C-SiC on the substrates. However, there was no significant difference between the layers grown on as-received and on re-polished substrates (which had smoother surfaces which should have favored 3C-SiC formation), the coverage of 3C-SiC compared to 6H-SiC was ~ 87 % in both cases. Thus, the surface smoothes does not influence noticeably the 3C-SiC nucleation. This observation suggests that during the temperature ramp up there is initial homoepitaxial 6H-SiC growth that creates perfectly on-axis regions (making large enough terraces for two dimensional nucleation instead of nucleation at steps which would replicate the 6H-SiC substrate), which is very important for the 3C-SiC formation [7,8].

Growth on annealed and silicon covered substrates showed less nucleation of 3C-SiC (~45 % only). The annealing is performed to create steps, which favor 3C-SiC growth during the VLS process [5]. In the VLS process, the 3C-SiC starts to nucleate at the terraces of these steps. However, in vapor phase growth, the steps enhance growth at the ledges and, thus, reproduction of the 6H-SiC substrate polytype, as proved by the lower percentage of 3C-SiC when using annealed substrates. Thus the growth mechanism in this case is step-controlled, and 3C-SiC is not likely to form heteroepitaxially on 6H-SiC.

On the other hand, almost pure 3C-SiC coverage was achieved on substrates with a VLS-grown 3C-SiC layer. About 99.9 % of the layer was 3C-SiC. The temperature ramp-up [9] is thus well matched with the 3C-SiC layer, and homoepitaxial growth occurs. The surface of the sublimation grown 3C-SiC layer was improved compared to the VLS.

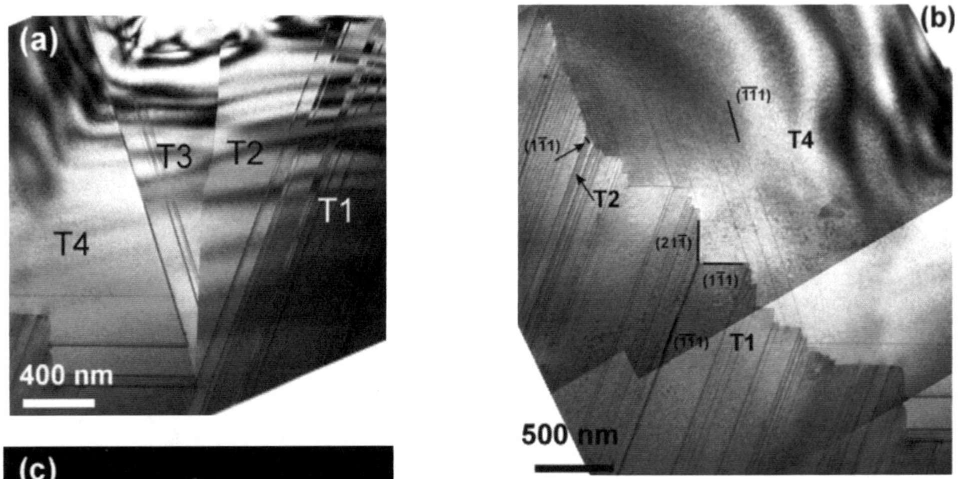

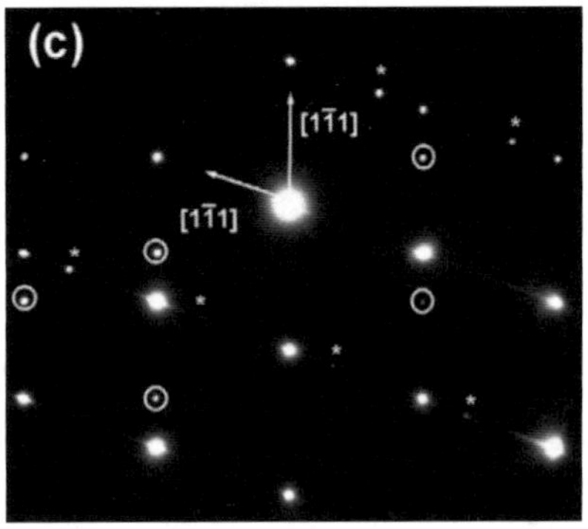

Figure 1. TEM micrographs of the layer grown on as-received 6H-SiC substrates. The images show the formation of (a) a fourfold twin complex, (b) interface between two twins with zigzag propagation where the facets follow the $(1\bar{1}1)$ and the $(21\bar{1})$ planes of the twins as deduced by the corresponding diffraction pattern in (c). The reflections due to the T2 and T4 twins are indicated by an asterisk * and a circle, respectively.

The microstructure and, consequently, the crystalline quality of the grown layers was investigated by TEM. The TEM micrographs for the growth (i) on as-received 6H-SiC substrates and (ii) on buffer layer of 3C-SiC, deposited prior the sublimation by VLS, are presented in figure 1 and figure 2, respectively. The following observations can be made: (i) when growth on 6H-SiC substrates was performed, the most important defects formed in the layers under investigation were twins. As it is shown in figure 1, multiple twinning appear along all $\{\bar{1}11\}$ planes. This leads to the formation of characteristic triangular defects as shown in figure 1a. They consist of four twins, denoted as T1-T4 in the figure and the twinning took place along the $\{\bar{1}11\}$ planes (T1/T2 and T3/T4 interfaces) and along the $(\bar{1}22)$ planes (T2/T3 interfaces). The $\{\bar{1}11\}$ twin interfaces were always observed to be coherent as clearly seen from the images. In addition, the multiple twinning lead to interfaces with zig-zag propagation, where facets along $(\bar{1}11)$ and $(\bar{2}11)$ planes are formed, which is seen in figure 1b and the corresponding diffraction pattern in 1c. The streaks in the diffraction pattern indicate the formation of SF or very thin slabs of micro-twins. In must be noted that the nucleation of the multiple fourfold complex, shown in figure 1a is not very well understood, but one can speculate by its appearance that it is predetermined by the surface pattern of the 6H-SiC substrate. In support of the latter suggestion we observed that (ii) in the layers grown when (111) 3C-SiC buffer was deposited before the sublimation epitaxy, the appearance of such multiple fourfold twins with perfectly coherent interfaces was not observed. In this case, due to the absence of selectivity from the seed, the forming twin interfaces were not coherent, which is shown in figure 2a. In addition, different domains misoriented about the $[01\bar{1}]$ zone axis were observed. This is clearly confirmed by the area which is out of contrast in figure 2a, denoted as A. For clarity, the appearance of such misoriented grains is also illustrated in figure 2b, where they are denoted by A1-A5. However, it is noteworthy that even if such defective areas were observed in the case of homoepitaxy, large areas with low defects density were also formed.

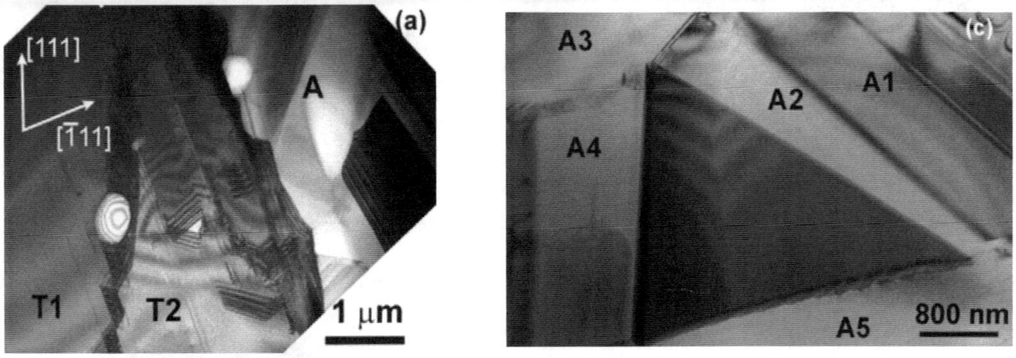

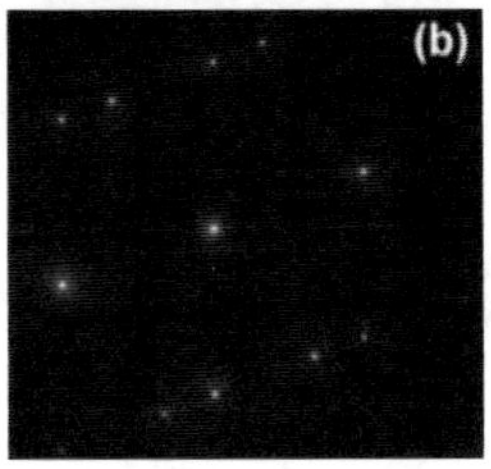

Figure 2. TEM micrographs of the layer grown on the VLS-grown (111) 3C-SiC seed. The images show the formation of (a) twins with incoherent interfaces along the $(1\bar{1}1)$ planes, confirmed by the diffraction pattern in (b). Image (c) shows the formation of grains misoriented by a few degrees about the $[01\bar{1}]$ axis denoted as A1-A5.

4. Conclusions

We found that almost 100% 3C-SiC could be obtained in homoepitaxial growth while conversion to 3C-SiC on 6H-SiC substrates was not fully possible. The initial nucleation in the latter case must be understood in detail regarding the two-dimensional nucleation, when 6H-SiC and 3C-SiC islands compete at the initial nucleation stage. Any steps present on the surface of the substrate will lower the formation probability of 3C-SiC.

The microstructural characterization showed the formation of coherent fourfold twins along all $\{\bar{1}11\}$ planes in the layers grown heteroepitaxially on 6H-SiC substrates. In the case of homoepitaxy, the multiple interfaces between the twins were incoherent and different domains were misoriented with respect to one another. However, it is noteworthy that even if such domains were formed, the defect density was quite inhomogeneous and large areas free of defects were seen.

Acknowledgement

This work was supported by FP6 Marie Curie Action - Research and Training Network - MANSiC - contract N°035735, Swedish Energy Agency and Ångpanneföreningen Research Foundation.

References

[1] Yoo W S and Matsunami H 1991 *J. Appl. Phys.* **10** 7124
[2] Syväjärvi M, Yakimova R, Tuominen M, Kakanakova-Georgieva A, MacMillan M F, Henry A, Wahab Q and Janzén E 1999 *J. Cryst. Growth* **197** 155
[3] Syväjärvi M, Ciechonski R R, Yazdi G R and Yakimova R 2005 *J. Cryst. Growth* **275** e1103–7
[4] Matsunami H and Kimoto T 1997 *Mat. Sci. Eng.* **R20** 125–60
[5] Soueidan M, Ferro G, Dazord J, Monteil Y and Younes G 2005 *J. Cryst. Growth* **275** e1011–6
[6] Soueidan M and Ferro G 2006 *Adv. Funct. Mater.* **16** 975
[7] Vasiliauskas R, Syväjärvi M, Beshkova M and Yakimova R 2009 *Mater. Sci. Forum* **615–617** 189–92
[8] Latu-Romain L, Chaussende D, Chaudouet P, Robaut F, Berthome G, Pons M and Madar R 2005 *J. Cryst. Growth* **275** 609
[9] Yakimova R, Yazdi G R, Sritirawisarn N and Syväjärvi M. 2006 *Mater. Sci. Forum* **527–529** 283

Journal of Physics: Conference Series **223** (2010) 012015 doi:10.1088/1742-6596/223/1/012015

Magnetron sputtering of polycrystalline LSMO/YBCO structures on sapphire substrates

B Blagoev[1,4], E Mateev[1], V Strbik[2], T Nurgaliev[1] and L S Uspenskaya[3]

[1]Emil Djakov Institute of Electronics, Bulgarian Academy of Sciences, 72 Tzarigradsko Chaussee, 1784 Sofia, Bulgaria
[2]Institute of Electrical Engineering, Slovak Academy of Sciences, Dubravska cesta 9, 84104 Bratislava, Slovak Republic
[3]Institute of Solid State Physics, Russian Academy of Sciences, 142432 Chernogolovka, Moscow Region, Russia

E-mail: bblagoev@ie.bas.bg

Abstract. $La_{0.7}Sr_{0.3}MnO_3$ (LSMO) films with different thickness were deposited directly on r-cut sapphire substrates by RF off-axis single magnetron sputtering with a view to produce a suitable buffer for growing HTS $YBa_2Cu_3O_{7-\delta}$ (YBCO) films on top. These LSMO layers were polycrystalline and showed a ferromagnetic behavior at room temperatures. The top YBCO films were sputtered in a DC off-axis double magnetron system. The polycrystalline structures thus produced were characterized and discussed.

1. Introduction

Interplay between superconductivity and ferromagnetism is an interesting and exciting phenomenon because of a variety exotic effects at the interface between these materials. Moreover, superconductor (SC) – ferromagnetic (FM) structures are of considerable practical importance. SC/FM layered structures could be used for tunable magnetic microwave devices [1–3], for dynamic modification of HTS critical parameters and for spin-polarized injection [4–7]. The oxide superconductor $YBa_2Cu_3O_{7-\delta}$ (YBCO) and the doped rare-earth manganese perovskite oxides (or manganites) $L_{1-x}A_xMnO_3$ (L = La, Pr,…; A = Ca, Sr, Pb,…) are promising candidates for preparing such structures.

The resistance of manganites decreases under an external magnetic field (CMR effect) due to double-exchange mechanism between 3*d*-electrons of Mn-ions together with a strong electron-phonon coupling (J-T polaron) [8–11]. The CMR effect gives possibility of using these materials in different microelectronic and spintronic applications [8, 9, 12–14]. The CMR effect is usually manifested at high magnetic fields (a few Tesla) in a narrow temperature range near the Curie temperature T_C, which limits the possibility for conventional applications [13, 15–17]. In ferromagnetic manganites, which possess grain-boundaries (polycrystalline materials or artificial grain boundaries), a significant magnetoresistance effect is observed at low magnetic fields (tens mT) and in a broad temperature range below T_C (the so-called low-field MR (LFMR) effect) [15–17]. The temperature of insulator-to- metal transition $T_{I\text{-}M}$ in single crystals is near T_C, whereas in polycrystalline material it has a much lower value [18–20].

[4] To whom any correspondence should be addressed.

The aim of this work is to grow manganite $La_{0.7}Sr_{0.3}MnO_3$ (LSMO) films on r-cut sapphire (Al_2O_3 or ALO) substrates by RF magnetron sputtering with a view to produce a suitable buffer for growing HTS $YBa_2Cu_3O_{7-\delta}$ (YBCO) films on top and double layer HTS/FM structures. It should be noted that ALO is characterized by a low dielectric constant $\varepsilon \sim 10$, low electric losses $tan\delta < 10^{-4}$, high heat conductivity and is considered as a very prospective substrate material for applications in nanoelectronics.

2. Experimental

The polycrystalline $La_{0.7}Sr_{0.3}MnO_3$ (LSMO) thin films were deposited by RF off-axis single magnetron sputtering on double-side polished 5×10×0.5 mm r-cut sapphire (Al_2O_3 or ALO) substrates supplied by Crystal GmbH. The sputtering process was carried out in Ar:O_2 (1:1) atmosphere at a total pressure of 5.3 Pa. The RF magnetron power and the substrate temperature were 30 W and 780°C, respectively. LSMO films with different thickness were prepared (10, 20, 40 nm – see table 1) by varying the deposition time. After the deposition procedure was finished, an in-situ annealing at 500°C of substrate temperature and 600 Torr of oxygen pressure for 30 min took place and the sample was cooled to room temperature (RT) (at rate ~ 15 °C/min). For one of the samples (sample 5) the annealing temperature was not held at 500°C but the sample was directly cooled to RT (table 1). Thus sample 5 was less oxygenated.

As a next step, the top YBCO films were sputtered by DC off-axis double magnetron system on one half of the surface of the LSMO-covered samples. All YBCO films were of 60 nm thickness except for the case of sample 4 where the thickness of the YBCO layer was 100 nm. The standard deposition conditions were used for in-situ growth of these YBCO films.

Table 1. Films deposition conditions.

sample	LSMO thickness, nm	YBCO thickness, nm	CHARACTERISTICS
1	40	60	Standard full oxidation of LSMO Standard YBCO thickness
2	10	60	Standard full oxidation of LSMO Standard YBCO thickness
3	20	60	Standard full oxidation of LSMO Standard YBCO thickness
4(!)	40	100	Standard full oxidation of LSMO **(!) Thicker YBCO**
5(!)	40	60	**(!) Less oxidation of LSMO** Standard YBCO thickness

The electrical resistance of the films was measured by a standard four-probe method from 300 K to 4 K. The presence of the ferromagnetic state in LSMO films was qualitatively confirmed by their response to the magnetic field. The magneto-optical visualization technique was used to investigate the magnetic inhomogeneities of LSMO samples.

3. Results and discussion

The LSMO films deposited on ALO substrates were polycrystalline and consisted of very small (possibly of nanometer scale) grains. These films were in the FM state at room temperature, which was confirmed by suspended samples being attracted by a permanent magnet. Moreover, an image of the edge area of the sample, obtained by the magneto-optic visualization technique (not shown here), demonstrated a brightness contrast due to the FM state of the sample. It is well known that magnetooptic (MO) visualization reveals the perpendicular component of the magnetic induction created by the magnetic defects and heterogeneity, FM domains and domain walls in the film. For example, several types of domain structures were observed by MO visualization in our epitaxial LSMO films grown on $LaAlO_3$ substrates [21]. It should be noted that MO visualization did not allow us to observe clearly the domain structure in LSMO films grown on ALO substrates. In our opinion, this is due to the spatial resolution of our MO method, which is not sufficiently high to resolve magnetic patterns on a nanometer scale, as the polycrystalline LSMO films deposited on ALO consist of nanometer-scale grains and the magnetic inhomogeneities in them could be expected to be on the nanometer scale, too.

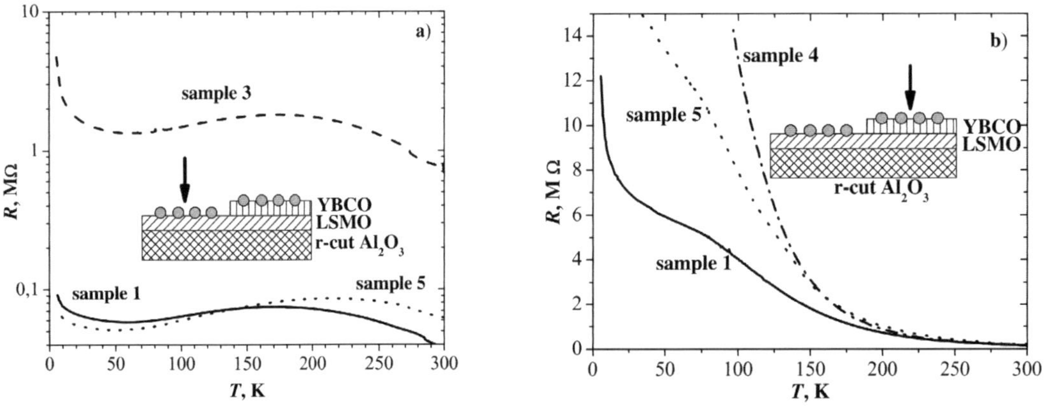

Figure 1. Electrical resistance of polycrystalline LSMO/YBCO structures: a) resistance of polycrystalline LSMO films on Al_2O_3 substrates; and b) resistance of the top polycrystalline YBCO films deposited on different LSMO underlayers. The contacts positions for the 4-point resistance measurement are shown by closed circles in the insets. The arrows show the areas measured.

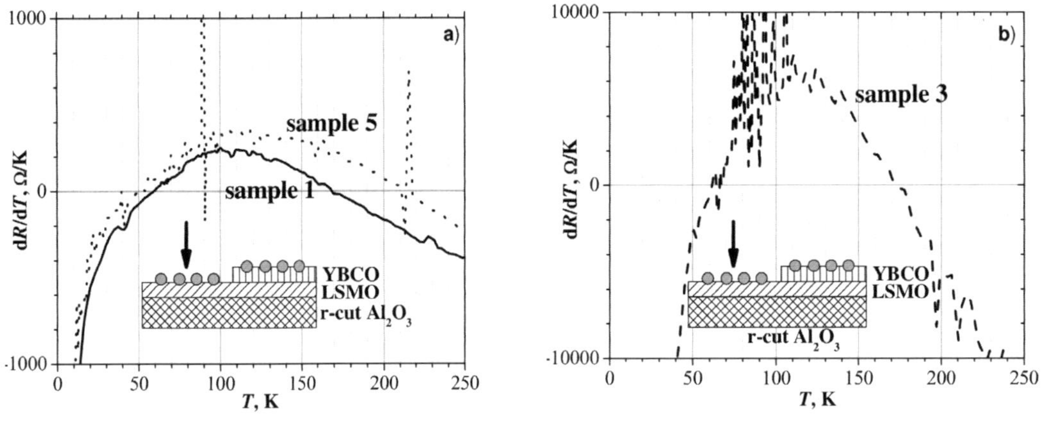

Figure 2. First derivative of the resistance of polycrystalline LSMO films with a) 40 nm and b) 20 nm thickness.

The resistance versus temperature (R *vs* T) dependences of these polycrystalline LSMO films are shown in figure 1a. These curves are typical for polycrystalline manganite films [18–20, 22]. In the high-temperature region, the R *vs* T dependence has a semiconductor-like behavior. At lower temperatures, an insulator-to-metal transition occurs and the resistance acquires a metallic character. Finally, an upturn of the low-temperature resistance is observed. The resistance is relatively high (in the MΩ range) which is in agreement with the results of other authors [22]. Higher values of the resistance (figure 1a) are observed for thinner films (< 40 nm). The resistance of the LSMO film with a thickness of 10 nm was too high for our measurement system and is not shown here. We assume that the grain-size of our films decreases as the film thickness is decreased in analogy with [23]. The films with a smaller grain-size possess a larger number of grain boundaries (which are a physical barrier for charge carriers) and thus demonstrate a higher resistance.

To locate the extrema of *R vs T* dependences (which correspond to the metal-insulator (M-I) peak and the upturn point of the resistance) more precisely, we calculated the first derivatives of the resistance d*R*/d*T* (figure 2). The M-I peak of sample 1 (standard deposition of 40 nm LSMO) is observed at $T_{I\text{-}M}$ =170 K and the upturn of the resistance, at 60 K (figure 2a). The resistance curve of sample 3 (standard deposition of 20 nm LSMO) shows a high level of "noise" and provokes a "noisy" derivative (figure 2b). $T_{I\text{-}M}$ of sample 3 is approximately 170 K and the upturn minimum is observed at about 65 K. Films with the same thickness, but different degree of oxidation, possess different metal-insulator transition $T_{I\text{-}M}$. It is interesting to note that the polycrystalline LSMO films with less oxidation show a higher $T_{I\text{-}M}$. The resistance peak and minimum of sample 5 (40 nm LSMO, less oxidation) are 210 K and 50 K, respectively (figure 2a).

The temperature dependences of the samples resistance following deposition of top YBCO layers are shown in figure 1b. All top YBCO layers were polycrystalline and not superconducting. Actually, the resistance in figure 1b is the effective resistance (R_{eff}) due to the combined action of the FM and HTS layers. The resistances of FM and HTS layers (R_{FM} and R_{HTS}, respectively) are connected in parallel: $\frac{1}{R_{FM}}+\frac{1}{R_{HTS}}=\frac{1}{R_{eff}}$. The effective resistance R_{eff} of the FM/HTS structures is highly nonlinear and has an insulating character. This rise in the resistance could be due to interfacial diffusion.

When obtaining thin film heterostructures, the interfacial properties are of crucial importance. In our polycrystalline heterostructures there exist two interfaces – ALO-LSMO and LSMO-YBCO. Polycrystalline LSMO (and $La_{1-x}Ca_xMnO_3$ or LCMO) films deposited on different substrate (Al_2O_3 and Si) were investigated in [14, 18, 19, 22, 24, 25]. The magnetic and transport properties of polycrystalline LSMO (LCMO) thick films (> 100 nm) depend more strongly on the interfaces between grains than on the substrate-manganite interface. The interfaces between grains are correlated with the deposition conditions and film thickness. Recently, S. N. Kale and co-workers [26] reported on the influence of strain on the structural and surface morphology of LSMO thin films formed on Si, Al_2O_3 (ALO) and $LaAlO_3$ (LAO) substrates. The strain induced in all the systems depends essentially on the lattice misfits and on the deposition conditions as well. If the lattice misfit is high, the polycrystalline growth with a large number of grain boundaries and smaller grain size is seen to be dominating. The distortions and strain in the films on ALO are the highest and mostly due to a structural as well as dimensional mismatch. The overall growth-mode for the films on either substrate can be considered as Volmer-Weber type in which the energy relaxation takes place via the formation of 3D islands to minimize the energy. The electronic structure and magnetism of a LSMO interface buried under $YBa_2Cu_3O_7$ capping layers of various thicknesses ≤ 8 nm was investigated by Stadler et al (1999, 2000) using X-ray magnetic circular dichroism (XMCD) and X-ray absorption spectroscopy (XAS) at the Mn L2,3 edge. A comparison of the data with spectra measured on a $La_{1-x}Sr_xMnO_3$ series (Abbate et al 1992) led to the conclusion that the cation stoichiometry at the interface is changed. A YBCO overlayer facilitates a cation displacement or interchange whose net effect is the out-diffusion of La ions: interchange of La with Ba ions or by exchange of Sr from deeper in the bulk with La at the interface. The XMCD signal is consistent with a progressive replacement of the ferromagnetic order by an antiferromagnetic structure [27, 28] which could lead to a rise in the resistance as well.

The reason why the polycrystalline YBCO layers in the polycrystalline FM/HTS structures are not superconducting could be the following. The polycrystalline bottom layer (LSMO) is very thin. Therefore, it consists of small grains and of many grain boundaries. The behavior of the resistance in polycrystalline manganites can be explained as being the result of the presence of different conduction channels. One type of channels is related to the intrinsic transport properties of the system and includes the grains with good intergrain contacts. The other type of channels includes the grains with "bad" contacts and reflects the effects of energy barriers. These inhibit metallic conduction at all temperatures due to poor connectivity and to effects of disorder and contamination in the structure [19].

The high resistance of our polycrystalline LSMO films is probably due to the presence of a great number of channels with poor connectivity between the grains, i.e., in our opinion, our LSMO layers possess poor conducting microareas: grain boundaries and physical barriers between the grains. The physical barriers appear because the grains are presumably not closely packed. The surfaces of the grains are contaminated and the Mn environment at the surface is not the same as that inside the grains [19]. Such poor conducting areas are amorphous and the superconducting YBCO crystalline phase can not be grown on them. To minimize the poorly conducting areas and to decrease the resistance of polycrystalline LSMO films on ALO substrates, we need further optimization of the film thickness (figure 1b) and of the annealing regime. A more accurate optimization of the technology, based on structural and electrical analyses of the samples, is necessary for producing polycrystalline LSMO/YBCO thin film structures with superconducting YBCO layers.

4. Conclusion

In summary, LSMO polycrystalline films and LSMO/YBCO polycrystalline bilayer structures were obtained on r-cut Al_2O_3 substrates by magnetron sputtering. The LSMO polycrystalline films were ferromagnetic, but they showed large values of the resistance, especially in the case of thinner films. The top YBCO layers were not superconducting because of the presence of contaminated poorly conducting areas on the LSMO surface. These areas are amorphous and thus hinder the epitaxial growth of the superconducting YBCO phase on them.

Acknowledgements

This work was supported by the National Science Fund at Ministry of Education and Science of Republic of Bulgaria under project F1503/05 and by the Slovak Grant Agency for Science (VEGA, No. 2/7125/27).

References

[1] Terakago M, Mine S, Sakatani T, Hontsu S, Nishikawa H, Nakamori M, Tabata H and Kawai T 2001 *Supercond. Sci. Technol.* **14** 1140–3
[2] Nurgaliev T, Miteva S, Jenkins A and Dew-Hughes D 2001 *IEEE Trans. Appl. Supercond.* **11** 446–9
[3] Nurgaliev T, Miteva S, Jenkins A and Dew-Hughes D 2003 *IEEE Trans. Microwave Theory and Technique* **51** 33–40
[4] Chen Z Y, Biswas A, Žutić I, Wu T, Ogale S B, Greene R L and Venkatesan T 2001 *Phys. Rev.* B **63** 212508-1–4
[5] Schoop U, Schonecke M, Thienhaus S, Herbstritt F, Klein J, Alff L and Gross R 2001 *Physica* C **350** 237
[6] Alverez G A, Wang X L, Puzzer T, Peleckis G and Dou S X 2009 *J. Appl. Phys.* **105** 07E326
[7] Wanchoo S K, John J, Bagwe V C, Pai S P, Narsale A M and Pinto R 2009 *J. Phys.: Conf. Series* **153** 012033
[8] Dörr K 2006 *J. Phys. D: Appl. Phys.* **39** R125–50
[9] Haghiri-Gosnet A-M and Renard J-P 2003 *J. Phys. D: Appl. Phys.* **36** R127–50
[10] Carbone G 2004 Structural and magnetic studies of strained thin films of $La_{2/3}Ca_{1/3}MnO_3$ *PhD Thesis* (Stuttgart Univ, Stuttgart, Germany)
[11] Nagaev E L 2001 *Elsevier Physics Reports* **346** 387–531
[12] Katsu H, Tanaka H and Kawai T 2000 *Appl. Phys. Lett.* **76** 3245–7
[13] Mathur N D, Burnell G, Isaac S P, Jackson T J, Teo B-S, MacManus-Driscoll J L, Cohen L F, Evetts J E and Blamire M G 1997 *Nature* **387**266–8
[14] Balcells L l, Carrillo A E, Martínez B, Sandiumenge F and Fontcuberta J 2000 *JMMM* **221** 224–30
[15] Gross R, Alff L, Büchner B, Freitag B H, Höfener C, Klein J, Lu Y, Mader W, Philipp J B, Rao M S R, Reutler P, Ritter S, Thienhaus S, Uhlenbruck S and Wiedenhorst B 2000 *JMMM* **211** 150–9
[16] Hwang H Y, Cheong S-W, Ong N P and Batlogg B 1996 *Phys. Rev.* B **77** 2041

[17] Paranjape M, Mitra J and Raychaudhuri A K 2003 *Phys. Rev.* B **68** 144409
[18] Ren Guang Ming, Yuan Song Liu and Tian Zhao Ming 2009 *Thin Solid Films* **517** 3748–51
[19] de Andrés A, García-Hernández M and Martínez J L 1999 *Phys. Rev.* B **60** 7328–34
[20] Yuan S L, Xia Z C, Liu L, Chen W, Zhao L F, Tang J, Zhang G H, Zhang L J, Cao H, Feng W, Tian Y, Niu L Y and Liu S 2003 *Phys. Rev.* B **68** 184423
[21] Uspenskaya L S, Nurgaliev T and Miteva S 2009 *J. Phys.* D **42** 185006
[22] García-Hernández M, Guinea F, de Andrés A, Martínez J L, Prieto C and Vázquez L 2000 *Phys. Rev.* B **61** 9549–52
[23] Zhao Y G, Cai W, Wu X S, Zhang X P, Wang K, Gao S N, Lu L and Zhu J 2005 *Appl. Phys.* A **81** 607–10
[24] Liu S M, Zhu X B, Yang J, Zhao B C, Sheng Z G, Song W H, Dai J M and Sun Y P 2004 *Elsevier Physica* B **353** 238–41
[25] Chen Yen-Hua, Ho Shu-Te and Wu Tai-Bor 2006 *Electrochemical and Solid-State Letters* **9** J27–30
[26] Kale S N, Mona J, Ganesan V, Choudhary R J and Phase D M 2009 Substrate dependent morphologies of self-assembled nanocrystalline manganite films: an atomic force microscopy study *Preprint Available at:* *http://publications.ictp.it*
[27] Ziese M 2001 *arXiv:cond-mat/0111263vl [cond-mat.str-el]*
[28] Ogale Satishchandra B 2005 *Thin films and heterostructures for oxide electronics* (New York: Springer Science) Chapter 5 pp 153–92

Journal of Physics: Conference Series **223** (2010) 012016 doi:10.1088/1742-6596/223/1/012016

Dilute GaAsN and GaInAsN grown by liquid phase epitaxy

M Milanova[1], G Koleva[1], R Kakanakov[1], P K Vitanov[2,5], Z Alexieva[2], E A Goranova[2], B Arnaudov[3], S Evtimova[3], C Barthou[4] and B Clerjaud[4]

[1]Central Laboratory of Applied Physics, Bulgarian Academy of Sciences, 59 St. Petersburg Blvd, 4000 Plovdiv, Bulgaria
[2]Central Laboratory of Solar Energy and New Energy Sources, Bulgarian Academy of Sciences, 72 Tzarigradsko Chaussee, 1784 Sofia, Bulgaria
[3]Faculty of Physics, St. Kl. Ohridski University of Sofia, 5 J. Bourchier Blvd, 1164 Sofia, Bulgaria
[4]Université Pierre et Marie Curie, Institut des NanoSciences de Paris, rue de Lourmel 140, 75015 Paris, France

E-mail: vitanov@phys.bas.bg

Abstract. Dilute III-nitrides, such as GaAsN and GaInAsN, are of considerable current interest both from a fundamental point of view and for applications in solar cells, GaAs-based long-wavelength photodetectors and diode lasers. The addition of nitrogen leads to material properties that deviate strongly from those expected for conventional III–V solid solutions.

The possibility was investigated to use liquid phase epitaxy to incorporate nitrogen in epitaxial GaAsN/GaAs and GaInAsN/GaAs heterostructures. The structures were grown from Ga- and Ga-In- melts containing powder GaN as a nitrogen source. The initial growth temperature was varied in the range 560°C - 660°C. The low temperature growth favors nitrogen incorporation in the epilayers. The optical transmission and photoluminescence spectra of a set of structures grown at different temperatures were studied showing ternary and quaternary dilute nitride solid solutions with nitrogen content about 0.2 at.%. The photoluminescence spectra show emission from localized nitrogen states as well.

1. Introduction

The III–V–N semiconductor solid solutions have attracted a great deal of attention because of their interesting physical properties and potential applications in optoelectronic devices and multi-junction solar cells. One technologically important feature of GaInAsN quaternary solid solutions is that they can have a perfect lattice match to a GaAs or InP substrate with an appropriate ratio of indium-to-nitrogen fractions, thus avoiding the growth problems encountered in strained layers, such as the ternary GaAsN and GaPN semiconductors.

Technologically, the incorporation probability of nitrogen in GaAs is very small and strongly depends on the growth conditions. GaAsN-based solid solutions and heterostructures are first grown by metaloorganic vapor-phase epitaxy (MOVPE) [1] and molecular-beam epitaxy (MBE) [2]. Only one group, Dhar et al. [3], has reported liquid-phase epitaxy (LPE) growth of these compounds. The advantages of the LPE are its simplicity, low cost and ability to produce materials of high quality.

[5] To whom any correspondence should be addressed.

In this paper, we show the possibility to use liquid phase epitaxy to incorporate nitrogen in epitaxial GaAsN/GaAs and GaInAsN/GaAs heterostructures and study the optical properties of GaAsN and GaInAsN LPE layers grown at different temperatures.

2. Experimental procedure

Epitaxial layers of GaAsN and GaInAsN, 0.2-0.8μm thick, were grown in a conventional horizontal sliding boat LPE reactor on (100) oriented n-type GaAs:Si substrates from Ga- or mixed Ga-In melt. Polycrystalline GaAs and powder GaN were used as an As and N sources, respectively. The N content in the melt was 1.5 at.% for all layers studied. Epitaxial layers were deposited at different initial temperatures in the range 560-660 °C at a cooling rate of 0.5 °C/min. The technological conditions of the as-grown samples are summarized in table 1. The layers were not subsequently annealed.

Table 1.Growth conditions of the samples investigated.

Sample	66 GaAs	68 GaAsN	70 GaAsN	72 GaAsN	74 GaInAsN	75 GaInAsN
T	660 °C	660 °C	595 °C	560 °C	570 °C	660 °C

The composition dependence of the effective band-gap of the layers was investigated by optical transmission spectroscopy. Transmission spectra were measured in the spectral range 850-2000 nm by a UV-3600 Shimadzu double-beam UV-VIS spectrophotometer. In order to increase the signal-to-noise ratio, the substrates of GaAsN/GaAs and GaInAsN/GaAs heterojunction samples as well as the GaAs reference sample were thinned down to 90 μm.

The CW photoluminescence (PL) spectra under excitation of a Coherent 488 nm 12 W Argon laser were measured at 5 K. They were analyzed by a Jobin-Yvon Spectrometer HR460 and a multichannel CCD detector (2000 pixel).

3. Results and discussion

3.1. Absorption spectra discussion

Figure 1 a), b) shows the transmission spectra of several GaAsN and GaInAsN layers respectively, compared with the reference GaAs sample 66.

It is seen that the absorption edge for all N-containing samples is red-shifted compared to the GaAs one. The red shift for the samples investigated increases when the growth temperature decreases. This

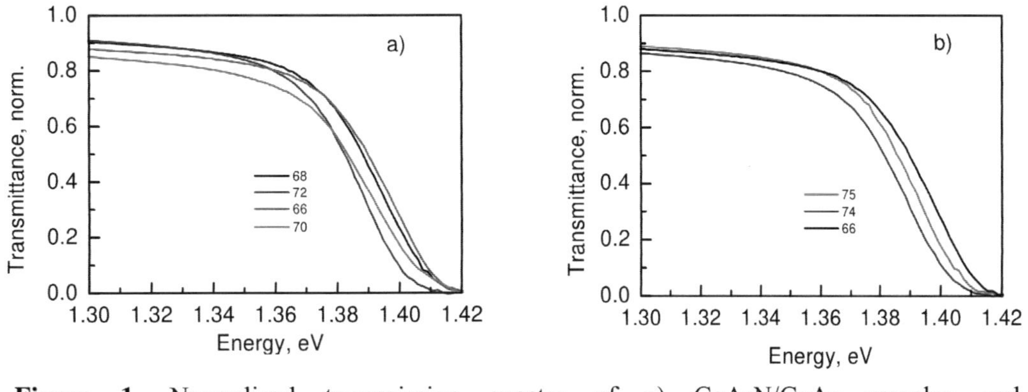

Figure 1. Normalized transmission spectra of a) GaAsN/GaAs samples and b) GaInAsN/GaAs samples.

Journal of Physics: Conference Series **223** (2010) 012016 doi:10.1088/1742-6596/223/1/012016

fact is in accordance with the effect of the growth temperature on the N concentration in GaAsN reported by Toivonen et al. [4], thus confirming the incorporation of nitrogen under our growth conditions. According to Tisch et al. [5], the red shift of the absorption spectra observed in figure 1 corresponds to a nitrogen content in the solid solution epitaxial layers near or less than 0.2 at.%.

3.2. PL spectra discussion

Figure 2 shows representative PL spectra for the samples listed in table 1.

First, narrow excitonic-like emission bands near 1.52 eV are recorded in the spectra of all N-containing layers and the reference GaAs layer, as well as at 1.49 eV for sample E75. Such narrow bands are typical for intentionally undoped GaAs of relatively good quality and are observed in similar N-containing GaAs-based structures, as published by Dhar et al.[3], Makimoto et al.[6] and Zhang et al. [7]. These bands have been explained, e.g. by Toivonen et al. [4], as due to free-to-impurity recombination.

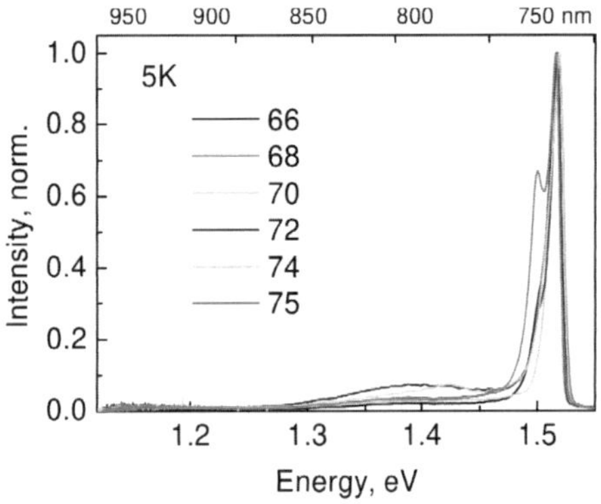

Figure 2. Low temperature PL spectra of all samples.

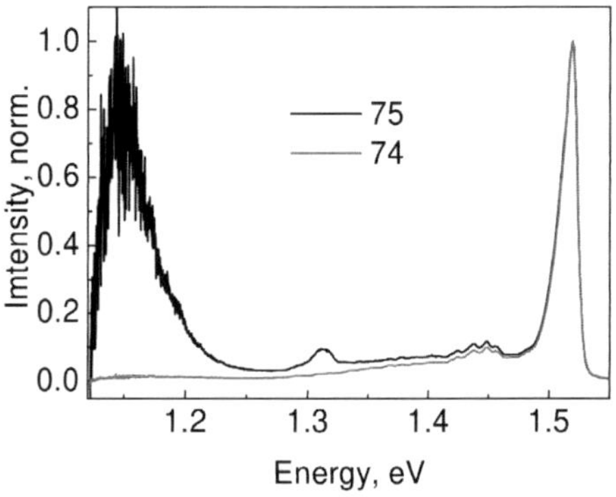

Figure 3. PL spectra of GaInAsN/GaAs samples.

A broad emission band at 1.3 - 1.5 eV is observed in the spectrum of the N-containing sample 72. The same band but with relatively lower intensity is observed in the spectra of the other N-containing samples 68-75 also shown in figure 2. This emission band is the only "individual" announcement of the N-content in the epitaxial layers studied and is an additional evidence for N presence in ternary or quaternary solid solutions, in accordance with the above discussed absorption spectra. Very weakly, near the noise level, this emission band is observable in the spectrum of the reference sample 66. For sample 70 the intensity of the 1.3 - 1.5 eV emission band is comparable with the intensity in the reference sample 66.

Figure 3 shows detailed emission spectra of the quaternary GaInAsN layers. We observe a set of 4 or 5 relatively narrow emission bands in the range 1.42 - 1.48 eV, shown in figure 3 for sample 74 and 75. These "peaks" grow up on the high-energy slope of the above discussed broad 1.3 – 1.5 eV emission band and support the assumption that the added In increases the optical activity of the nitrogen in the GaAs-based epilayers. Similar sets of 5 peaks in the same spectral region are reported by Makimoto et al.[6] and Zhang et al. [7];they are assigned - simultaneously with the peaks at 1.518 eV - to an excitonic activity of localized nitrogen states - a free exciton at 1.518 - 1.514 eV and an exciton bound on nitrogen as an isoelectronic impurity and on N-N pairs [6]. These excitonic lines are manifested at N-concentration in the layers up to 1.5×10^{19} cm^{-3}. They disappear over this limit and are displaced by the wider emission band at 1.425 eV, which is assumed to belong to the GaAsN solid solution. In our case we observe the localized nitrogen states luminescence simultaneously with the "solid solution-related" 1.3 – 1.5 eV emission band.

Bands at 1.32 eV are observed in the emission spectra of GaInAsN layers - figure 3 black line. Similar emission bands are reported by Sprutte et al. [8] in QWs of GaInAsN and by Wang et al. [9] in

GaAsSbN, associated with potential fluctuations and N-clusters of more than two N atoms. The lowest-energy emission band observed in the GaInAs layer is at 1.15 eV - sample 74, figure 3. Similar bands are reported also by Toivonen et al. [4] and by Fan et al. [10] for QW structures.

4. Conclusion

Dilute nitride - GaAsN and GaInAsN - 0.2 – 0.8 μm thick layers are grown on GaAs substrates by low-temperature LPE using powder GaN as a nitrogen source. The heterostructures are investigated by optical transmission and photoluminescence spectroscopy. The absorption edge for all N-containing samples is red shifted compared to the GaAs one corresponding to nearly 0.2 at.% nitrogen in the solid solution. The PL spectra at 5 K show – except the GaAs-related excitonic emission bands – several emission bands at 1.3 - 1.5 eV and at 1.15 eV, attached to the ternary and quaternary solid solutions, respectively. A set of 4 or 5 relatively narrow emission bands in the range 1.42 - 1.48 eV in the PL spectra of GaInAsN layers is an evidence that the added In increases the optical activity of the nitrogen in the GaAs-based layers and that the nitrogen acts in localized states as well.

Acknowledgments

This work was supported by the National Sciences Fund of the Ministry of Education and Science of Bulgaria under Grant D 03 369/06 and INTAS-ref. N 061000017-8536.

References

[1] Kurtz S, Reedy R, Barber G D, Geisz J F, Friedman D J, McMahon W E and Olson J M 2002 *J. Crystal Growth* **234** 318
[2] Auvray L, Dumont H, Dazord J, Monteil Y, Bouix J, Bru-Chevallier C and Grenouillet L 2002 *Materials Science in Semiconductor Processing* **3** 505
[3] Dhar S, Halder N, Mondal A and Bansal B 2005 *Semicond. Sci. Technol.* **20** 1168
[4] Toivonen J, Hakkarainen T, Sopanen M and Lipsanen H 2000 *J. Cryst. Growth* **221** 456
[5] Tisch U, Finkman E and Salzman J 2002 *Appl. Phys. Lett.* **81** 463
[6] Makimoto T, Saito H, Nishida T and Kobaiashi N 1997 *Appl. Phys. Lett.* **70** 2984
[7] Zhang Y, Mascarenhas A, Geisz J F, Xin H P and Tu C W 2009 *Phys. Rev.* B **63** 085205
[8] Sprutte S G, Coldren C W, Harris J S, Wampler W, Krispin P, Ploog K and Larson M 2001 *J. Appl. Phys.* **89** 4401
[9] Wang W J, Su F H, Ding K, Li G H, Yoon S F, Fan W J, Wisaksono S and Ma B S 2006 *Phys. Rev.* B **74** 195201
[10] Fan W J, Yoon S F, Cheah W K, Loke W K, Ng T K, Wang S Z, Liu R and Wee A 2004 *J. Cryst. Growth* **268** 470

Thermal solar absorber made of diamond-like carbon thin films

S S Tinchev, P I Nikolova and Y T Dyulgerska[1]

Emil Djakov Institute of Electronics,
Bulgarian Academy of Sciences, 72, Tzarigradsko Chaussee, 1784 Sofia, Bulgaria

E-mail: petranka@ie.bas.bg

Abstract. Solar photo-thermal collectors were prepared from pure diamond-like carbon (DLC) thin films on aluminum substrates. By only changing the deposition conditions (different bias voltage) of the DC PECVD process, coatings were fabricated consisting of five sublayers with the necessary optical parameters. As expected, absorption better then 0.9 was achieved already for submicrometer absorber thickness. Accelerated thermal ageing tests performed at 200°C and 300°C in air showed that the reflectance changes only slightly. Long operating life can, therefore, be expected of such absorbers.

1. Introduction

Solar absorbers are the most important part of solar collectors. An ideal solar absorber should be spectrally selective, i.e. it should absorb the solar radiation in the whole solar spectrum (350 ÷ 2500 nm) and should have a minimal thermal emission at ~ 8 μm for low temperature solar collectors [1, 2]. Existing absorbers in solar collectors for photo-thermal conversion are usually produced by different chemical and electrochemical processes [3]. A typical example is the so-called "black chrome" [4], which has become one of the most widely used solar absorber material.

More recently, the interest has risen in physical vapor deposited coatings (mostly by sputtering) [5]. These processes avoid wastes and are highly reproducible. Such coatings consist of different materials, e.g. aluminum oxide, titanium or chromium oxynitride [6]. Among them, metal containing amorphous hydrogenated carbon is very promising [7]. These coatings are in fact nanostructured materials containing clusters with dimensions on the nanometer length scale. By changing the metal content (usually chromium [8]) during deposition, one can change the optical properties of the coating. Such solar absorbers based on a multilayer of a-C:H/Cr nanocomposites are produced on a large scale by the "Ikarus coatings", Germany [9]. An advantage of the a-C:H/Cr is the absence of grain boundaries, which may be susceptible to corrosion. Therefore, for such coatings a life time of more then 25 years is expected, although oxidation of the metal particles could decrease the absorber life time. Deposition of such a-C:H/Cr coatings, however, is a complicated process, whilechromium as a heavy metal is environmentally harmful and should be avoided.

Recently, we succeeded in fabricating a DLC solar absorber made without metal inclusions [10]. In our coatings, the necessary variation of the optical properties is achieved by simply changing the bias voltage during DC PECVD deposition. A typical design of our solar absorber is shown in figure 1. It consists of 5 sublayers deposited on an aluminum substrate at 2.5 kV; 2 kV; 1.5 kV; 1 kV and 0.5 kV.

[1] To whom any correspondence should be addressed.

The first sublayer is softer then the others, with small mechanical stress and good adhesion to the substrate. The following sublayers are deposited with decreasing bias voltages, which make them more optically transparent and harder. The top sublayer is hard and plays also the role of a protective coating of he whole absorber.

The necessary thickness of such an absorber has been estimated earlier [11]. Using the measured absorption coefficient and assuming ~ 95% absorption, we have found that the absorber film thickness should be at least 0.6 μm. In this work, these conclusions were proved experimentally and an accelerated ageing test was carried out, which predicts long life behavior of our absorbers.

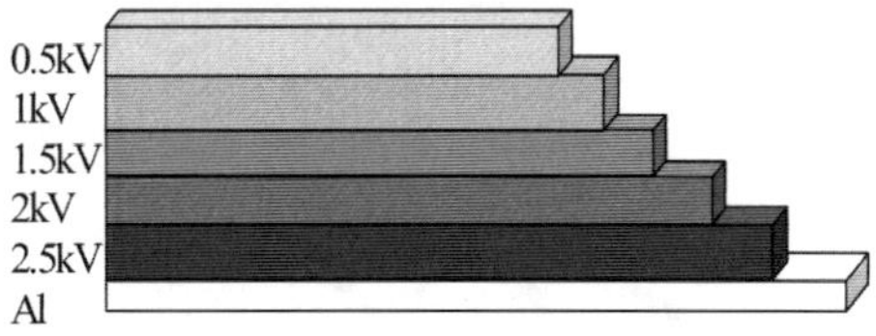

Figure 1. Schematic representation of a thermal solar absorber.

2. Experimental

Thin films of diamond-like carbon were deposited in DC a PECVD reactor [10]. Benzene vapors diluted with argon were chosen as a precursor. Both silicon and aluminum substrates were used for the experiments. Aluminum was chosen as a substrate material because of its good adhesion to the DLC and its high thermal conductivity. The DLC films deposited on aluminum are very smooth with roughness about of 0.5 nm as the AFM image shows (figure 2).

To reduce further the reflectance from the absorber surface, rough Al substrates were also used. The roughness of such a substrate was about 4 μm, just as needed to obtain small, broadband and omnidirectional reflection from the absorber surface in the visible optical range.

Three sets of absorbers were fabricated with 6 min, 12 min and 18 min deposition time of each of the 5 sublayers. The total absorber thickness, measured by a surface profilometer, were found to be $d_1 = 0.88$ μm; $d_2 = 1.76$ μm and $d_3 = 2.41$ μm, respectively. The reflectance R(λ) of the complete absorber was measured by a Shimadzu 3600 UV-VIS-NIR spectrometer equipped with an integrating sphere in the wavelength range from 350 nm to to 2500 nm to estimate the total solar absorption α .

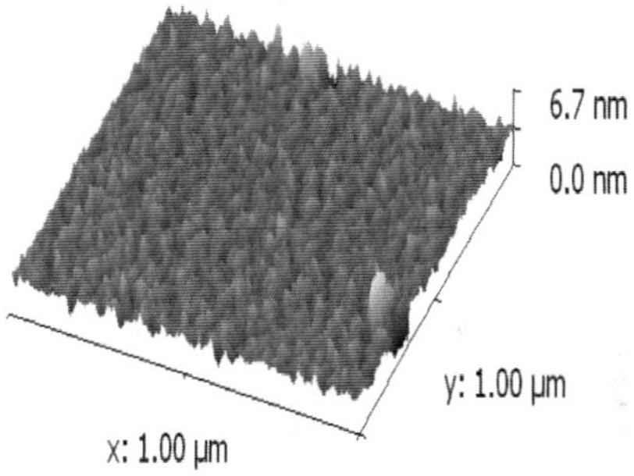

Figure 2. Atomic force micrograph of absorber with thickness d_1= 0.88 μm on Al substrate.

In order to simulate accelerated ageing [12] the absorbers were subjected to high-temperature treatment at 200°C for 10 h and at 300°C for 10 h in air. After the absorbers were exposed to these temperatures, the reflectance was measured again and compared with the values prior to the thermal treatment.

The solar absorption α of the samples before and after thermal ageing treatment was calculated using the relation:

$$\alpha = \frac{\int_{0.39\mu m}^{2\mu m} [W(\lambda)(1 - R(\lambda))] d\lambda}{\int_{0.39\mu m}^{2\mu m} W(\lambda) d\lambda},$$

where *W(λ)* is the solar radiation intensity at different wavelength for air mass 1.5 (ISO9845-1:1992).

The degradation of the diamond-like carbon coatings on silicon substrates after annealing at 300°C was monitored also by Raman spectroscopy with a laser operating at wavelength of 633 nm.

3. Results

Figure 3 shows the reflectance spectra of the 3 fabricated absorbers. All have spectrally selective properties: reflection below 10 % at solar maximum radiation and increasing reflectance toward 2500 µm. Interference pattern are observed in all samples, whose position depends obviously on the film thickness. The absorber 1 (d_1) with 5 sublayers deposited at 6 min each shows a minimal reflectance of 3.7 % at 780 nm. As expected, the three absorbers (d_1, d_2, d_3) have sufficient absorptance: $\alpha_1 = 0.9150$, $\alpha_2 = 0.9163$ and $\alpha_3 = 0.9292$.

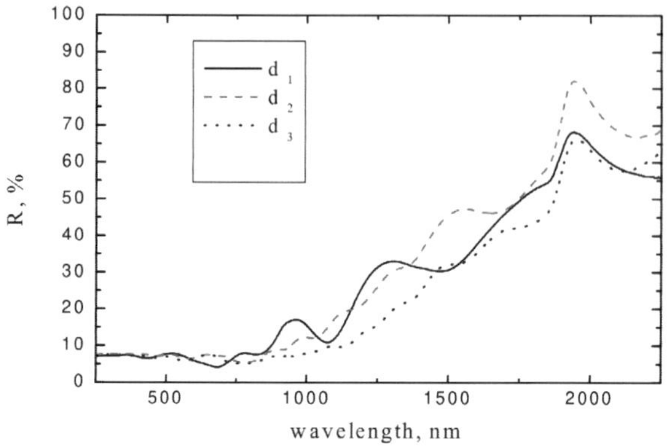

Figure 3. Total reflectance of the three mulilayered absorber coatings with different thicknesses before tempering.

Figure 4 shows the reflectance spectra changes after the heat treatment at 200°C for 10 h – figure 4a and at 300°C for 10 h – figure 4b. No significant changes can be seen in the spectra. Slight shifts of the interference pattern were only observed connected probably with small films thickness changes due to carbon oxidation.

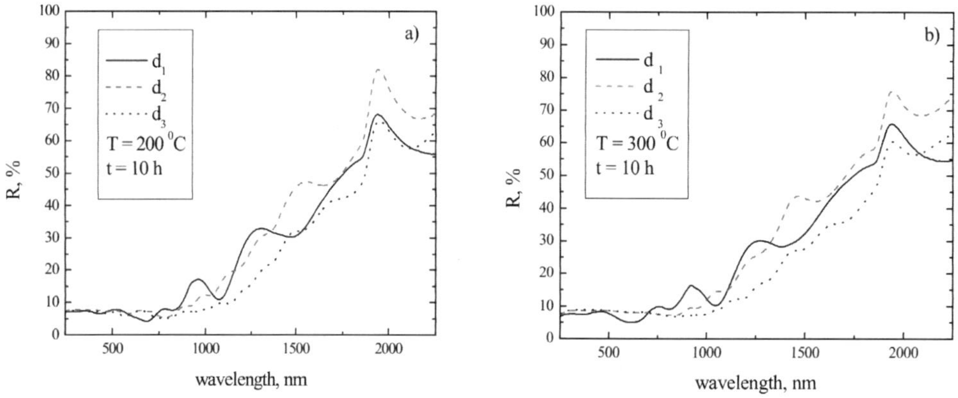

Figure 4. Reflectance spectra after heat treatment at: a) 200 °C for 10 h; b) 300 °C for 10 h.

As seen in figure 5, the positions of the reflectance maxima differ slightly, i.e. there is no considerable temperature influence on the optical thickness.

A possible degradation of the coatings caused by the accelerated ageing was monitored by Raman spectroscopy [13]. Figure 6 shows the Raman spectra of the tree absorbers after annealing at 300°C. The spectra were almost identical with two broad maxima: the so-called G-peak at about 1580 cm^{-1} and D-peak at about 1350 cm^{-1}. These peaks are well known to originate from sp^2-bonded graphite-like regions and from disordered carbon, respectively. It is also well known that by analysing their position and width one can determine the size of the sp^2-bonded regions [14]. The narrow peaks, as well the increased D-peak of our samples are indication for the existence of graphite-like regions in the absorbers.

In our coatings we did not observe any shift in the position of both peaks from their typical positions. These Raman spectra, however, should be interpreted with caution, because the samples are not homogeneous in depth and some modification of the samples by the laser beam power was noticed during the Raman measurements. The modification is caused due to the high absorption of the samples at the laser wavelength of 633 nm, which falls in the middle of the solar spectrum. However, this is the best demonstration of the excellent properties of our solar absorbers. In single films one cannot observe any modification because their absorption is not so high.

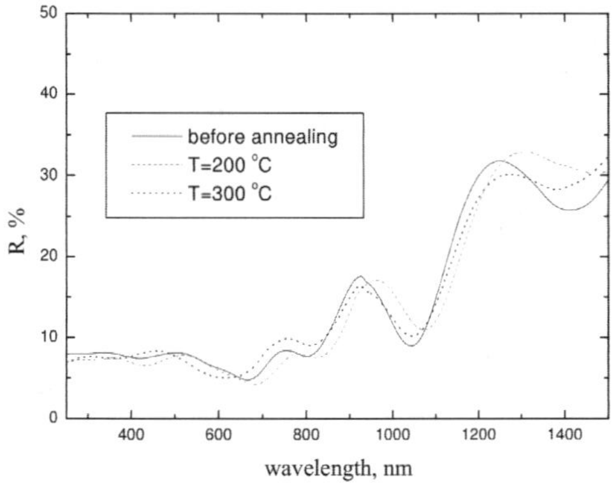

Figure 5. Comparative representation of the reflectance spectra before and after thermal treatment of the absorber with thickness d_1.

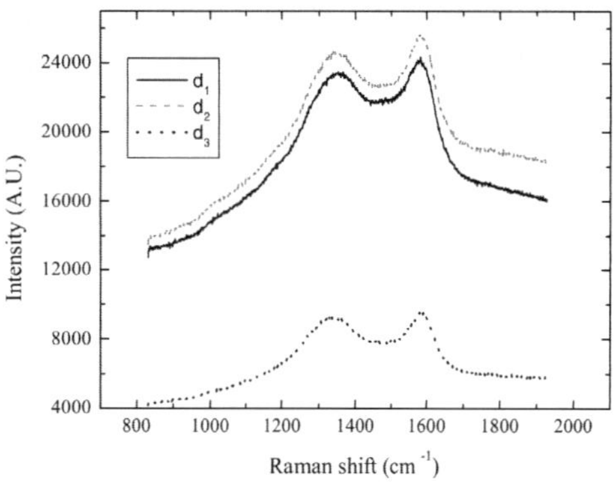

Figure 6. Raman spectra of the absorbers after annealing at 300˚C.

The calculated absorptions of all tree absorbers are presented in the table 1. There is only a small difference between absorbers with different thickness, although absorber 2 is twice as thick, and absorber 3 is three times as thick as absorber 1. This confirms our estimation that submicrometer thickness is already sufficient to achieve the necessary absorption. In all cases absorption better then 90 % was achieved. No significant changes were observed in all samples after accelerated ageing treatment at 200°C and 300°C.

Table 1. Solar absorption of 3 mulilayered absorber coatings with different thicknesses.

Absorption α	before annealing	after T = 200°C	after T = 300°C
α_1 (d_1)	0.9150	0.9123	0.9132
α_2 (d_2)	0.9163	0.9223	0.9211
α_3 (d_3)	0.9291	0.9156	0.9183

4. Discussion

In the experiments described, no substantial difference was found in the absorption of the three sets of samples, although the thickness of absorber 2 exceeds that of absorber 1 by factor of 2, and that of absorber 3, by a factor of three. The thinnest one (0.88 μm) has already the necessary absorption and should be preferred because of the lower emittance, a (probably) better adhesion and the lower cost. In addition, only small changes were found in α_1 after a high-temperature test (table 1). Furthermore, the interference maxima in reflection measured after the annealing (figure 5) show small shifts only, indicating that the optical thickness of the sample is not decreased considerably.

5. Conclusions

In conclusion we would like to point out that we succeeded in fabricating a solar absorber made of diamond-like carbon thin films without metal inclusions consisting of 5 sublayers deposited on an aluminum substrate with a total thickness of 0.88 μm. Accelerated thermal ageing performed at 200°C and 300°C in air showed that there are not substantial differences in the absorption and the structure of the DLC layers. Therefore long-life behavior of our absorber should be expected.

Acknowledgements

The work was supported by the Bulgarian National Science Fund under Project NNP E-04-01.

References

[1] Bogaerts W F and Lampert C M 1983 *J. Mater. Sci.* **18** 2847
[2] Granqvist C G and Wittwer V 1998 *Sol. Energy Mater. Sol. Cells* **54** 39
[3] Bostrom T, Westin G and Wackelgard E 2007 *Sol. Energy Mater. Sol. Cells* **91** 38
[4] Lee T K, Kim D H and Auh P C 1993 *Sol. Energy Mater. Sol. Cells* **29** 149
[5] Nunes C, Teixeira V, Collares-Pereira M, Monteiro A, Roman E and Martin-Gago J 2002 *Vacuum* **67** 623
[6] Barshilia H, Selvakumar N, Rajam K and Biswas A 2008 *Sol. Energy Mater. Sol. Cells* **92** 1425
[7] Gampp R, Oelhafen P, Gantenbein P, Brunold S and Frei U 1998 *Sol. Energy Mater. and Sol. Cells* **54** 369
[8] Milde F, Dimer M, Hecht Ch, Schulze D and Gantenbein P 2000 *Vacuum* **59** 825
[9] Oelhafen P and Schuler A 2005 *Solar Energy* **79** 110
[10] Tinchev S, Nikolova P, Dyulgerska Y, Danev G and Babeva Tz 2005 *Sol. Energy Mater. Sol. Cells* **86** 421
[11] Nikolova P, Tinchev S and Alexieva Z 2009 *J. Optoelectr. Adv. Mater. - Symposia* **1** 327
[12] Carlsson B, Moller K, Kohl M, Frei U and Brunold S 2000 *Sol. Energy Mater. Sol. Cells* **61** 255
[13] Baek S, Parks S-J, Lee K-R and Seok C-S 2008 *Mater. Sci. Engin.* A **483–484** 245
[14] Ferrari A and Robertson J 2000 *Phys. Rev.* B **61** 14095

16th International Summer School on Vacuum, Electron, and Ion Technologies (VEIT 2009) IOP Publishing
Journal of Physics: Conference Series **223** (2010) 012018 doi:10.1088/1742-6596/223/1/012018

Morphology and gas sensing properties of as-deposited and thermally treated doped thin SnO_x layers

B Georgieva[1,3], D Nihtianova[2], J Pirov[1] and I Podolesheva[1]

[1]Acad. J. Malinowski Central Laboratory of Photoprocesses,
Bulgarian Academy of Sciences, Acad. G. Bonchev Str., Bl.109, 1113 Sofia, Bulgaria
[2]Central Laboratory of Mineralogy and Crystallography,
Bulgarian Academy of Sciences, Acad. G. Bonchev Str., Bl.107, 1113 Sofia, Bulgaria

E-mail: biliana@clf.bas.bg

Abstract. Thin layers intended for gas sensors are prepared by vacuum co-evaporation of TeO_2 and Sn. The as-deposited layers consist of a nanosized oxide matrix and finely dispersed dopants (Te, Sn, TeO_2 or SnTe, depending on the atomic ratio $R_{Sn/Te}$). In order to improve the characteristics of the layers they are additionally doped with platinum. The gas sensing properties are strongly dependent on the atomic ratio $R_{Sn/Te}$, as well as on the structure, composition and surface morphology. The as-deposited layers with $R_{Sn/Te}$ 0.8 are highly sensitive humidity sensors working at room temperature. Thermally treated Pt-doped layers with $R_{Sn/Te}$ 2.3 are promising as ethanol sensors.

With the aim of obtaining more detailed knowledge about the surface morphology, structure and composition of layers sensitive to different environments, various techniques - TEM, SAED, SEM, EDS in SEM and white light interferometry (WLI), are applied. It is shown that all layers with $1.0 > R_{Sn/Te} > 2$, as-deposited and thermally treated, exhibit a columnar structure and a very smooth surface along with the nanograined matrix. The thermal treatment causes changes in the structure and composition of the layers. The ethanol-sensitive layers consist of nanosized polycrystalline phases of SnO_2, Sn_2O_3, Sn_3O_4 and TeO_2. This knowledge could help us understand better the behaviour and govern the characteristics of layers obtained by co-evaporation of Sn and TeO_2.

1. Introduction

Resistive gas sensors based of metal oxide semiconductors, such as SnO_2, have been widely studied during the last decades mostly due to their high sensitivity, fast response, small size, low cost and low power consumption [1]. To resolve the problem with their low selectivity, doping with different additives, such as noble metals or other oxides, have been used [2]. Since the morphological and electrical parameters of the sensors are strongly affected by the preparation conditions, different methods have been investigated for their preparation with the aim to improve further their characteristics.

We developed a method for preparing thin SnO_2 layers by thermal vacuum co-evaporation of Sn and TeO_2 [3]. During the co-deposition, both substances react and nanograined amorphous SnO_2 layers doped with a dispersed phase of Te, Sn, TeO_2 or SnTe (depending on the atomic ratio $R_{Sn/Te}$) are formed. Thermal treatment causes changes in the structure, surface morphology and composition. As we have

[3] To whom any correspondence should be addressed.

already reported, at $R_{Sn/Te}$ 0.8, the as-deposited layers possess good humidity sensing properties at room temperature [4]. At higher $R_{Sn/Te}$ values (2.3), and after thermal treatment and Pt-doping, the layers show high sensitivity towards ethanol vapors at substrate temperatures 120 and 240 °C [5].

The aim of our work was to use different techniques (WLI, TEM, SAED, SEM and EDS in SEM) to obtain more detailed information about the surface morphology, structure and composition of the humidity- and ethanol-sensitive layers, as well as about the changes the Pt-doped layers with $R_{Sn/Te}$ 2.3 undergo upon thermal treatment which makes them sensitive to ethanol. This knowledge could help us manage the characteristics of layers obtained by co-evaporation of Sn and TeO_2.

2. Experimental

Tin oxide layers were deposited under vacuum (10^{4} Pa) on stationary substrates using thermal vacuum co-evaporation of Sn and TeO_2 from independent Knudsen-type cells. During the co-evaporation, the condensation rate of each substance was controlled by a quartz oscillator. The additional doping element (Pt) was deposited on the sensing layers in a separate vacuum cycle. The atomic ratio $R_{Sn/Te}$ of the humidity sensitive layers was ≈0.8, and that of the ethanol-sensitive, ≈ 2.3. Thermal treatment was carried out up to 360 °C with a step of 40 °C with a 15-min delay at each temperature, before or after doping with platinum.

The layer thickness was controlled by a Talystep profilometer (Rank Taylor Hobson), and the $R_{Sn/Te}$, by EDS in SEM. Transmission electron microscopy (bright and dark field) and SAED were used to observe the morphology and structure of the layers and SEM, to study their growth profile. The samples intended for SEM were evaporated on glass substrates to a thickness of about 800 nm. Samples suitable for direct imaging in TEM were prepared on Cu grids by removing the layers from the glass substrate using vapor-phase hydrogen fluoride (HF) treatment and floating off in water. WLI (White Light Interferometric Profiler MicroXAM S/N 8038) was used to determine the surface roughness. The resolution of the interferometer was 1–2 nm in vertical (z) and 1-3 µm in lateral (x, y) direction. For this investigation, the as-deposited amorphous ($R_{Sn/Te}$ ≈ 0.8) and the thermally treated ($R_{Sn/Te}$ ≈ 2.3) Pt-doped layers were prepared on glass substrates.

To test the layers as gas sensors, Cr comb-like electrodes were patterned using a photolithographic technique [6]. The layers were exposed to humidity and ethanol vapors in a test chamber equipped with temperature- and RH-controllers. The electrical resistance R [Ohm] of undoped and Pt-doped samples was measured at room temperature as a function of the relative humidity (RH [%]) or as a function of the substrate temperature in the presence of ethanol vapours. The apparatus used was a multichannel ohmmeter (consisting of a resistor-connector-unit and a multichannel analog-to-digital converter PCI 6024E), product of National Instruments. The data acquisition and processing, as well as the control of the ohmmeter were computerized, using Lab View software.

3. Results and Discussion

3.1. Structure and composition

Figure 1 shows the profile of as-deposited (a) and thermally treated (b) layers with thickness d = 800 nm and $R_{Sn/Te}$ ≈ 2.3. As it could be seen, the amorphous as-deposited layer exhibits a columnar structure. After thermal treatment, the shape of the columns changes, obviously due to structural transformations of the layer. Along with the nanograined matrix, this columnar structure indicates a large free surface which is important for the gas sensing properties. Analogous results have been obtained earlier for layers with $R_{Sn/Te} < 1$.

The surface profile morphology of as-deposited (a) and thermally treated Pt-doped (b and c) layers is shown in figure 2. The root-mean-square (rms) values for as-deposited amorphous layers with $R_{Sn/Te}$ 0.8 and for layers with $R_{Sn/Te}$ 2.3, doped with Pt before or after the thermal treatment respectively, are given in the same figure. Regardless of the difference between the rms-values of layers doped with Pt before and after thermal treatment, the surfaces of both as-deposited amorphous and thermally treated samples are generally very smooth.

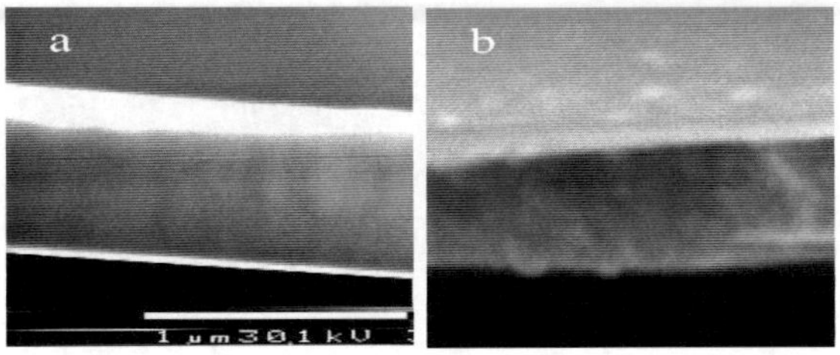

Figure 1. SEM images of the growth profile of as-deposited (a) and thermally treated (b) layers with $R_{Sn/Te}$ 2.3 and d = 800 nm.

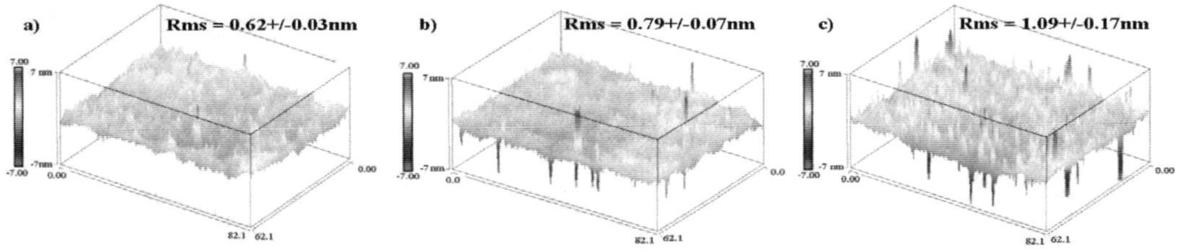

Figure 2. Surface profile obtained using a WLI surface profiler of: a) as-deposited amorphous layers with $R_{Sn/Te}$ 0.8; b) layers with $R_{Sn/Te}$ 2.3, Pt-doped before thermal treatment and c) layers with $R_{Sn/Te}$ 2.3, Pt-doped after thermal treatment.

The TEM micrographs of thermally treated SnO_2 layers with $R_{Sn/Te}$ 2.3, Pt-doped before thermal treatment, are shown in figure 3. It can be clearly seen that upon heating the layers become crystalline and consist of a mixture of Sn oxides and TeO_2. It should be noted that the crystalline layers do not show sensitivity to humidity; this is why we do not comment the changes layers with $R_{Sn/Te} < 1$ undergo upon thermal treatment.

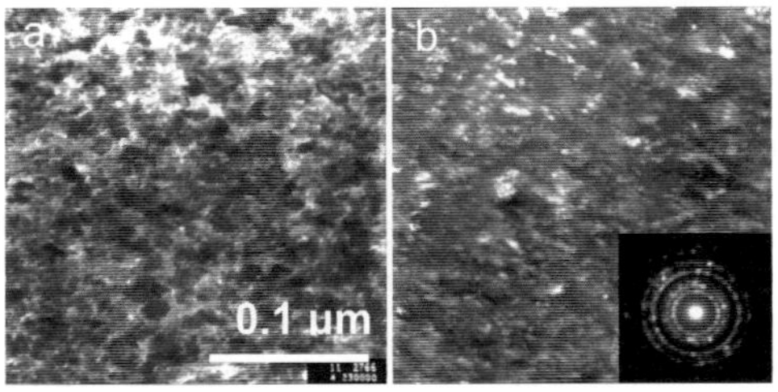

Phases:

TeO_2	Paratellurit	PDF 42-1365
SnO_2	Cassiterite	PDF 41-1445
Sn_2O_3		PDF 25-1259
Sn_3O_4		PDF 20-1293

Figure 3. TEM images: a- bright field; b- dark field and SAED pattern of a layer with $R_{Sn/Te}$ 2.3, Pt-doped before thermal treatment.

3.2. Gas sensing properties

3.2.1. Humidity sensing properties. In our previous investigations [4], we have shown that the amorphous layers with $R_{Sn/Te} \approx 0.8$ are very sensitive to the ambient relative humidity at room temperature. The response is very fast, the recovery period, very short and no hysteresis effects are observed. Accelerated ageing tests have shown that the layers exhibit a good long-term stability. However, at low humidity levels the resistance of the sensing layers is too high. One of the ways to overcome this problem is additional doping, e.g. with some metals (in our case with Pt). The effect of 2-3 at.% Pt on the response curve is shown in figure 4. As it could be seen, doping with platinum leads to a decrease of the resistance with one order of magnitude compared with the undoped layers. Therefore, a substantial resistance decrease, without any changes in sensitivity, could be attained through optimal doping.

3.2.2. Ethanol sensing properties. Without Pt-doping, the crystalline layers with $R_{Sn/Te} \approx 2.3$ show no sensitivity to ethanol, while introduction of Pt makes them sensitive to ethanol vapors. The sensitivity depends on the layer preparation conditions, on the substrate temperature, etc. Since, however, these results will be the object of another publication they will not be discussed further here.

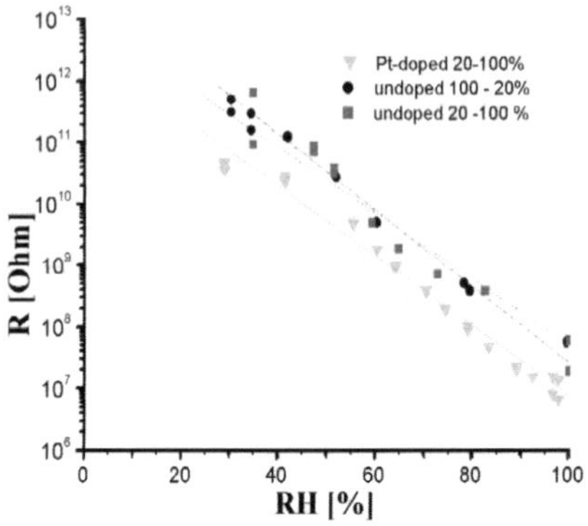

Figure 4. R [Ohm] as a function of RH [%] of amorphous as-deposited layers undoped and doped with Pt.

4. Conclusions

Using a method developed in our laboratory, doped SnO_2 thin layers were synthesized with desired composition and structure which govern the sensing properties. The as-deposited amorphous layers with $R_{Sn/Te} \approx 0.8$ are very sensitive to ambient humidity at room temperature but at low humidity levels their resistance is too high. It was shown that this problem could be overcome through Pt-doping. The study of the morphology and composition by various techniques revealed that, along with the nanograined matrix, the layers exhibit a columnar structure indicating a large free surface and that the surface of all layers is very smooth. Furthermore, it was shown that the thermal treatment of the layers with $R_{Sn/Te} \approx 2.3$ leads to the formation of nanocrystalline phases of SnO_2, Sn_2O_3, Sn_3O_4 and TeO_2. These layers, additionally doped with Pt, possess good sensing properties to ethanol.

As the layer preparation method is compatible with the conventional microelectronics technologies, after optimisation the sensors can be integrated on one chip with field-effect transistors or with entire subsystems for data processing.

Acknowledgements

The authors express their appreciation to "Centre for Industrial and Engineering Optics and FOCAS, DIT", Dublin, Ireland, for the surface measurements and to Dr. Ts. Babeva for the fruitful discussions and comments. This work was supported by the Bulgarian Ministry of Education and Science under Contract DO 02-148/16.12.2008.

References

[1] Yamazoe N, Sakai G and Shimanoe K 2003 *Catalyst Surveys from Asia* **7** 63
[2] Rumyanceva M, Safonova O, Boulova M, Ryabova L and Gashkov A 2003 *Russian Chem. Bulletin, International Edition* **52** 1217
[3] Podolesheva I, Dimov V, Tarassov M, Platikanova V and Konstantinov I 1998 *J. Vac. Sci. Technol.* A **16** 2608
[4] Georgieva B, Podolesheva I, Pirov J and Platikanova V 2005 *J. Optoelectr. Adv. Mater.* **7** 2595
[5] Georgieva B, Podolesheva I and Spasov G 2010 *Solid State Phen.* **159** 133
[6] Mednikarov B and Sahatchieva M 2005 *J. Optoelectr. Adv. Mater.* **7** 1415

16th International Summer School on Vacuum, Electron, and Ion Technologies (VEIT 2009) IOP Publishing
Journal of Physics: Conference Series **223** (2010) 012019 doi:10.1088/1742-6596/223/1/012019

Nanolaminated TiN/Mo_2N hard multilayer coatings

I N Martev[1,4], D A Dechev[1,2], N P Ivanov[1,2], Ts D Uzunov[2] and E P Kashchieva[3]

[1]Emil Djakov Institute of Electronics,
Bulgarian Academy of Sciences, 72 Tzarigradsko Chaussee, 1784 Sofia, Bulgaria
[2]Faculty of Engineering and Education, Technical University of Sofia,
59 Bourgasko Chaussee, 8800 Sliven, Bulgaria
[3]Department of Physics, University of Chemical Technology and Metallurgy,
8 Kl. Ohridski Blvd., 1756 Sofia, Bulgaria

E-mail: inmartev@ie.bas.bg

Abstract. The paper presents results on the synthesis of hard multilayer coatings consisting of titanium nitride and molybdenum nitride thin films with thickness of several nm. The TiN and Mo_2N films were successively deposited by reactive DC magnetron sputtering. These multilayer structures were investigated by Auger electron spectroscopy (AES), transmission electron microscopy (TEM), selected area electron diffraction (SAED), X-ray diffraction (XRD), cross-section scanning electron microscopy (CSSEM) and cross-section electron probe microanalysis (CSEPMA). The mechanical properties of the multilayer coatings, namely, hardness, Young's modulus and the coefficient of plastic deformation were measured. The adhesion was evaluated by the Rockwell-C-impact test. Coatings with different total thickness were examined with respect to adhesion to substrates of tool materials.

1. Introduction

Multilayer coatings of alternatively ordered thin films of transition metal nitrides with nanoscale thickness are known to have hardness superior than that of a single film. Different physical mechanisms have been proposed to explain this effect. The decisive role has been assigned to the phenomena occurring at the interface region of the thin composing films. The dislocation blocking at this region is considered as the most important mechanism governing the properties of the coatings. The idea of composite hardness introduced in [1] may be used only as an overall suggestion for explanation of the advantageous effect observed. A deeper understanding of the nature of the processes at the interfaces is needed in order to produce better coatings and for a more complete utilization of the hard nanolaminated structures. The thickness of the interfaces usually increases at elevated deposition temperatures due to interdiffusion of the elements involved. This effect causes undesirable deterioration of the coating properties and can be avoided to a great extend by keeping the deposition temperature as low as possible. Multilayer coatings of different materials deposited at relatively low temperatures by magnetron sputtering and other physical vapor deposition (PVD) methods possess high-quality mechanical properties [2-5]. Multilayers consisting of pairs with varying thickness period showed that the best mechanical characteristics were obtained at a period of about 10 nm or more [6-8].

[4] To whom any correspondence should be addressed.

The TiN/Mo_2N multilayer system has not yet been well studied although some preliminary investigations performed have shown the potential for its utilization [9]. In this work we investigated in detail the compositional, structural etc. characteristics of hard TiN/Mo_2N multilayer nanolaminated structures. The mechanical properties and adhesion of the coatings were of primary interest.

2. Experimental

The TiN/Mo_2N multilayer structures (MLS) were deposited by reactive DC magnetron sputtering. The deposition was carried out in a high vacuum system with pure Ti and Mo magnetron targets assembled on water cooled holders. Magnetron discharges in an Ar and a mixed Ar/N_2 gas atmosphere were utilized for the two process regimes: pretreatment of the substrate surface by high-pressure discharge prior to deposition and multilayer structure growth. The substrates temperature during the deposition was maintained at 360 ^oC. Substrates of high-speed steel (marked as R18) and of a sintered hard alloy of TiC and WC with Co as a binding element (marked as KM1) were used. The process parameters are shown in tables 1 and 2.

Several deposition series of MLS with different thickness were grown: 2×5 TiN/MoN MLS (total thickness of 115 nm), 2×10 TiN/MoN MLS (total thickness of 230 nm) and 2×60 TiN/MoN MLS (total thickness of about 1300 nm).

Table 1. Substrates pretreatment parameters.

Discharge gas	Discharge pressure, Pa	Discharge voltage, V	Discharge current, A	Substrate temperature, ^0C	Treatment time, s
Argon	4.0	1200	0.1	300 - 320	600

Table 2. Film deposition parameters.

Film Material	Process pressure, (N_2/Ar), Pa	N_2 partial pressure, Pa	Discharge Voltage, V	Discharge current, A	Deposition temperature, ^0C
TiN	1.2×10^{-1}	4.5×10^{-3}	460	1.0	360
Ta_2N	1.2×10^{-1}	4.0×10^{-2}	580	0.8	360

Auger electron spectroscopy (AES) and X-ray diffraction (XRD) were utilized to investigate the compositional and structural characteristics of the TiN films. These investigations were performed in previous experiments in order to optimize the process parameters to achieve the strongest adhesion of the TiN films to substrates [10] as the deposition runs of MLS started with a TiN film. The AES analysis was performed in a derivative mode by a cylindrical mirror analyzer with a relative resolution of 0.6 %. A primary electron beam (3 keV, 5 μA) at normal incidence to the sample surface was used. Surface cleaning was carried out with a focused Ar ion beam (current density approx. 60 $\mu A/cm^2$; ion energy of 3 keV). X-ray diffraction patterns were obtained using a Philips X-ray diffractometer in the Bragg-Brentano geometry with Cu-Kα radiation. The data were collected for different aperture sizes, scanning steps and scanning times to achieve a maximum clarity of the analysis.

The microstructure and the crystallization trend of the films were studied by transmission electron microscopy (TEM) and selected area electron diffraction (SAED) using a Philips 400 electron microscope. The films were deposited on fresh fractured surfaces of NaCl crystals.

Cross-section scanning electron microscopy (CSSEM) and cross-section electron probe microanalysis (CSEPMA) were performed on the 2×60 TiN/MoN ML structure for qualitative investigation of the elemental content of the multilayer structure as well as to evaluate its the thickness.

The mechanical properties were measured by a Fischerscope H100 nanotester loaded with 20 mN.

The adhesion of MLS with different thickness to the substrates was measured by Rockwell-C Indentation Test.

3. Results and discussion

The composition of the TiN films was evaluated by AES analysis. Problems of peak overlapping of the N(KLL) and $Ti(L_3M_{23}M_{23})$ Auger transitions as well as Auger peak shape changes due to chemical effects make the interpretation of the Auger spectra difficult. In this study we used a comparative analysis with data obtained from a previous AES analysis of titanium nitride films implemented in direct and derivative modes [11]. The composition of the films was evaluated with an uncertainty of about 10 %. Figure 1 presents an AES spectrum recorded after the removal of contaminants, mainly carbon, from the film surface. The composition of the film was evaluated as near stoichiometric.

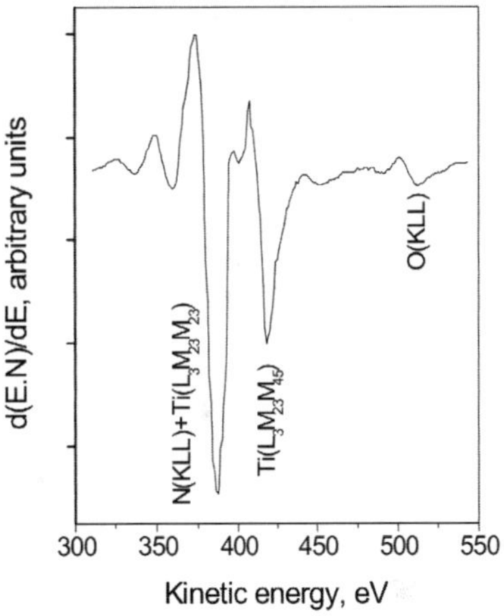

Figure 1. AES spectrum of TiN film.

The phase composition and crystallographic features of the TiN films were investigated by XRD. Figure 2 shows the diffraction pattern of a film deposited at optimized conditions over a broad range of 2θ, from 5 to 95 degrees, using a Ni filter, a step of 0.03 degrees and a scanning time of 0.5 s/point. This spectrum reveals the presence of δ-TiN (as a dominant phase) and ε-Ti_2N phases in the film (together with peaks originating from the substrate). This XRD analysis also shows that the films grow preferentially along the (200) crystallographic direction of the *fcc* (face-centered cubic) δ-TiN phase. The lattice parameter of the δ-TiN phase evaluated from the main δ-TiN (200) peak is 0.425 nm which is slightly higher than that of bulk stoichiometric titanium nitride (0.424 nm [12]).

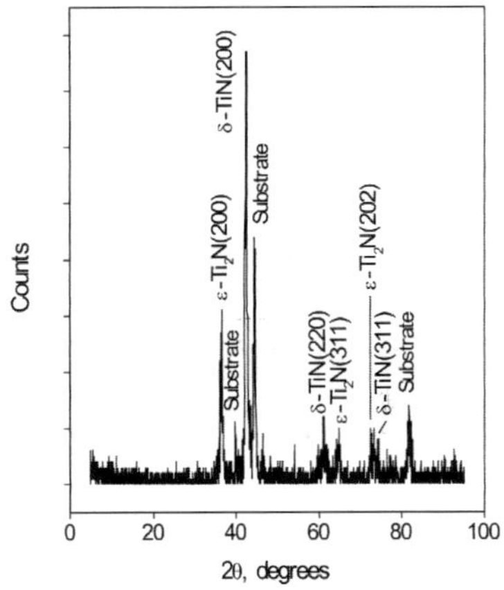

Figure 2. XRD spectrum of TiN.

The TEM observations show that the TiN films are relatively homogeneous with a certain tendency for agglomeration and chain formation. This appears in combination with the uniformly distributed nano-crystals with size varying in the narrow range between 30 and 50 nm. This mono disperse structure corresponds to a polycrystalline character (figure 3). As revealed by the SAED patterns, there exists some tendency towards texturing that could be assigned to the chain formation registered. The Mo_2N films are more homogeneous and also possess a polycrystalline mono-dispersed structure. In this case, the size of the uniformly distributed nano-crystals vary from 50 to 100 nm (figure 4). The SAED patterns confirm the presence of a crystalline Mo_2N phase.

Figure 5 presents the scanning electron microscopy image of cross-section of the 2×60 TiN/Mo_2N MLS deposited on a Si substrate together with the electron probe microanalysis of the same cross section. The thickness of the nanolaminated structure was precisely measured. The EPMA pictures give a quantitative idea for the elemental content of the structure, namely, X-ray emission from a Si substrate and from Ti and Mo atoms superimposed on a large background.

The mechanical properties of the nanolaminated multilayer structures grown on substrates of tool materials are shown in table 3. The same characteristics of the substrates are presented in the table for comparison. The MLS coatings hardness is about 30 % higher than that of single TiN films and 20 higher than that of bulk TiN.

The adhesion was measured for MLS coatings with different thickness, from 100 to 1350 nm. The test shows that for thicknesses of at least 250 nm the adhesion is strong. Figures 6 and 7 present pictures of the test for coatings on steel R18 and hard sintered alloy KM1 with thickness of 120, 240 and about 1400 nm respectively for 5, 10 and 60 pairs of TiN and Mo_2N films.

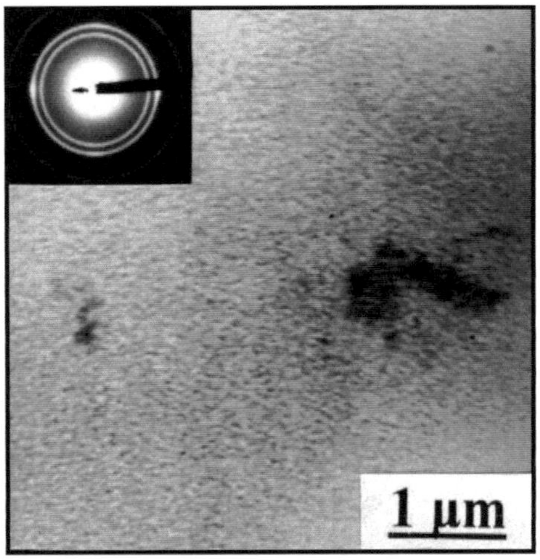

Figure 3. TEM micrograph and SAED patterns of TiN thin film.

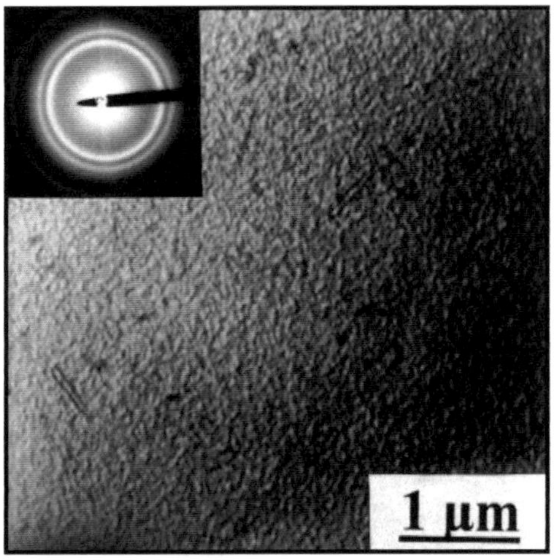

Figure 4. TEM micrograph and SAED patterns of Mo_2N thin film.

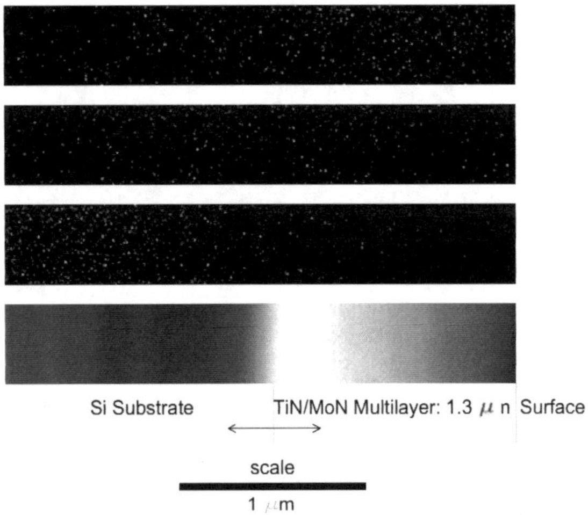

Figure 5. SEM & EPMA Images of Cross Section of 60x2 TiN/MoN Nanolaminated multilayer structure.

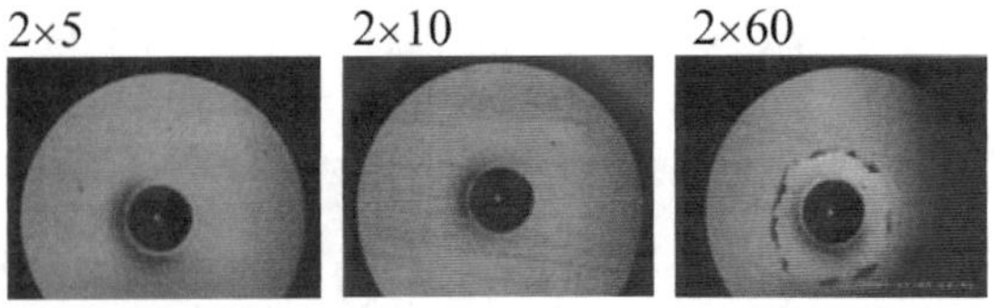

Figure 6. Rockwell-C Indentation Test of adhesion: TiN/Mo_2N MLS on R18.

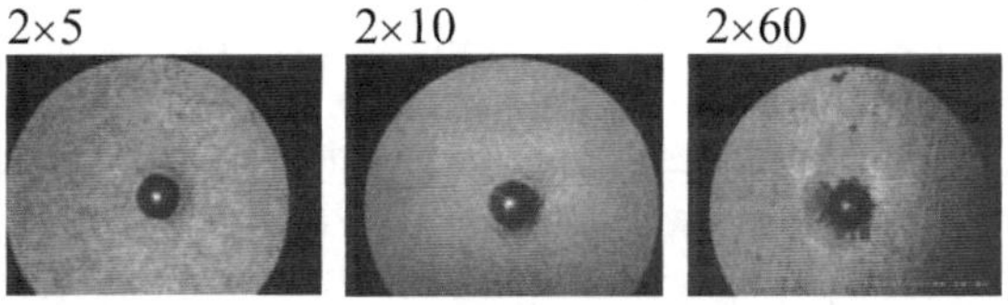

Figure 7. Rockwell-C Indentation Test of adhesion: TiN/Mo_2N MLS on KM1.

Table 3. Mechanical properties of nanolaminated 60x2 TiN/Mo_2 MLS on R18 and KM1.

Sample	Hardness, GPa	Indentor penetration depth, μm	Young's modulus, GPa	Full energy, nJ	Energy of plastic deformation, %	Loading force, mN
60×2 TiN/MoN MLS on R18	24.1	0,181	224	2,08	37,4	20
60×2 TiN/MoN MLS on KM1	25.9	0,178	366	1,82	43,3	20
R18 without coating	10.3	0,295	172	2,53	63,15	20
KM1 without coating	13.4	0,243	237	1,91	59,04	20

4. Conclusions

Hard multilayer coatings consisting of successively deposited TiN and Mo_2N films with nanoscaled thickness were deposited at relatively low temperatures on substrates of tool materials. This two-step PVD process was successfully utilized to grow structures with high mechanical characteristics. Thick multilayer structures with thickness of 1.3 μm of cubic TiN and cubic Mo_2N films showed much higher hardness compared with that of the substrate tool materials (twice as high) as well as to that of single TiN and Mo_2N films (more than 40 %). These favorable features of the synthesized TiN/Mo_2N nanolaminated multilayer coatings make them very promising for application as hard and wear resistant coatings, biocompatible coatings etc.

Acknowledgment: This work was sponsored by the Bulgarian National Science Fund at the Ministry of Education and Science under contract F-1514/05.

References

[1] Kim S H, Baik Y J and Kwon D 2004 *Surf. Coat. Technol.* **187**/1 47-53
[2] Hubler R 2002 *Surf. Coat. Technol.* **158-159** 680-4
[3] Ma C-H, Huangl J-H and Chen H 2004 *Thin Solid Films* **446**/2 184-93
[4] Tentardini E K and Hubner R 2001 *Nucl. Instr. Meth. Phys. Res.* B **157** 626-9
[5] Carraso C, Segars L, Benavente B and Vergara V 2004 *J. Mater. Proc. Technol.* **147** 371-6
[6] Yaoa S H, Kaob W H, Suc Y L and Liuc T H 2005 *Mater. Sci. Engin.* A **392**/1-2 380-5
[7] Zhou Z, Rainforth W M, Rother B, Ehiasarian A P, Hovsepian P Eh and Munz W-D 2004 *Surf. Coat. Technol.* **183**/2-3 275-82
[8] Yang Q and Zhao L R 2003 *Surf. Coat. Technol.* **173**/1 58-66
[9] Nordin M, Larsson M and Hogmark S 1998 *Surf. Coat. Technol.* **106**/2-3 234-41
[10] Martev I N, Dechev D A, Ivanov N P, Uzunov Ts D and Kashchieva E P 2008 *J. Physics: Conf. Series* **113** 012025
[11] Vignes J L, Langeron J P, Grigirov G I, Martev I N and Stoyanova M V 1991 *Vacuum* **42** 151-3
[12] Toth L E 1971 *Transition Metal Carbides and Nitrides* (Academic press New York)

Preparation of wire structured ZnO films by electrochemical deposition

H Nichev[1], M Sendova-Vassileva[1], P Andreev[1], D Dimova-Malinovska[1,3] and K Starbova[2]

[1]Central Laboratory of Solar Energy and New Energy Sources,
Bulgarian Academy of Sciences, 72 Tzarigradsko Chaussee, 1784 Sofia, Bulgaria
[2] Central Laboratory of Photoprocesses, Bulgarian Academy of Sciences,
Acad. G. Bonchev Str., Block 109, 1113 Sofia, Bulgaria

E-mail: doriana@phys.bas.bg

Abstract. This article reports the synthesis of wire structured ZnO by electrochemical deposition on a glass substrate covered by thin film of ITO or SnO_2 doped with F (SnO_2:F) and on a carbon tissue. The influence of the structure of the transparent conductive films on the structural properties of the ZnO films obtained was studied by SEM and Raman spectroscopy. The results show that the films have a columnar structure with grain size of about 200-500 nm. The films deposited on the SnO_2:F coated glass substrate have better structural properties. Pillar structure is observed on the surface of the fiber of the carbon tissue.

1. Introduction

Crystalline pillar structures such as ZnO wires deposited on a substrate have received increasing attention due to potential applications in solar cells, nanoscale electronics, optical and sensing devices, etc. [1, 2]. Different physical and chemical methods have been applied for the synthesis of ZnO thin films with wire structure [3]. Since, however, the preparation conditions determine to a great extent their properties, their fabrication with predetermined size and perfect morphology is still a challenging task.

In this work preliminary results about the preparation of ZnO films with pillar structure by electrodeposition on different substrates: ITO or SnO_2 doped with F (SnO_2:F) coated glass and carbon tissue are reported. The influence of the type of the substrates on the surface morphology and the structural properties of the deposited ZnO films are investigated.

2. Experimental

ZnO films were deposited by a electrochemical process using a three-electrodes potentiostatic system with saturated calomel electrodes as reference electrodes. $In_xSn_{1-x}O_2$ (ITO), SnO_2 doped with F (SnO_2:F) coated glass substrates and carbon tissue (for bio-medical applications) were used as substrates and working electrodes. The ZnO films were deposited from an aqueous solution containing $ZnCl_2$ 5×10^{-3} M, KCl 0.1 M, with pH = 4.00 and using an air flow. The electrochemical deposition took a place at 60^0C. The deposition process was carried out potentiostatically at a potential of –0.9 V vs. the saturated calomel electrode.

[3] To whom any correspondence should be addressed.

The surface morphology and the thickness of the deposited films were determined using scanning electron microscopy (SEM) and the structure of the deposited ZnO films was studied by Raman spectroscopy. The Raman spectra were recorded by a Horiba Jobin Yvon LabRAM HR800 spectrometer using a 1800 l/mm grating and a HeNe laser for excitation. The samples were placed under the 100X achromatic objective of a BX41 microscope and measured in back scattering configuration.

3. Results and discussion

Figures 1 a and b show SEM micrographs of the surface morphology of the ITO deposited on a glass substrate (a) and of the ZnO wire structured film deposited on an ITO coated glass substrate. The cross sectional images of the ZnO films grown on the ITO surface taken at 0^0 and 30^0 are demonstrated in figures 2 a and b, respectively. The ITO film is about 1 μm thick. It exhibits columnar grains with a diameter of less than 200 nm grown with their c-axis perpendicular to the substrate surface. The interface between the ITO and the grown ZnO film is rough. The ZnO film has a columnar structure with well-defined grains with size between 200 and 500 nm. The surface of ZnO films exhibits high porosity.

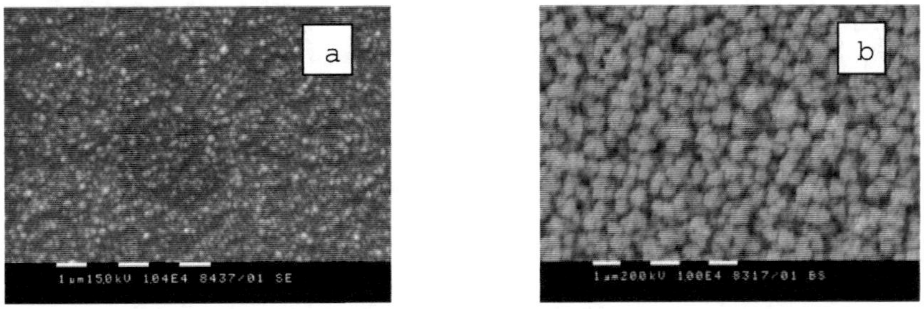

Figure 1. SEM images of the surface of the ITO layer on a glass substrate (a) and of a ZnO wire structured film on an ITO coated glass substrate (b).

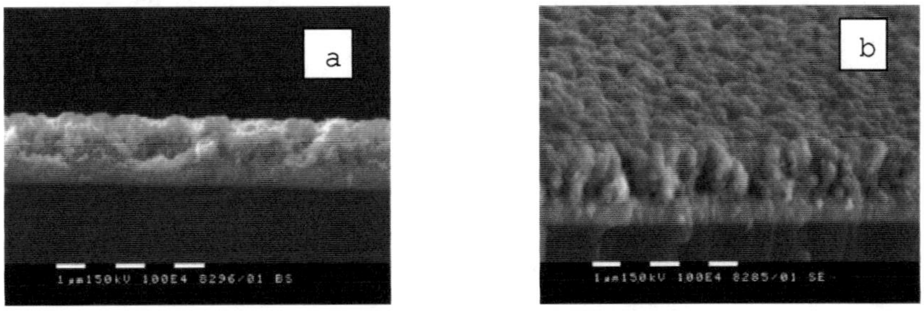

Figure 2. SEM cross-section taken at 0° (a) and at 30° (b) of a ZnO layer on ITO coated glass substrate (b).

Figure 3 a and b demonstrate the surface morphology of the SnO_2:F and the electrodeposited ZnO film on the SnO_2:F coated glass substrate, respectively. The SnO_2:F film surface shows tetrapodlike morphology with a relatively uniform grain size (figure 3 a). Cross-section SEM micrographs of a ZnO film deposited on a SnO_2:F /glass substrate taken at 0° and at 30° are displayed in figures 4 a and b, respectively. The interface between SnO_2:F and the ZnO film grown on it is very homogeneous and relatively smooth compared with that in the case of using an ITO coated glass substrate. The ZnO film is porous, has a columnar structure with grain diameter of about 200-500 nm as in the case of the film grown on ITO coated glass (figure 1 b and figure 2 b), but is denser.

Figure 3. SEM images of the surface of SnO_2:F (a) and of ZnO wire structured film on a SnO_2:F coated glass substrate.

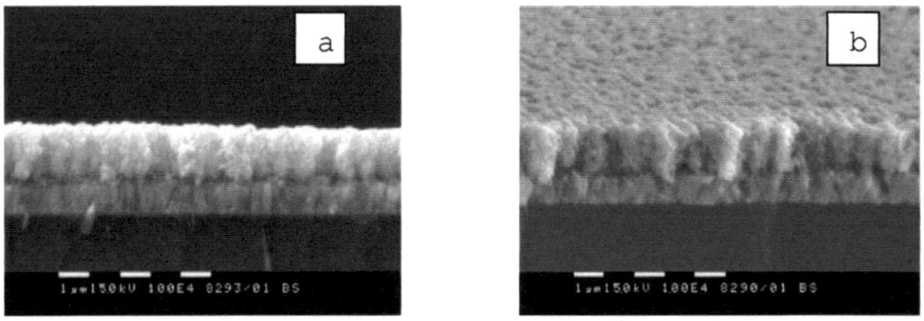

Figure 4. SEM cross-section taken at 0° (a) and at 30° (b) of a ZnO layer on SnO_2:F coated glass substrate.

The structure of the ZnO film deposited was studied by Raman spectroscopy. Raman spectra of the ZnO films deposited on ITO and SnO_2:F coated glass substrate are shown in figures 5 and 6, respectively. For comparison, the Raman spectra of the ITO and SnO_2:F films on glass substrate are given as well. The strongest SnO_2 band at 634 cm^{-1} (A_{1g}) [18] is clearly visible in all spectra. Typical bands for ZnO at 437 cm^{-1} (E_2), 407 cm^{-1} (E_1), 574 cm^{-1} (A_1) are observed [19]. The Raman bands in the spectrum of the ZnO film deposited on SnO_2:F coated glass substrate are better expressed and their Full Width at Half Maximum (FWHM) is smaller which is an indication for better structural quality of

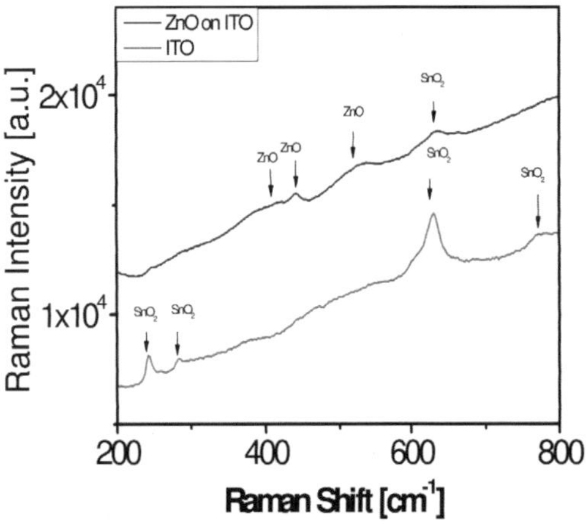

Figure 5. Raman spectra of a ZnO film deposited on an ITO coated glass substrate (a). For comparison the Raman spectrum of an ITO film deposited on a glass substrate is given too (b).

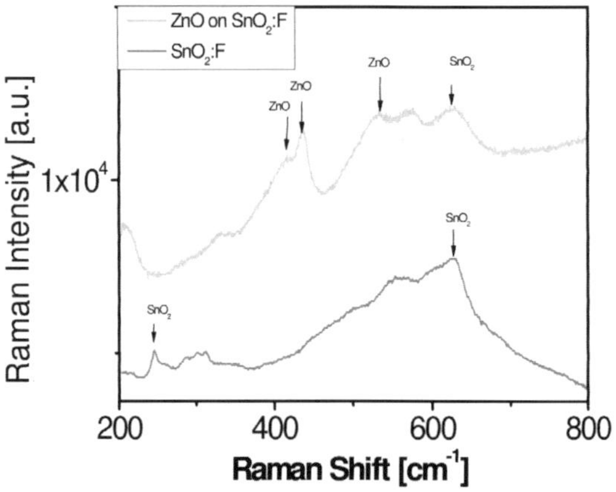

Figure 6. Raman spectra of a ZnO film deposited on a SnO_2:F coated glass substrate (a). For comparison, the Raman spectrum of a SnO_2:F film deposited on a glass substrate is given too (b).

the film. This conclusion is confirmed by the observations from SEM micrographs. It is reasonable to assume that the tetrapode structure and the flatter surface of the SnO_2:F films create conditions for growth of ZnO with better structural quality.

ZnO pillars were deposited on the surface of a carbon tissue used for bio-medical applications. The SEM images of the deposited ZnO pillars are demonstrated in figure 7. The pillars have a size of about 200-500 nm and growth perpendicular to the surface of the tissue fibers.

Figure 7. SEM of ZnO pilars on a carbon tissue surface. Note the different magnification of the two pictures.

4. Conclusion

We studied the structural properties of wire structured ZnO films deposited by electrochemical process using a three-electrodes potentiostatic system with saturated calomel electrodes as reference electrodes on two types of glass substrates coated with ITO and SnO_2:F, respectively. The possibility to deposit ZnO with pillar structure using carbon fibered tissue as a substrate is demonstrated as well. The structure was investigated by SEM and Raman spectroscopy. It was shown that the structure of the substrates influences the structure of the electrodeposited ZnO films. The films have a columnar structure with grain size of about 200-500 nm. Denser ZnO films with better structural quality were deposited using the SnO_2:F coated glass as a substrate. The specific tetrapod structure and more homogeneous surface morphology of the SnO_2:F films created more favorable conditions for ZnO films growth.

Acknowledgements

This work is funded by the Bulgarian National Science Fund (projects DO02-207/2008 and DO02-167/2008).

References

[1] Schropp R E I, Li H, Franken R H, Rath J K, van der Werf C H M, Schüttauf J W A and Stolk R L 2008 *Thin Solid Films* **516** 6818-23
[2] Yan J-F, Lu Y-M, Liang H–W, Liu Y-Ch, Li B-H, Fan X-W and Zhou J-M 2005 *J. Crystal Growth* **280** 206-12
[3] Huang M H, Wu Y, Feick H, Tran N, Weber E and Yang P 2001 *Adv. Mater.* **13** 113
[4] Djurisic A B and Leung Y H *Small* 2006 **2** 944-49
[5] Sun Z-P, Liu L, Zhang L and Jia D-Z 2006 *Nanotechnology* **17** 2266-72
[6] Zhang Y, Huang Y, He J, Dai Y, Wang S and Zhou C 2005 *Mater. Res. Soc. Symp. Proc.* **879 E**, 771-4
[7] Song J, Wang X, Liu J, Liu H, Li Y and Wang Z 2008 *Nanoletters* **8** 203-6
[8] Dimova-Malinovska D, Tzenov N, Tzolov M and Vassilev L 1998 *Mater. Sci. Eng.* **B52** 59-62
[9] Lee J-H, Ko K-H and Park B-O 2003 *J. Crystal Growth* **247** 119-25
[10] Nanto H, Minami T, Shooji S and Tanaka S 1984 *J. Appl. Phys.* **55** 1029-103
[11] Fahoume M, Maghfoul O, Aggour M, Hartiti B, Chraibi F and Annaoui E 2006 *Solar Energy Materials & Solar Cells* **90** 1437-44
[12] Pauzauskie P J and Yang P **2006** *Nanowire photonics, Materials Today* **9** 36-40

[13] Kong X, Ding Y and Wang Z 2004 *J. Phys. Chem.* **B 108** 570-76
[14] Dai Y, Zhang Y, Li Q K and Nan C W 2002 *Chem. Phys. Lett.* **358** 83-6
[15] Lim Y S, Park J W, Hong S-T and Kim J 2006 *Mat. Sci. Engin.* **B 129** 100-3
[16] Gui Y, Xie C, Zhang Q, Hu M, Yu J and Weng Z 2006 *J. Crystal Growth* **289** 663-6
[17] Chenga H, Chengb J, Zhangb Y and Wanga Q-M 2007 *J. Crystal Growth* **299** 34-6
[18] Abello L, Bochu B, Gaskov A, Koudryavtseva S, Lucazeau G and Roumyantseva M 1998 *J. Solid State Chem.* **135** 78-85
[19] Damen T C, Porto S P S and Tell B 1966 *Phys. Rev.* **142** 570-4

Journal of Physics: Conference Series **223** (2010) 012021 doi:10.1088/1742-6596/223/1/012021

Mirror coatings for eye-safe laser generation

T K Tenev[1], I K Miloushev and R A Peyeva

Georgi Nadjakov Institute of Solid State Physics,
Bulgarian Academy of Sciences, 72 Tzarigradsko Chaussee, 1784 Sofia, Bulgaria

E-mail: tenev@issp.bas.bg

Abstract. Eye-safety is of paramount importance in the use of laser-based instruments. Eye safe output at 1.54 micron is very important for various applications such as robot vision, range finding and laser radars in civil and military use. Nonlinear crystals have been used in state-of-the-art rangefinders for providing eye-safe output in the 1550 ± 50 nm band. We present our results on designing a back and a front mirror for eye-safe laser generation at a wavelength of 1538 nm. The back mirror is designed for high transmission at 1067 nm and high reflection at 1351 and 1538 nm. The front mirror is designed for high transmission at 1067 nm, high reflection at 1351 nm and partial reflection at 1538 nm. The optimized design is achieved using titanium dioxide and silicon dioxide layers. In the experiment, the coatings are fabricated by electron beam evaporation on BK7 glass substrates. The characteristics of the coated samples are compared with the results of the calculations, showing good correspondence. The laser mirrors so prepared passed the laser damage and durability tests.

1. Introduction

"Eye safe" lasers are very important for different applications in optical communications and lidar techniques.

Eye hazards caused by laser radiation depend on the wavelength, laser power, exposure time and type of tissue. Excessive exposure to visible radiation causes photochemical and thermal retina damage. According to the American standard ANSI Z136.1-1986, the wavelength of 1.5 μm is treated as safe for direct eye exposure to the beam for energy densities 100 times larger than for 10.6 μm (CO_2 laser) and 2×10^5 times larger than for 1.06 μm (Nd:YAG laser). According to the standards developed by International Electrotechnical Committee, the application of laser devices that do not satisfy the safety requirements will be soon considerably limited.

At present, the methods developed of eye protection against laser radiation have to do with the application of "eye-safe" lasers that emit radiation outside the spectral transmission region of eye, i.e. outside the range 400-1400 nm.

The rapid development of Raman-active media has made stimulated Raman scattering (SRS) in crystals a promising method for wavelength conversion of laser radiation [1-3]. One especially interesting approach among the SRS systems is the self-stimulated Raman laser that achieves laser emission and SRS action in one crystal.

In the past few years, all-solid-state self-stimulated Raman lasers have been reported in materials such as Nd:KGd$(WO_4)_2$ [4,5], Nd:PbWO_4 [6], Yb:KGd$(WO_4)_2$ [7] and Yb:KY$(WO_4)_2$ [8].

---

[1] To whom any correspondence should be addressed.

The $KGd(WO_4)_2$ (KGW) crystal is popular for optically active components due to its suitability as a rare-earth host material, its high third-order nonlinear susceptibility, high thermal conductivity, and strong Raman conversion properties. KGW has prominent Raman modes at 768 and 901 cm^{-1} arising from the tungsten-oxygen sublattice in the crystal which are useful for wavelength conversion in the near infrared spectrum [9]. Compact and efficient lasers have been constructed using passive resonator Q-switching [10-14]. In this case, the laser mirrors are common for all elements and must cover the spectral requirements for several spectral lines.

In this paper we report the design and fabrication of back and front (output) mirrors coatings on BK7 glass substrate to be used for self-stimulated Raman laser generation in Nd:KGW Various specifications of back and front mirrors were found in the scientific literature and in the technical catalogues. The back mirror should have high transmission at the wavelength 1067 nm, together with high reflection at the lasing wavelength 1351 nm and at the SRS wavelength 1538 nm. For the front mirror one needs high transmission at the wavelength 1067 nm; high reflection at the lasing wavelength 1351 nm and partial reflection (about $30 \pm 10\%$) at the SRS wavelength 1538 nm for extracting a portion of the energy through the mirror. High transmission of both mirrors at 1067 nm is necessary to prevent lasing at this wavelength and promote lasing at 1351 nm used for SRS conversion to 1538 nm.

2. Results and discussion

Optimization with different starting number of layers and specified transmission (reflection) at the wavelengths of interest is performed using the software package TFCalc (Spectra Software, Inc). The selection of the coating materials is based on their refractive index values, physical stability characteristics, their transparency in the desired wavelength regions and interface compatibility. The commonly used high index coating materials are TiO_2, Ta_2O_5, HfO_2, ZrO_2 and the low index coating material is SiO_2. TiO_2 is chosen as a high index and SiO_2, as a low index coating material since this combination has a maximum refractive index difference and yields the widest high-reflection zone. Increasing the number of layers increases the cumulative stress and sometimes limits the efficiency of the device. Random errors in the deposited layer thicknesses lead to discrepancies between the calculated and the implemented spectral characteristics. Therefore, it is better to use a minimum number of layers in the design. The characteristics desired (transmission/reflection) were designed by optimizing the thicknesses of the high and low refractive index layers. The refractive indices and dispersions of the materials used were determined beforehand and used in the optimization. The starting designs were mirrors at a reference wavelength of 1450 nm with different number of layers (between 15 and 30). The reflectivity desired at each wavelength of choice was added as discrete targets and the respective thickness optimization was carried out. Appropriate sensitivity and yield analysis was then carried out to check the feasibility of the solutions obtained. The optimized design was achieved with seventeen and nineteen layers of alternate layers of TiO_2 and SiO_2 for the back and front mirror, respectively.

The mirror coatings were fabricated by using electron-beam gun evaporation system in a Leybold A700QE vacuum coating unit. Local high absorption generally occurs due to the loss of film stoichiometry and can be minimized by depositing the film in an oxygen atmosphere. Initially, the system is evacuated to a vacuum better than 5×10^{-6} mbar. The evaporation took place at the working vacuum of 5×10^{-5} mbar for SiO_2 and 3.5×10^{-4} mbar for TiO_2. The rate of evaporation in the case of SiO_2 was 0.8 nm/s and 0.4 nm/s for TiO_2. The thicknesses of the layers were controlled by a quartz microbalance system Inficon XTC-2. The sample holders were fixed to a dome that was rotated with respect to the central point of the coating chamber for better uniformity. The substrate temperature during deposition was maintained at 300^0C within a tolerance of $\pm 10^0$C.

The theoretically designed index profile for the back mirror is given in figure 1; the corresponding transmittance curve (dashed line) is given in figure 2. As shown in figure 2, the theoretically achieved transmission is 99.8% at 1067 nm, 0.33% at 1351 nm (the lasing wavelength) and 0.38% at 1538 nm (SRS wavelength). Similarly, the theoretical index profile for the front mirror is given in figure 3, while the corresponding design transmittance is given in figure 4 (dashed line). As is evident from figure 4, the theoretically achieved transmission is 99.5% at 1067 nm, 1.0% at 1538 nm and 70% at 1538 nm.

The experimental results for the back and the front mirrors are also shown in figure 2 and figure 4 (solid lines) (as measured by a Perkin-Elmer Lambda 950 spectrophotometer). The achieved transmission for the back mirror is T=97.9% at 1067, T=0.38% at 1351 and T=0.34% at 1538 nm. The experimentally measured transmittance of the front mirror is T=97.5% at 1067, T=1.2% at 1351 and T=67% at 1538 nm.

The theoretical and experimental transmission curves are in close agreement There are, however, some discrepancies in the experimentally achieved transmission values compared to the theoretical values due to the absorption and scattering losses of the columnar microstructure and to the dispersion of the coating materials deposited. At 1067 nm, the transmittance measured depends also on the quality of the AR coating on the second surface of the BK glass substrate.

The coated samples were tested to withstand five 10 ns pulses of 1500 MW/cm^2 laser radiation at 1067 nm in a single point. The samples are tested for repeatability of the characteristics and durability in view of regular commercial production. The coated samples passed the adhesion and abrasion tests.

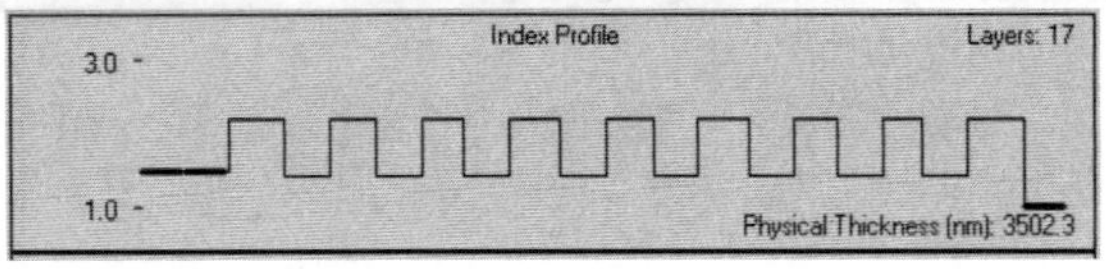

Figure 1. Back mirror: Refractive index vs. physical thickness of medium.

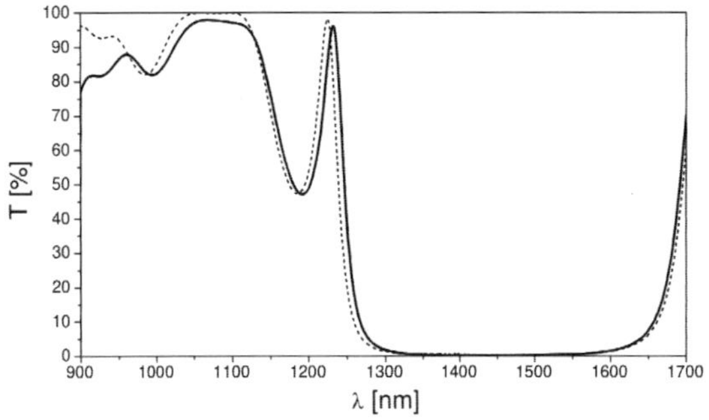

Figure 2. Back mirror transmittance vs. wavelength: calculated transmittance curve (dashed line) and the corresponding measured transmittance curve (solid line).

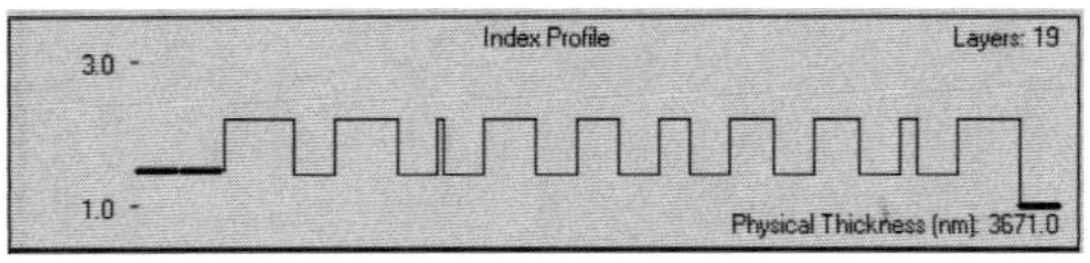

Figure 3. Front mirror: Refractive index vs. physical thickness of medium.

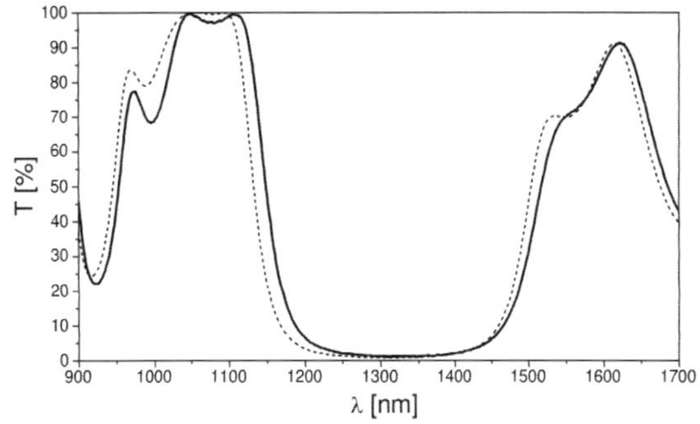

Figure 4. Front mirror transmittance vs. wavelength: calculated transmittance curve (dashed line) and the corresponding measured transmittance curve (solid line).

3. Conclusion

We designed and fabricated back and front (output) mirrors coatings for eye-safe self-stimulated Raman laser generation in a Nd:KGW passively Q-switched intracavity Raman laser. Laser mirrors are common for all laser elements and must cover the requirements for three spectral lines simultaneously. The optimized designs were achieved with a minimum number of layers in order to avoid cumulative stress and errors in fabrication. The spectral characteristics of the samples fabricated were compared with the results of the calculations, showing good correspondence. In preliminary experiments with such a Raman laser, pulses with 30-ns duration and 8-mJ energy were generated.

Acknowledgments
The experiments are carried out in the frame of a research project between ISSP-BAS and "Optix" Co., Panagyurishte. We acknowledge the contribution of "Optix" Co. team for manufacturing and testing the samples.

References
[1] Cerny P, Jelinkova H, Zverev P and Basiev T T 2004 *Prog. Quantum Electron.* **28** 113
[2] Pask H M 2003 *Prog. Quantum Electron.* **27** 3
[3] Zverev P G, Basiev T T and Prokhorov A M 1999 *Optical Materials* **11** 335
[4] Grabtachikov A S, Kuzmin A N, Lisinetskii V A, Orlovich V A, Ryabtsev G I and Demidovich A A 1999 *Appl. Phys. Lett.* **75** 3742
[5] Findeisen J, Eichler H J and Peuser P 2000 *Opt. Commun.* **181** 129
[6] Chen W, Inagawa Y, Omatsu T, Tateda M, Takeuchi N and Usuki Y 2001 *Opt. Commun.* **194** 401
[7] Lagatsky A A, Abdolvand A and Kuleshov N V 2000 *Opt. Lett.* **25** 616
[8] Grabtachikov A S, Kuzmin A N, Lisinetskii V A, Orlovich V A, Demidovich A A, Danailov M B, Eichler H J, Bednarkiewicz A, Strek W and Titov A N 2002 *Appl. Phys.* B **75** 795
[9] Major A, Aitchison J S, Smith P W E, Langford N and Ferguson A I 2005 *Opt. Lett.* **30** 421
[10] Ustimenko N S and Gulin A V 2002 *Quantum Electron.* **32** 229
[11] Abazadze A Yu, Zverev G M, Kolbatskov Yu M and Ustimenko N S 2004 *Quantum Electron.* **34** 20
[12] Huang Jianhong et al. 2007 *Opt. Lett.* **32** 1096-98
[13] Grabtchikov A S, Kuzmin A N, Lisinetskii V A, Orlovich V A, Demidovich A A, Yumashev K V, Kuleshov N V, Eichler H J and Danailov M B 2001 *Optical Materials* **16** 349
[14] Ustimenko N S and Zabotin E M 2005 *Instruments and Experimental Techniques* **48** 239 (Translated from Pribory i Tekhnika Eksperimenta 2005 **2** 112)

Journal of Physics: Conference Series **223** (2010) 012022 doi:10.1088/1742-6596/223/1/012022

Effect of the substrate surface topology and temperature on the structural properties of ZnO layers obtained by plasma enhanced chemical vapour deposition

S Kitova[1] and G Danev

Acad. J .Malinowski Central Laboratory of Photoprocesses,
Bulgarian Academy of Sciences, Acad. G. Bonchev Str., Bl.109, 1113 Sofia, Bulgaria

E-mail: skitova@clf.bas.bg

Abstract. In this work thin ZnO layers were grown by metal-organic PECVD (RF 13.56 MHz) on Si wafers. Zn acetylacetonate was used as a precursor and oxygen as oxidant. A system for dosed injection of the precursor and oxidant into the plasma reactor was developed. The influence of the substrate surface topology and temperature on the structural properties of the deposited layers was studied. ZnO and graphite powder dispersions were used to modify the silicon wafers before starting the deposition process of the layers. Some of the ZnO layers were deposited on the back, unpolished, side of Si wafers. Depositions at 400 °C were performed to examine the effect of the substrate temperatures on the layer growth. The film structure was examined by XRD and SEM. The results show that all layers are crystalline with hexagonal wurtzite structure. The crystallites are preferentially oriented along the c-axis direction perpendicular to the substrate surfaces. ZnO layers deposited on thin ZnO seed films and clean Si surface exhibit well-developed grain structures and more c-axis preferred phase with better crystal quality than that of the layers deposited on graphite seed layer or rough, unpolished Si wafer.

1. Introduction

Zinc oxide is a unique material with its high chemical and thermal stability, high mechanical strength, high piezoelectric coefficient, high exciton binding energy (60 meV) and wide band gap (3.37 eV) [1, 2]. In the last years, nanostructured ZnO, such as nanotubes, nanowires, nanoribbons, nanofibres and etc. have attracted extraordinary attention due to their potential application in chemical sensors, light-emitting diodes, field emitters, solar cells, nanogenerators and other optoelectronic devices [3, 4], However, the performance of these devices is strongly affected by the material properties and, therefore, numerous attempts have made to improve the crystallinity of ZnO layers. So far, various deposition techniques, such as thermal evaporation, pulsed laser deposition, sputtering, sol-gel deposition, template assisted and solution processes, molecular beam epitaxy, metal-organic chemical vapor deposition MOCVD, have been used in attempts to grow highly crystalline ZnO layers on substrates of practical importance, such as sapphire and Si wafers (for reviews see [3, 4] and the reference cited therein). Among them, MOCVD can offer high growth efficiency, large area uniformity and suitability for growing components containing volatile elements [5].

[1] To whom any correspondence should be addressed.

The plasma enhanced MOCVD (PECVD) technology is one of the most attractive for synthesis of ZnO layers at low and moderate substrates temperatures and minimal impact on them. Diethylzinc and dimethylzinc are usually applied as precursors in the PECVD process [1, 6]. However, these precursors are less desirable because of their high toxicity and flammability. Zinc acetylacetonate ($Zn(acac)_2$) should be a good candidate for use in low temperature fabrication of crystalline ZnO layers because of its non-toxicity, non-explosiveness and non-flammability, low vaporization temperatures (250 °C), low thermal decomposition temperatures (150-500 °C), high vapor pressure [6]. However, ($Zn(acac)_2$) is rarely used as a precursor in PECVD.

In our previous paper we have studied the effect of the PECVD technological parameters, using $Zn(acac)_2$ and O_2 as reactants, on the composition, structural and optical properties of ZnO layers grown at low substrate temperature (~40 °C) on glasses, optical grade polycarbonates and Si wafers [7]. The results showed that the layers obtained at the highest RF power of 2000 W and O_2 gas ambient pressure (about 300 mTorr) are amorphous, with high optical transmission in the visible region and E_g of 3.36 eV, which is very close to that of the bulk ZnO material. It is known that among the important parameters for producing high-quality crystalline ZnO films are the deposition temperature and the substrate surface [3, 4]. A selective growth of high-quality ZnO nanorods on Si substrates at 800 - 900 °C was demonstrated using a pattern thin film of ZnO as a seed layer [8].

In this paper, the effect was studied of the temperature and the surface topology of the substrate on the structural properties of ZnO layers deposited on Si wafers. The layers were grown by PECVD, using $Zn(acac)_2$ and O_2 as reactants. Depositions were performed at 400 ºC to examine the effect of temperature on the growth. To facilitate the formation of ZnO nuclei, two different very thin seed layers of graphite and ZnO were spin coated prior to the ZnO deposition. In order to study the influence of the surface roughness on the layer growth, some of the ZnO layers were deposited on the back, unpolished, side of Si wafers.

2. Experimental

The ZnO layers were grown in PECVD equipment (GENUS 8720). The gas plasma was excited by a 13.56 MHz RF generator. The RF power was 2000 W. Oxygen and $Zn(acac)_2$ ($Zn(C_5H_7O_2)_2.xH_2O$) (Sigmas-Aldrich) were used as precursors. The Zn source was sublimated by heating up to 140 °C and carried by nitrogen to the growth chamber. A system was developed and attached to one of the chamber windows to enable dosed injection of the precursor to the plasma reactor. The stable precursor temperature resulted in a stable precursor flow rate, even if we could not directly measure it. The O_2 gas was introduced separately into the reaction chamber. All gas flows were controlled by mass flow meters controllers at a constant flow rate. Chamber pressure of 250 mTorr and gas ratio O_2/N_2 of 4:1 were kept constant during the deposition process. The substrates were heated to 400 ºC. The deposition time was typically 20 min. The thickness of the layers was about 1.2 μm measured by a Talystep type profilometer.

The substrates used to deposit the layers were Si wafers. Preliminarily cleaning of the substrate surface is very important to grow ZnO layers with good crystal quality and to enhance ZnO nucleation. For this purpose, the Si wafers were dipped in a 10% dilute HF acid solution and then rinsed in deionized water and isopropyl alcohol. Two different very thin seed films were formed by spin coating (1000 rpm, 60 s) of graphite powder dispersion (1 vol.%) in isopropyl alcohol and ZnO powder dispersion (1 vol.%) in isopropyl alcohol.

The crystal structure of the layers was characterized by X-ray diffraction (XRD) using a Philips (PW 1710) apparatus with Cu-Kα radiation separated by a focusing graphite monochromator. The alignment of the diffractometer is in the symmetrical Bragg-Brentanno focusing geometry. Scherrer's equation was used to estimate the average crystallite sizes, without making the corrections for the instrumental geometrical broadening of the peaks and stripping the α1 component from the α2 component radiation. The value of λ was taken as 0.154 nm. The surface morphology of the layers was examined by a scanning electron microscope (SEM Philips 515).

3. Results and discussion

X-ray diffraction patterns of the layers deposited at 400 ºC on Si wafers, ZnO seed film; C seed film; and unpolished side of a Si wafer (samples 1 - 4) are presented in figure 1. The XRD results show that all ZnO layers are polycrystalline. Three peaks at 31.7°, 34.3° and 36.2° are observed. These peaks can be indexed to the $(10\bar{1}0)$, (0002) and $(10\bar{1}1)$ planes of the hexagonal wurtzite structure of the ZnO crystal with cell parameters a = 3.25 Å and c = 5.22 Å, which agree well with the values reported in JCPDS card # 89-1397 [9].

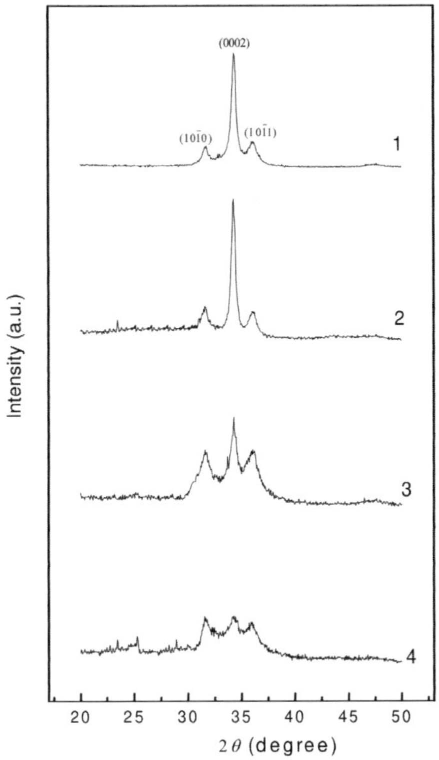

Figure 1. XRD patern of layers deposited at 400 ºC on substrates: 1) Si wafer; 2) ZnO seed film; 3) C seed film; 4) unpolished side of Si wafer.

Figure 2. Top-view SEM images of the same layers as in figure 1.

The results show that the crystallites are preferentially oriented along the c-axis direction perpendicular to the substrate surface, as the (0002) reflection is greatly enhanced with respect to the usual $(10\bar{1}1)$ maximum reflection of ZnO. In order to quantify the intensity ratio of the (0002) crystal orientation, $\alpha_{(0002)}$ was defined as:

$$\alpha_{(0002)} = \frac{I_{(0002)}}{I_{(0002)} + I_{(10\bar{1}0)} + I_{(10\bar{1}1)}}, \quad (1)$$

where $I_{(0002)}$, $I_{(10\bar{1}0)}$ and $I_{(10\bar{1}1)}$ denote the intensities of the (0002), $(10\bar{1}0)$ and $(10\bar{1}1)$ planes, respectively. $\alpha_{(0002)}$ is found to be 0.63, 0.69, 0.49 and 0.46 for sample 1 to 4. The full width at half maximum (FWHM) is 0.51, 0.47, 0.76 and 1.92 respectively. The highest value of $\alpha_{(0002)}$ = 0.69 and the narrow FWHM of 0.47 found for the layers deposited on a ZnO seed film (sample 2) are indications for a more c-axis preferred phase with better crystal quality than that in the rest of the samples. The lower values of $\alpha_{(0002)}$ and the increased FWHM values found for layers deposited on a graphite seed film and on a rough Si surface (samples 3 and 4) point to a more randomly oriented polycrystalline wurtzite phase with smaller size. Based upon Scherrer's equation, the average crystallite size of samples 3 and 4 is 11.1 nm and 4.3 nm, while that of samples 1 and 2 is 16.3 and 17.8 nm, respectively.

Top-view SEM images of samples 1 - 4 are shown in figure 2. The images show the well-developed grain structures of layers deposited on clean Si and a ZnO seed film, and the denser, without well defined grains, smoother surface of layers deposited on a C seed film and rough unpolished Si.

Most probably, the preferred orientation observed of deposited polycrystalline layers and their related structure are connected with the so called preferential nucleation (PO). For the case of ZnO, c-axis PO nucleation has been considered to arise because of the lowest surface free energy of this plane among the hexagonal crystallographic planes [10]. Therefore, it could be assumed that initial nuclei with predominant c-axis PO were formed in the ZnO seed film and at the onset of the growth of the layer; their PO have determined the PO of the layers obtained. On the contrary, initial nuclei with random orientation dominate on the substrates with the C seed film and on the rough unpolished side of the Si wafer, which have many active centers with different energy on crystal imperfections. The results obtained indicate that these nuclei cause the growth of also a-axis phases (with $(10\bar{1}0)$ and $(10\bar{1}1)$ planes) and, as a whole, to the smaller grain size of the layer obtained.

4. Conclusions

Crystalline ZnO layers were produced by PECVD at 400 °C on Si substrates with different topography. The raw material was $Zn(acac)_2$ which is non-toxic and easy to handle. The XRD results indicate that all layers exhibit a hexagonal wurtzite structure. The crystallites are preferentially oriented along the c-axis direction perpendicular to the substrate surfaces. It was found that ZnO layers deposited on thin ZnO seed films and clean Si have well developed grain structures with more c-axis preferred phase and better crystal quality than that of the rest of the samples. Due to this characteristic, the layers deposited on ZnO seed layers and Si are potential materials for preparing chemical and biosensors where an inherently large surface to volume ratio of structured materials is an important prerequisite for enhanced sensitivity.

Acknowledgements

The authors thank Prof. G. Schmahl, Dr. J. Thieme and Dr. B. Niemann from Roentgenphysik-Uni of Goettingen-Germany for the gift of the PECVD installation, to Sc.D. I. Tomov for carrying out the XRD analysis, and to Dr. K.Starbova for the SEM observations.

This study was supported by National Science Fund Grant # D002-207.

References

[1] Ozgur U, Alivov Y, Liu C, Teke A, Reshchikov M, Dogan S, Avrutin V, Cho S and Morkoc H 2005 *J. Appl. Phys.* **98** 041301
[2] Wang X, Song J and Wang Zh 2007 *J. Materials Chemistry* **17** 711
[3] Mende L and Driscoll L 2007 *Materialstoday* **10** 40
[4] Lu J, Chang P and Fan Zh 2006 *Mater. Sci. Eng.* R**52** 49
[5] Ma Y, Du G, Yin J, Yang T and Zhang Y 2005 *Semicond. Sci. Technol.* **20** 1198
[6] Karpina V, Lazorenko V, Lashkarev C, Dobrowolski V, Kopylova L, Baturin V, Pustovoytov S, Karpenko A, Erimin S and Lytvyn P 2004 *Cryst. Res. Technol.* **39** 980
[7] Minchev M, Kitova S and Danev G 2009 *JOAM* **11** 1312
[8] Conley J, Stecker L and Ono Y 2006 *J. Electronic Mater.* **35** 795
[9] *Powder Diffraction File 89-1397 for hexagonal Zinc oxide* (2001 JCPDS-International center for Diffraction data)
[10] Kajikawa Y 2006 *J. Crystal Growth* **289** 387

Optical properties of magnetron sputtered Tb^{3+} ions containing thin dielectric films for thin film solar cells applications

M Sendova-Vassileva[1,4], K Baumgartner[2], O Angelov[1], B Holländer[3], D Dimova-Malinovska[1] and R Carius[2]

[1]Central Laboratory of Solar Energy and New Energy Sources, Bulgarian Academy of Sciences, 72 Tzarigradsko Chaussee, 1784 Sofia, Bulgaria
[2]Institut für Energieforschung 5, Forschungszentrum Jülich GmbH, D-52425 Jülich, Germany
[3]Institut für Bio- und Nanosysteme 1, Forschungszentrum Jülich GmbH, D-52425 Jülich, Germany

E-mail: marushka@phys.bas.bg

Abstract. Results are presented on the deposition and characterization of thin SiO_2 and Al_2O_3 films containing Tb^{3+} ions developed for application as spectral converters. The films are prepared by RF magnetron co-sputtering. The photoluminescence (PL) is measured at room temperature using the 488 nm line of an Ar laser. The dependence is studied of the PL intensity on the Tb concentration in the film. It is found that the intensity exhibits a maximum at about 1 at.%. Annealing studies are performed on SiO_2:Tb using two different methods to improve the PL intensity. In both regimes of annealing, the best results are obtained at 650 - 700°C. After treatment at this temperature the Tb PL increases 2.5 - 3 times.

1. Introduction

One of the ways proposed for increasing the efficiency of solar cells is the application of spectral converters [1]. By absorbing light with a shorter wavelength and emitting at a longer wavelength they can transform the solar spectrum to a spectral region better suited to the absorption characteristics of the solar cell materials. For the implementation of this idea, semiconductor nanoparticles [2] or rare earth ions [3] could be used.

Solar cells with glass on the front side have a poor blue and near UV response due to absorption in the glass and in the doped layer. This poor response of so-called superstrate solar cells could be enhanced by spectral conversion of the solar spectrum. Tb^{3+} ions embedded in a transparent film deposited on the front of the glass could convert light from the UV region into the range of 400 – 600 nm [4]. This light would pass through the glass and be utilized by the thin film Si based solar cell deposited on its back.

In this report we present a study of the optical properties of thin films of SiO_2 and Al_2O_3 containing Tb^{3+} ions deposited by magnetron co-sputtering. The dependence was investigated of the

[4] To whom any correspondence should be addressed.

photoluminescence intensity in the above range on the Tb concentration and on heat treatment of the films.

2. Experimental

Thin films were deposited by magnetron co-sputtering using a SiO_2 or an Al_2O_3 target. Pieces of Tb foil of different areas were placed inside the erosion zone of the target. Silica, polished Si and sapphire substrates were used without intentional heating. The substrates were placed in the center of the sample holder, which was parallel to the target at a distance of 12 cm from it. The area of the Tb pieces (i.e., the Tb concentration) and the sputtering power, *P*, (i.e., the deposition rate) were varied. The Tb concentration was measured by Rutherford backscattering spectroscopy (RBS). The thickness of the films was determined by a profilometer.

The photoluminescence (PL) was measured at room temperature by using the 488 nm line of an Ar laser at a power of 2 mW and a nitrogen-cooled CCD camera attached to a single monochromator.

Annealing was performed to improve the PL intensity of the SiO_2:Tb films employing two different methods: (i) 30 min in vacuum in a tube furnace and (ii) rapid thermal annealing (RTA) for 5 min in nitrogen atmosphere in the temperature interval 600 – 900°C. The PL spectra after the annealing were recorded with a double monochromator and a photomultiplier. The 488 nm laser beam was chopped at 85 Hz and the PL signal was detected by a lock-in amplifier.

3. Results and discussion

The photons of the 488 nm Ar laser line excite the Tb^{3+} ions resonantly from the $^7F_6$ ground state into the $^5D_4$ level [4]. The radiative transitions to the $^7F_{3,4,5}$ levels were recorded. Figure 1 shows a typical PL spectrum of a SiO_2:Tb thin film. This spectrum was spectrally corrected for the experimental setup.

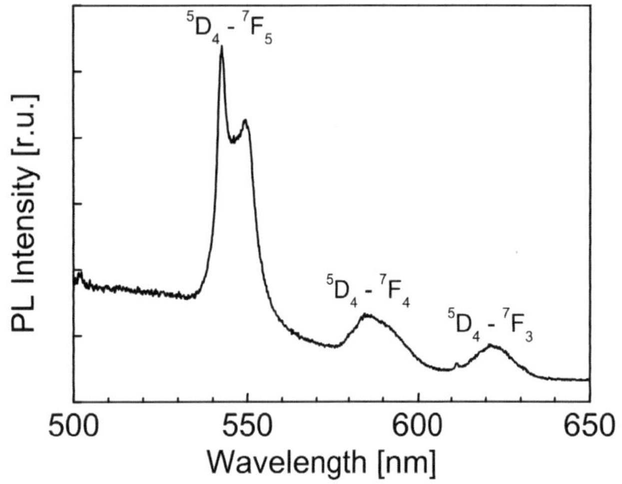

Figure 1. PL spectrum of a 220 nm thin SiO_2:Tb film on a Si substrate containing 1.1 at.% Tb.

Three characteristic bands were observed and identified as the following 4f-4f transitions: $^5D_4 - {}^7F_5$ (~ 546 nm), $^5D_4 - {}^7F_4$ (~ 586 nm) and $^5D_4 - {}^7F_3$ (~ 622 nm). These emission lines appeared on a broad background which could be related to the luminescence of defects inside the SiO_2 matrix.

Different series of samples were sputtered with SiO_2 or Al_2O_3 targets on different kinds of substrates and at two different sputtering powers of 150W and 180W. The intensity of the 542 nm PL band, normalized against the film thickness, is plotted against the area of Tb foil used during the sputtering of the SiO_2:Tb films in figure 2, where the different series of samples are shown with different symbols and the substrate and power used are shown. The inset in figure 2 shows the dependence of the Tb concentration on the Tb foil area for most of the same samples, which can be approximated by a linear function. Figure 3 exhibits similar data for the Al_2O_3:Tb films. In this case all the samples were sputtered at one and the same power of 150 W. The PL intensity, normalized against the film thickness, exhibits a maximum at about 1 at.% of Tb for the SiO_2 and apparently for the Al_2O_3 matrix. At higher Tb content concentration quenching of the PL sets in [5,6]. It was found that the PL intensity of the thin films on Si substrates was higher than on silica substrates. This could be caused by the different optical properties, which was not considered in the data evaluation. The silica substrates are transparent, whereas the Si substrates lead to interference effects because of the greater refractive index.

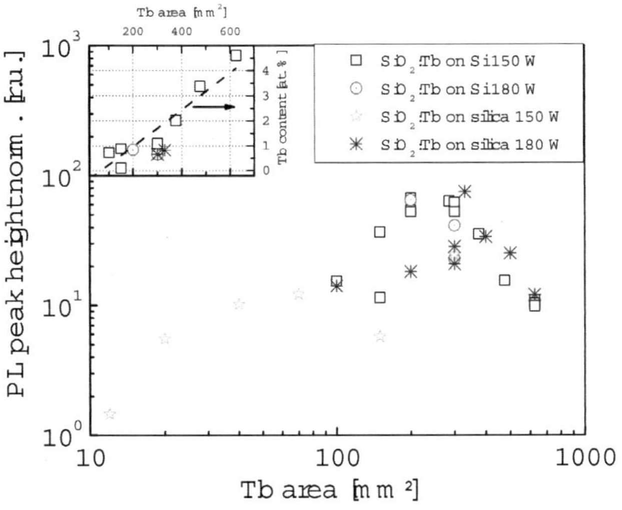

Figure 2. Dependence of the PL intensity from thin SiO_2:Tb films on the Tb area. The deposition conditions are indicated in the figure. The intensities are normalized against the film thickness. The Tb area corresponds to the Tb concentration in the films (see inset).

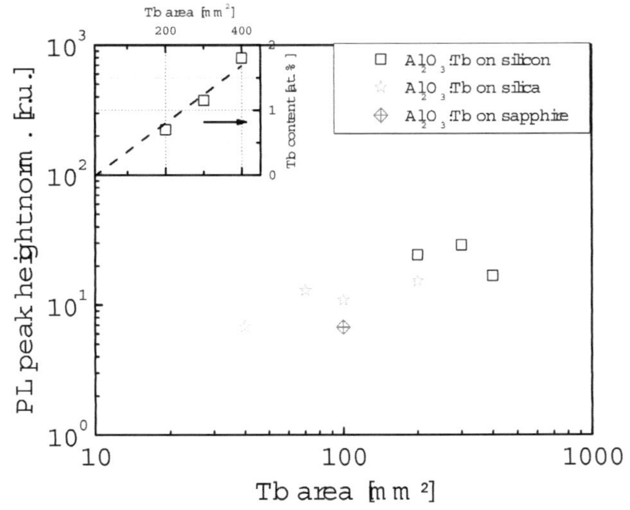

Figure 3. Dependence of the PL intensity from thin Al_2O_3:Tb films on the Tb area. The substrates used are indicated in the figure. The intensities are normalized against the film thickness. The Tb area corresponds to the Tb concentration in the films (see inset).

Further, the dependence was investigated of the PL intensity of the SiO_2:Tb samples on the annealing temperature using the two different annealing methods described above. The sample used for the experiment was a 260 nm thin SiO_2:Tb film on a Si substrate, which was sputtered with a Tb area of 150 mm² (Tb concentration of 0.89 at.%). The ratio of the PL intensity in the 542 nm band after the annealing to that before the treatment is given in table 1. The best annealing conditions for thin SiO_2:Tb films were found to be in the temperature range 650 – 700°C.

In figure 4 the influence of the two annealing methods is compared for 700°C. Annealing in vacuum for 30 min leads to an increase of the background luminescence, which could be related to defects inside the SiO_2 matrix, because of possible outgassing (figure 4 (a)). The samples annealed by

RTA in a nitrogen atmosphere for 5 min did not show this effect (figure 4 (b)).

Similar annealing could be used for processing the Tb^{3+} containing films for solar cell applications. After the treatment the substrates with the dielectric film on the front side can be used for depositing thin film solar cells on the back side.

Table 1. Dependence of the PL intensity of the SiO_2:Tb samples on the annealing temperature (T).

Annealing T [°C]	600	650	700	800	900
PL ratio – vacuum	2.1	2.4	2.2	1.9	1.5
PL ratio - RTA	2.4	2.5	2.7	2.5	1.8

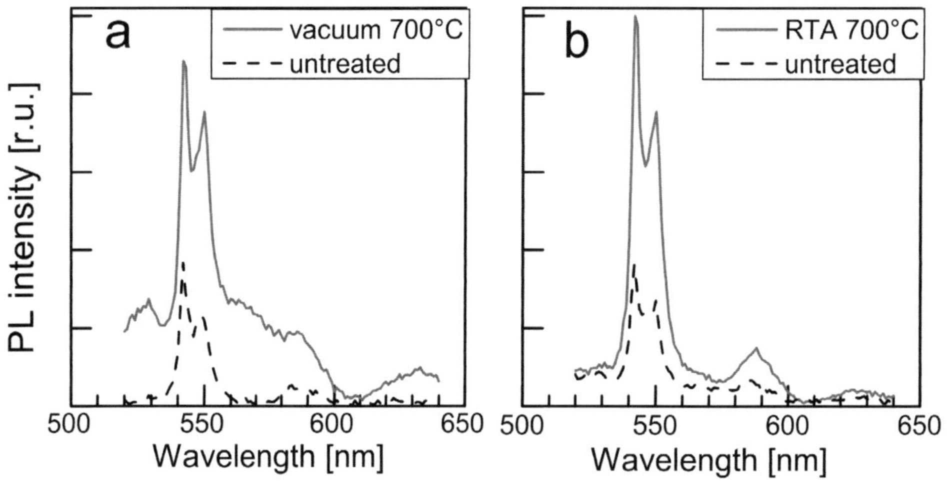

Figure 4. PL spectra of a sample with 260 nm thin SiO_2:Tb film on a Si substrate, sputtered with a Tb area of 150 mm². The sample is annealed (a) in a tube furnace for 30 min in vacuum and (b) by RTA for 5 min in nitrogen both at 700°C.

4. Conclusion

Thin films of SiO_2 and Al_2O_3 containing Tb^{3+} ions were prepared by magnetron co-sputtering. They exhibited photoluminescence in the visible between 500 and 650 nm which makes them suitable for application as spectral converters in thin film silicon based superstrate solar cells. The maximum PL intensity was observed for samples containing about 1 at.% Tb. At higher concentrations quenching sets in. Annealing in vacuum in a tube furnace and by RTA in nitrogen atmosphere was applied in an effort to increase the PL intensity. The RTA at 700°C gave the best results – a close to threefold increase of the PL intensity after annealing and no additional background luminescence. Such treatment could be applied to the substrate with the Tb^{3+} containing film on it before the solar cell deposition.

Acknowledgements

Part of this work is funded by the German Ministry of Education and Science (BMBF) in the project NANOVOLT (#03SF0322A).

References

[1] Maruyama T and Kitamura R 2001 *Sol. Energy Mater. Sol. Cells* **69** 207
[2] van Sark W G J H M, Meijerink A, Schropp R E I, van Roosmalen J A M and Lysen E H 2005 *Sol. Energy Mater. Sol. Cells* **87** 395
[3] Nakata R, Hashimoto N and Kawano K 1996 *Jpn. J. Appl. Phys.* **35** L90
[4] Reisfeld R and Jorgensen C K 1977 *Lasers and Excited States of Rare Earths* (Berlin: Springer) p 140
[5] Reisfrld R and Jorgensen C K 1987 *Handbook on the Physics and Chemistry of Rare Earths* **9** ed K Gschneider and L Eyring (Amsterdam: Elsevier) p 1
[6] Amekura H, Eckau A, Carius R and Buchal C 1998 *J. Appl. Phys.* **84** 3867

Corrosion properties of nitride, oxide and multilayer coatings on stainless steel and titanium-based substrates

A Zykova[1,5], V Safonov[2], J Walkowich[3], R Rogowska[3] and S Yakovin[4]

[1]Institute of Surface Engineering, 4 Zalutinskaya Str., Kharkov, Ukraine
[2]National Science Center "Kharkov Institute of Physics and Technology",
1 Academicheskaya Str., Kharkov, Ukraine
[3]Institute for Sustainable Technologies, National Research Institute,
6/10 K. Pulaskiego Str., Radom, Poland
[4]Department of Physical Technologies, Kharkov National University,
31 Kurchatov Ave., Kharkov, Ukraine

E-mail: zykov@bi.com.ua

Abstract. A comparative analysis is made of the parameters of oxide, nitride and multilayer coatings, such as TiN, deposited by means of Arc-PVD, oxide Al_2O_3 films deposited by magnetron sputtering (MS), and of multilayer TiN/Al_2O_3 on stainless steel (1H18N9) and a titanium-based material (Ti4Al6V). The corrosion examinations of anodic polarization by potentiodynamic method, Tafel and Stern curves and also impedance method at NaCl and SBF solutions were presented. The best corrosion resistance characteristics are exhibited by TiN/Al_2O_3 multilayer coatings on both stainless steel and titanium substrates.

1. Introduction

Various modern techniques are widely used for multilayer (including nano-structured) coatings deposition: electrochemical oxidation, plasma spray, modified Arc-physical vapor deposition (PVD) with separation, magnetron sputtering (MS) and ion-beam assisted deposition (IBAD). Novel multilayer coatings are applied in different industrial areas thanks to their high hardness and wear resistance properties [1]. However, many such applications require high stability in an aggressive and corrosive environment. Acquiring detailed information on the mechanical and corrosion properties of multilayer coatings depending on the deposition method and the specific technological conditions is very important for future successful application in real processes and devices.

Corrosion is one of the major processes that cause problems when metals and alloys are used as implants in the body [2]. Corrosion of implants in the aqueous medium of body fluids takes place via electrochemical reactions [3], so that it is necessary to understand the electrochemical principles that are most relevant to the corrosion processes. The body fluid environment may well decrease the fatigue strength of the metal implant and enhance the release of iron, chromium, nickel, titanium ions, which have been found to be powerful allergens and carcinogens [4]. The presence of titanium in the surrounding tissues of these implants in the form of titanium compounds and subsequent failure of implants due to fatigue, stress corrosion cracking and poor wear resistance have been reported [5,6].

[5] To whom any correspondence should be addressed.

Journal of Physics: Conference Series **223** (2010) 012024 doi:10.1088/1742-6596/223/1/012024

Release of metal ions into the tissues adjacent to the implants results in accumulation of harmful products in the tissues and internal organs of animals [7]. This may also be detrimental to the bone attachment and further bone growth on the implant surface.

2. Materials and methods

The substrates for deposited coatings were stainless steel (1H18N9) and a titanium-based material (Ti4Al6V) samples. The substrates were ultrasonically cleaned consecutively in acetone, ethanol and deionized water and then dried.

The TiN coatings were deposited by means of Arc- PVD. The main parameters of the process were described in our previous study [7, 8]: reactive atmosphere N_2, deposition pressure 8 10^{-2} Pa, arc current 100 A, substrate bias – 150 V, deposition temperature 593°K, thickness of deposited coatings 2 - 5 μm. TiN (MS), Al_2O_3 (MS) and multilayer coating deposition were performed in a high-vacuum pumping system with base pressure about 10^{-5} mBar. Details on the magnetron and the ion source in the sputtering chamber were given in [9]. The coatings thickness, adhesion properties, hardness and elastic modulus were evaluated by standard techniques [9].

A comparative analysis was made of the corrosion parameters of multilayer coatings types such as TiN (Arc)/Al_2O_3 and TiN (MS)/Al_2O_3. This was performed using Potentiostat PARSTAT 2263 (AMETEK, USA) equipment and included anodic polarization by applying the potentiodynamic method in the potential range 1.0 V - +2.0 V at a scanning rate of 1 mV/s, Tafel 0.050 V – +0.050 V and Stern 0.020 V – +0.020 V range curves and the impedance method in the frequency range 100 kHz – 10 MHz in 0.5 N NaCl and SBF (NaCl - 8,035, $NaHCO_3$ - 0,355, KCl - 0,225, K_2HPO_4 $3H_2O$ - 0,231, $MgCl_2$ $6H_2O$ - 0,311, $CaCl_2$ - 0,292, Na_2SO_4 - 0,072 at pH = 7.4 and temperature of 37°C) solutions. The samples were immersed in the electrolyte and the potential was monitored as a function of the time until the potential reached a steady value. In the solutions corrosion occurs in the form of anodic dissolution, namely, uniform removal process for the substrates and corrosive electrolyte pore penetration for the ceramic coatings.

Electrochemical impedance spectroscopy (EIS) is a powerful analytical technique, which can provide a large amount of information on the corrosion reactions, the mass transport and electrical charge transfer characteristics of coated materials in various solutions. The impedance spectrum reflects dielectric behavior, oxidation-reduction reactions and mass migration which are determined by the electrical and chemical properties of the corrosion medium and the electrode materials. Over the frequency bandwidth of interest, the impedance are presented in various ways by both the Nyquist and Bode plots. The EIS spectra describe the electrical charge transfer kinetics and details on the physical and electrochemical corrosion characteristics of the substrate/coating interface. EIS measurements were performed in 0.5 N NaCl and SBF solutions. Platinum wire and Ag/AgCl were used as counter and reference electrodes, respectively, in the frequency ranges 100 kHz – 10 MHz at constant (5 mV) amplitude and (250 mV) initial potential for all measurements. The impedance parameters $|Z|$, polarisation resistance Rp and capacitance C were calculated from Nyquist and Bode plots. The surface topography before and after the corrosion test was investigated by SEM (HITACHI, Japan) and AFM (Quesant Instrument Corporation, USA).

3. Results and discussion

The best mechanical parameters were obtained in the case of nitride coatings on both SS and Ti substrates and for multilayer nitride/oxide coatings [9]. The anodic polarization corrosion tests using the potentiodynamic technique in the potential range 1.0 V - +2.0 V at a scanning rate of 1 mV/s are presented in figure 1 a) for SS, SS/TiN, SS/Al_2O_3, SS//TiN(Arc)/Al_2O_3, SS/TiN(MS)/Al_2O_3 and in figure 1 b) for Ti, Ti/TiN, Ti/Al_2O_3, Ti//TiN(Arc)/Al_2O_3, Ti/TiN(MS)/Al_2O_3 coatings in 0.5 N NaCl solutions and by means of Tafel 0.050 V - +0.050 V and Stern 0.020 V - +0.020 V range curves, The impedance spectra of all samples were recorded before and after polarization to evaluate the performance of the coatings under equilibrium conditions and after the onset of the corrosion process, respectively.

The Nyquist impedance plot was obtained from the real ($Z_{re} = R_s+R_p/(1+\omega^2R_p^2C_d^2)$) and the imaginary ($Z_{im} = \omega R_p^2C_d /(1+\omega^2R_p^2C_d^2)$) impedance at different frequencies to determine the charge-transfer kinetics (R_s is the electrolyte solution resistance, R_p is the polarization resistance and C_d is the capacitance at the interface).

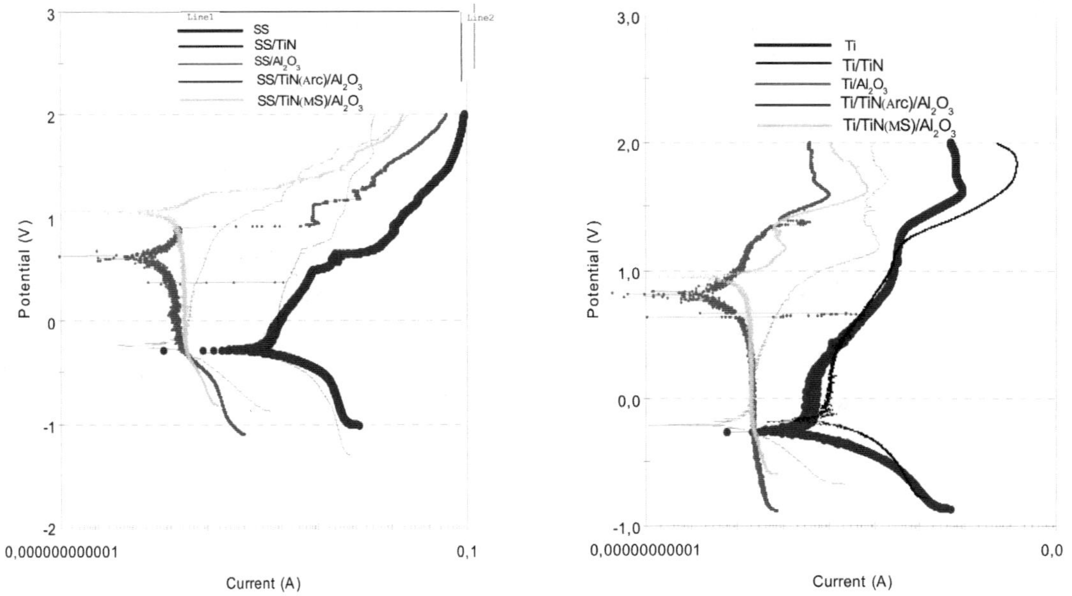

a) b)

Figure 1. Anodic polarization curves for a) SS, SS/TiN, SS/Al_2O_3, SS//TiN(Arc)/Al_2O_3, SS/TiN(MS)/Al_2O_3 and b) Ti, Ti/TiN, Ti/Al_2O_3, Ti//TiN(Arc)/Al_2O_3, Ti/TiN(MS)/Al_2O_3 coatings in 0.5 N NaCl solutions.

Figure 2, a) and b) show the Nyquist plots for SS, SS/Al_2O_3 and for Ti, Ti/Al_2O_3 coatings in SBF solution. The data show that coating deposition improves charge-transfer kinetic at counter electrode-electrolyte interfaces. The surface with ceramic oxide coatings has strong capacitive response due to their electrically inert properties and high dielectric constants. The surface topography before and after corrosion tests was investigated by SEM and AFM. The data for the coatings in Figure 3 further confirms the main results inferred from the polarization curves and EIS measurements.

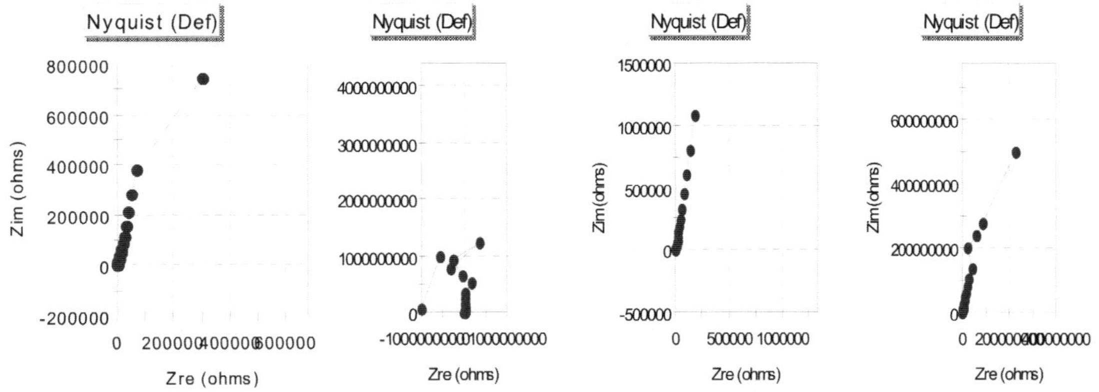

a) b)

Figure 2. Nyquist plots for a) SS, SS/Al_2O_3 and b) Ti alloy, Ti/Al_2O_3 in SBF solution.

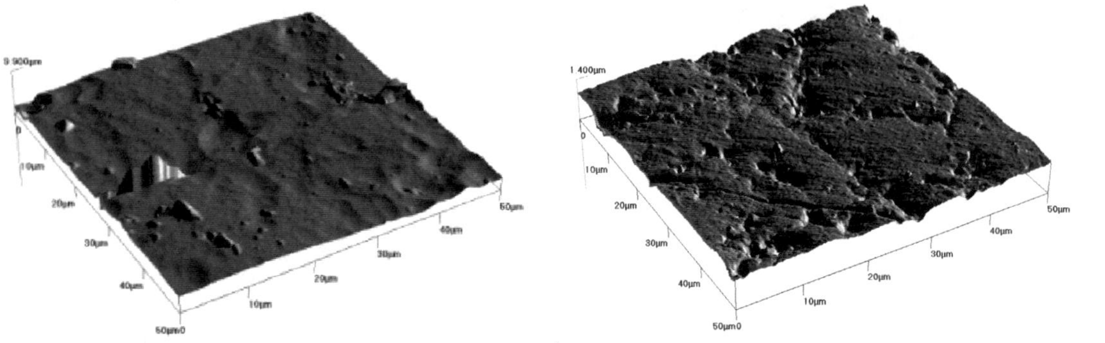

a)

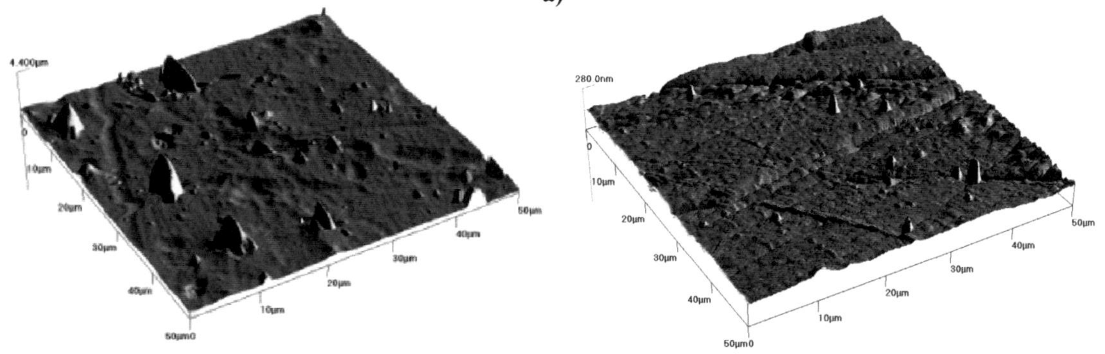

b)

Figure 3. Surface topography and failure after corrosion tests (0.5 N NaCl) by AFM: a)Ti/TiN (Arc)/Al_2O_3 and Ti/TiN (MS)/Al_2O_3 and b) SS/TiN (Arc)/Al_2O_3 and SS/TiN (MS)/Al_2O_3 coatings.

4. Conclusion

The results show that the best corrosion resistance characteristics in NaCl and SBF solutions were exhibited by the multilayer coatings TiN/Al_2O_3 both on stainless steel and titanium substrates. However, the nitride corrosion resistance is lower on 1H18N9 than on Ti4Al6V and lower than that obtained for oxide coatings. The formation of TiN interlayers results in improvement of the mechanical parameters and corrosion resistance characteristics of multilayer coatings TiN/Al_2O_3 deposited by Arc –PVD and magnetron sputtering.

References

[1] Serro A P, Completo C, Colaco R, dos Santos F, Lobato da Silva C, Cabral J M S, Araujo H, Pires E and Saramago B 2009 *J. Surface&Coating Technol.* **203** 3701-7
[2] Eisenbarth E, Velten D, Schenk-Meuser K, Linez P, Biehl V, Duschner H, Breme J and Hildebrand H F 2002 *Biomolecular Engin.* **19** 243-9
[3] Dion L, Bordenave F, Lefebvre R and Boreille C V 1994 *J. Mater. Sci.: Mater. Medicine* 18-24
[4] Kim D J, Lee M H, Lee D Y and Han J S 2000 *J. Biomedical Mater. Res.* **53** 438-43
[5] Ratner B D 1993 *J. Biomedical Mater. Res.* **27** 837-50
[6] Velten D, Biehl V, Aubertin F, Valeske B, Possart W and Breme J 2002 *J. Biomedical Mater. Res.* **59** 18-28
[7] Zykova A V, Luk'yanchenko V V and Safonov V I 2005 *J. Surface & Coating Technol.* **200** 90-93
[8] Zykova A, Safonov V, Luk'yanchenko V, Walkowicz J, Rogovska R and Yakovin S 2008 *J. Phys.: Conf. Series* **113** 012029
[9] Zykova A, Safonov V, Luk'yanchenko V, Walkowicz J, Rogovska R and Yakovin S 2009 *Problems of Atomic Sci. & Technol. Series: Plasma Physics* **15** 156-8

Mass-sensitive TiO_2-based acoustic sensor element

V Georgieva[1], N Donkov[2,5], D Spassov[1], I Jordanova[3] and A Grechnikov[4]

[1]Georgi Nadjakov Institute of Solid State Physics,
Bulgarian Academy of Sciences, 72 Tzarigradsko Chaussee, 1784 Sofia, Bulgaria
[2]Emil Djakov Institute of Electronics,
Bulgarian Academy of Sciences, 72 Tzarigradsko Chaussee, 1784 Sofia, Bulgaria
[3]Faculty of Physics, St. Kl. Ohridski University of Sofia,
5 J. Bouchier Blvd., 1164 Sofia, Bulgaria
[4]Vernadsky Institute of Geochemistry and Analytical Chemistry,
Russian Academy of Sciences, 19 Kosygin Str., 119991 Moscow, Russia

E mail: nikolaidd@abv.bg

Abstract. The sorption properties with respect to gaseous NH_3 of e-beam evaporated TiO_2 layers were investigated by a quartz crystal microbalance (QCM) allowing the detection of nanograms of sorbed mass. The TiO_2 (249 nm thick) layers were deposited on 16 MHz resonators. The as-deposited TiO_2 layers were characterized by ellipsometry, X-ray diffraction (XRD) and X-ray photoelectron spectroscopy (XPS). The measurements of the QCM-TiO_2 system's sorption properties were carried out on a laboratory setup allowing operation in dynamic and static modes under strict control of the process parameters which were recorded and processed automatically by an electronic measuring system. The sorption properties of TiO_2 were determined using the frequency shift (ΔF) of the QCM in NH_3 with concentrations ranging from 10 ppm to 1000 ppm. The time of sorption and de-sorption were defined; the maximal QCM frequency shifts for each NH_3 concentrations were measured and the sorbed masses were calculated. It was determined that ΔF increases from 8 Hz at 10 ppm to 216 Hz at 1000 ppm, which corresponds to sorbed masses of 2.78 ng and 75.19 ng, respectively. The results show that the QCM-TiO_2 system could very successfully be used as a sensor element for NH_3 detection in the range 10 - 1000 ppm.

1. Introduction
Titanium dioxide films have a wide range of applications, such as dielectric capacitors films; in photovoltaic devices; in photocatalysis [1], etc. Recently, the CO, NO_x, O_3 and SO_2 sensing properties of TiO_2 have been demonstrated [2]. The aim of this paper is to investigate the gas sensing properties of e-beam evaporated TiO_2 in view of NH_3 sensor application.

2. Experimental details
The TiO_2 films were deposited by e-beam evaporation of TiO_2 powder (initial pressure of 7×10^{-4} Pa, working pressure of 3×10^{-3} Pa and deposition rate of 50 nm/min). The thickness and the refractive index of the TiO_2 films were determined using a Rudolph 435 ellipsometer, λ = 632.8 nm. The surface

[5] To whom any correspondence should be addressed.

morphology of the TiO_2 deposited on the resonators was studied using SEM (Philips 515). The structure and the composition of the films were determined by XRD and XPS. An URD6 diffractometer with Cu-K_α radiation, $\lambda = 1.54178$ Å and a flat graphite monochromator were used in the XRD analysis. The diffraction patterns were registered in two geometries: Bragg-Brentano (B-B) and asymmetric grazing angle incidence mode with a fixed angle of incidence of 2^0. The XPS analysis was carried out on an ESCALAB Mk II (VG Scientific) spectrometer with an Al K_α X-ray source (1486.6 eV). The photoelectron line of carbon C 1s at 285.0 eV was used as an energy reference. The sorption properties of TiO_2 to NH_3 were investigated by the QCM method. The QCM structures were formed on an AT-cut quartz plate ($\phi = 8$ mm) with Au electrodes ($\phi = 5$ mm and thickness 60 nm), and had a resonance frequency of 16 MHz. Some of the equivalent circuit parameters such as the static capacity (C_0) and the dynamic resistance (R_q) were measured with a selective level meter; while others parameters such as the dynamic inductance (L_q) and the quality factor (Q) were calculated. The sorbed mass of NH_3 (Δm) was calculated by the Sauerbery equation. The experimental setup and the methodology of measurements were described in detail in [3]. The main stages of the measurement process were as follows: 1) purging of the structures with dry air to reach saturation and fix the initial frequency, F_0; 2) creating a certain NH_3 concentration; 3) reaching saturation at frequency - F_s (frequency of saturation for sorption time, t_s); 4) purging of the structures with dry air to restore the initial frequency due to complete NH_3 desorption. The desorption time, t_d, was also measured. The studies of the QCM – TiO_2 system were performed in the interval 10 - 1000 ppm NH_3. The NH_3 concentration in the test chamber was controlled by accurate measurements of the NH_3 flow and the diluting gas using MFCs, and by means of indicator tubes. The measurements were performed at a temperature of 26°C maintained with an accuracy of ±0.01° C.

3. Results and discussion

The results of the ellipsometric and SEM studies of the TiO_2 films indicated a thickness of 249 nm, refractive index of 2.167, and a smooth and uniform surface morphology. The main QCM parameters were determined as $R_q = 5\ \Omega$, $L_q = 2.798$ mH, $Q = 55544$, indicating high quality of the structures. The process of TiO_2 film evaporation worsens the QCM parameters as a result of mass loading. This leads to an increase in Rq to 6.7Ω and a decrease in Q by about 20%. Despite these changes in the QCM parameters, their values remain satisfactory. The diffraction pattern registered in the grazing mode (figure 1) shows a halo in the 2θ angles up to 40°. These points to the presence of an amorphous-like disordered phase. There are a number of diffraction peaks superimposed on the amorphous halo. These peaks correspond to a polycrystalline β-TiO_2. One of the peaks could be related to γ-TiO_2. For the latter phase, however, only one peak is observed which makes its identification somehow ambiguous. The amorphous halo and the peak associated with γ-TiO_2 are absent in the diffraction pattern registered in the B-B geometry (figure 2). Since in this geometry the X-ray path through the film is considerably shorter than in the case of grazing beam incidence, the phases with lower concentration in the film produce diffraction peaks whose intensity is too low to be distinguished from the background. Thus, the films have a predominantly polycrystalline structure consisting mainly of β-TiO_2 and, possibly, with a negligible amount of γ-TiO_2. An amorphous phase is also present and its content is less that that of the polycrystalline phase. Figure 3 presents the Ti 2p and O 1s XPS regions of the spectra which indicate that the films are not stoichiometric, since the Ti 2p line differs significantly from the characteristic doublet of stoichiometric TiO_2 [4]. The deconvolution of the Ti 2p spectrum shows clearly the presence of a Ti-suboxide state, namely Ti^{3+}. Applying a standard fitting procedure for the Ti^{4+} 2p doublet yields binding energies of the Ti^{4+} $2p_{3/2}$ and Ti^{4+} $2p_{1/2}$ peaks of 458.6 eV and 464.1 eV, respectively, as well as a FWHM of 3.4 eV. Similarly to the Ti^{3+} photoelectron line, the Ti^{3+} $2p_{3/2}$ and Ti^{3+} $2p_{1/2}$ peaks are positioned at 456.1 and 461.6 eV, respectively, while the FWHM was found to be 2.5 eV. The energy separation between $2p_{3/2}$ and $2p_{1/2}$ states of both doublets is 5.5 eV, and the binding energy shift between Ti^{4+} and Ti^{3+} doublet was 2 eV.

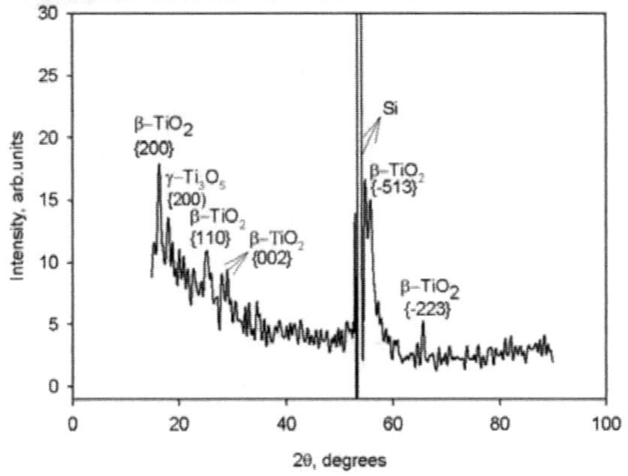

Figure 1. XRD of TiO_2 taken under asymmetric grazing beam geometry at incidence angle of 2^0.

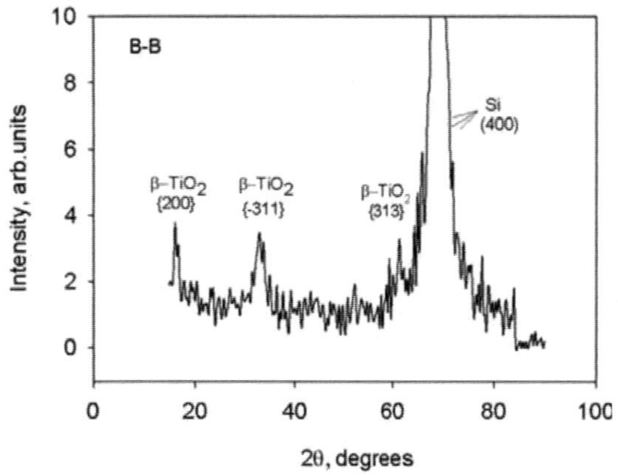

Figure 2. XRD of TiO_2 taken under symmetric Bragg-Brentano (B-B) geometry.

These values are in good agreement with the data reported by other authors for Ti^{4+} and Ti^{3+} oxidation states [5]. The Ti-suboxide concentration estimated from the spectra deconvolution is 28 at.%. The O 1s line (figure 3b) consists of a single peak, slightly asymmetric towards the binding energies. It is well deconvoluted into 2 Gaussian peaks: a main one at 530.7 eV (FWHM of 0.8 eV) and a second low intensity peak at 532.2 eV (FWHM of 1.2 eV). The peak at 530.7 eV is representative of 4-fold coordination oxygen in the bulk of TiO_2 [6]. The asymmetry of the O 1s line of TiO_2 is usually interpreted as the effect of dissociative adsorption of hydroxyl groups on the film's surface [7]. Therefore the smaller peak at 532.2 eV is attributed to surface hydroxyl groups. The FTC characteristics were taken for all NH_3 concentrations in the range studied. Figure 4 presents a typical FTC of a QCM with TiO_2 film at a 100 ppm NH_3 concentration. Initially, the sensor is operated in dry air flow for surface cleaning purpose as well as for providing a reference point. It is seen that the QCM frequency quickly saturates at the constant value F_0. Subsequently, a flow of NH_3 is added to the carrier gas to create 100 ppm of NH_3. The QCM frequency decreases as NH_3 is sorbed until a dynamic equilibrium after 44 s is reached between the sorbed and desorbed molecules in the mass-sensitive film. The steady-state frequency F_s at this stage is recorded. Finally, the system is purged in dry air for about 72 s after which the QCM frequency increases back to its initial value of F_0. The FTC curves measured allow the determination of the main parameters of each QCM- TiO_2 systems as a function of the NH_3 concentration. These are: the sorption and desorption times, the rates of the two processes, the maximal

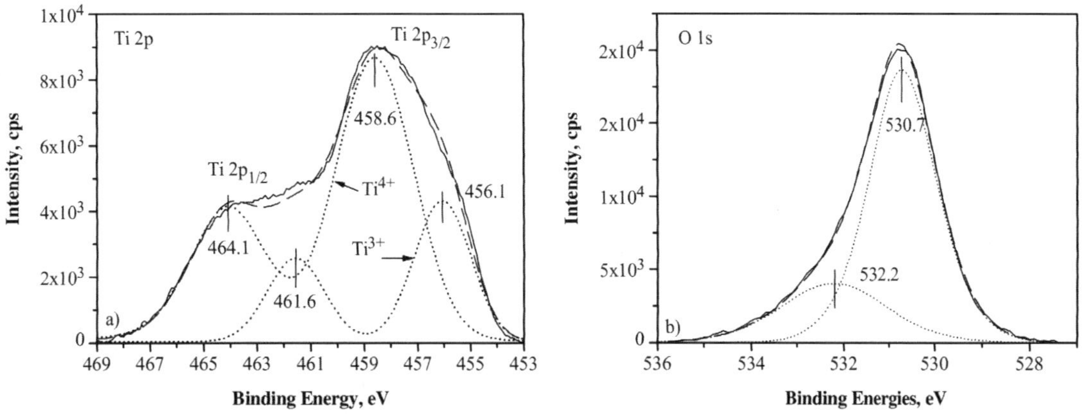

Figure 3. Ti 2p a) and O 1s b) XPS spectra of evaporated TiO_2 films. Solid lines – experimental spectra, dashed and dotted lines – deconvolution.

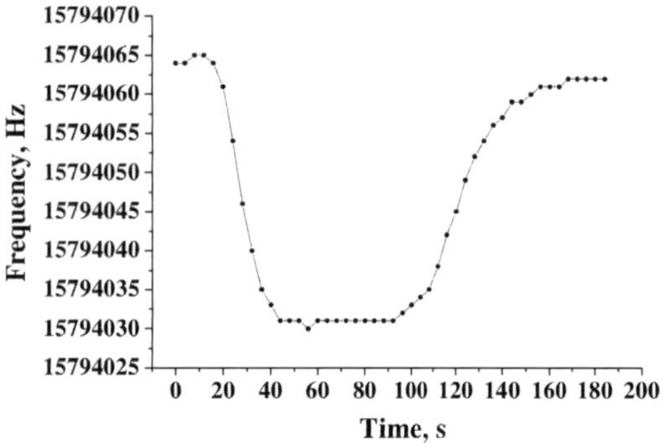

Figure 4. FTC of QCM with TiO_2 film at 100 ppm NH_3.

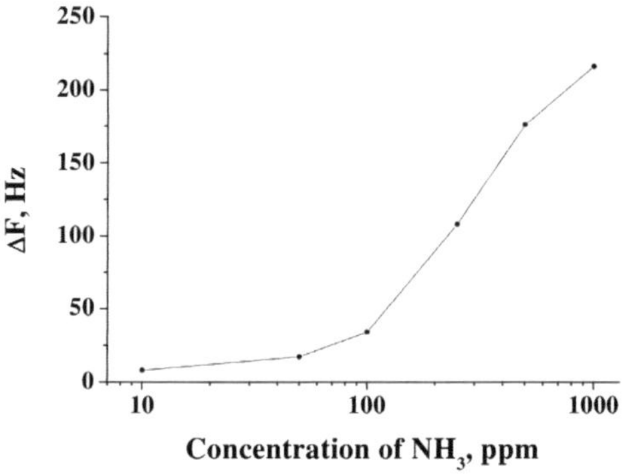

Figure 5. Frequency shift of QCM with TiO_2 film vs. NH_3 concentration.

frequency shift (ΔF) used for calculating the sorbed mass (Δm). The time t_s rises significantly - from 28 to 580 s - as the NH_3 concentration is increased from 50 ppm to 1000 ppm. In the experiments, t_s at NH_3 exceeding 100 ppm was determined as a frequency change in the range of 0.2 Hz for 2 s. The desorption time of the QCM is also strongly affected by the concentration of the test gas. The desorption process runs more slowly at higher concentrations - from 48 s at 10 ppm NH_3 it rises almost tenfold to 440 s at 500 ppm. We also observed that t_d is mainly determined by the purging gas flow, which was kept constant. Figure 5 illustrates the dependence of ΔF calculated as the difference between F_0 and F_s on the NH_3 concentration. Values of ΔF of 8 Hz at 10 ppm NH_3 to 216 Hz at1000 ppm NH_3 were measured. The magnitude of the frequency shift determines the sensitivity of the system. The sorbed mass calculated using the Sauerbery equation range from 2.78 ng at 10 ppm to 75.19 ng at 1000 ppm.

4. Conclusion

The analyses performed indicate that the TiO_2 films are not stoichiometric. The structure is predominantly polycrystalline (β- and γ-TiO_2) with a presence of amorphous phase. The films are sensitive to gaseous NH_3 in the concentration interval from 10 ppm to 1000 ppm. The technology of the TiO_2 films formation is fully compatible with QCM preparation. The QCM-TiO_2 system exhibits fast response and repeatability, which is necessary for the development of NH_3 detection sensors.

Acknowledgements

The work was supported in part by the Bulgarian National Science Fund at the Ministry of Education and Science under grant NT3/06.

References

[1] Gyorgy E, Socol G, Axente E, Mihailescu I.N, Ducu C and Ciuca S 2005 *Appl. Surface Sci.* **274** 429-33
[2] Comini E, Guidi V, Ferroni M and Sberveglieri G 2004 *Sensors and Actuators* B **100** 41-6
[3] Georgieva V, Stefanov P,. Spassov L, Raicheva Z, Atanassov M, Tincheva T, Manolov E and Vergov L 2009 *J. Optoelectr. Adv. Mater.* in press
[4] Wagner D C, Riggs W M, Daris L E and Moulder J F 1979 *Handbook of X-ray Photolectron Spectroscopy* ed G E Muilenberg (Perkin-Elmer Corp Eden Praire MN)
[5] Boffa A B, Galloway H C, Jacobs P W, Benitez J J, Batteas J D, Salmeron M, Bell A T and Somorjai G A 1995 *Surf. Sci.* **326** 80-92
[6] Moulder J F, Stickle W F, Sobol P E and Bomben K D 1995 *Handbook of X-ray photoelectron spectroscopy – A reference book of standard spectra for identification and interpretation of XPS data* (Minnesota USA: Physical Electronics Inc.) pp 44-7
[7] Wang L-Q, Baer D R, Engelhard M H and Shultz A N 1995 *Surf. Sci.* **344** 237-50

16th International Summer School on Vacuum, Electron, and Ion Technologies (VEIT 2009) IOP Publishing
Journal of Physics: Conference Series **223** (2010) 012026 doi:10.1088/1742-6596/223/1/012026

Investigation of the uncertainties of the estimated optical constants and thickness of very thin semiconductor layers

P Gushterova[1], B Hristov and P Sharlandjiev

Central Laboratory for Optical Storage and Processing of Information, Bulgarian Academy of Sciences, Acad. G. Bonchev Str., Bl. 101, 1113 Sofia, Bulgaria

E-mail: pgushterova@dir.bg

Abstract. We recently developed an approach to the estimation of the complex permittivity (ε) and thickness (d) of very thin layers using measurements of their transmittance, front-side reflectance and back-side reflectance. The approach is based on a limited expansion of the Abelès characteristic matrix elements and is especially designed for characterization of very thin layers. In this paper we investigate the uncertainties of the estimated real part of ε (ε_1), the imaginary part of ε (ε_2) and d of semiconductor thin layers that are due to the methodological error and the experimental uncertainties in the optical quantities measured. It is shown that the effect of the uncertainties in the measurable quantities is significantly stronger than that of the methodical error and increases considerably with the decrease of the ratio d/λ (λ being the wavelength). An efficient two-step procedure is proposed to reduce this effect. First, we supply a criterion for determination of d with the lowest uncertainty from the ensemble of estimations constructed on a wavelength by wavelength basis. This is crucial for the next step: the estimation of ε_1 and ε_2. The approach proposed ensures estimation of ε_1, ε_2 and d with the highest accuracy, limited only by the methodological error.

1. Introduction

Optically isotropic layers with a physical thickness (d) of less than $\lambda/30$ (λ being the wavelength) are key components for many applications in X-ray optics [1], microelectronics [2] and photonics [3].

The complex permittivity (ε) and d of thin layers are routinely estimated by reflectance (R) and/or transmittance (T) measurements. To evaluate d and ε, minimization techniques are most often applied [4, 5, 6]. These techniques could be divided in two broad groups: derivative-free (Simplex, direct search) methods and derivative (gradient, Levenberg–Macquardt) methods, which have several features in common: they are iterative; they need an initial guess for the model parameters; they may result in a local minimum, rather than in a global one. Using constraints on the parametric values helps but does not solve all the difficulties. Moreover, the results on very thin absorbing layers are often unreliable due to the strong correlation between ε and d [7] and the parameters obtained are susceptible to small experimental uncertainties in the optical quantities measured [8]. This is why the development of new methods for determination of ε and d, best adapted to this concrete optical problem, is still of great practical and theoretical interest.

[1] To whom any correspondence should be addressed.

In several papers, we have reported a straightforward approach to the estimation of the real (ε_1) and imaginary (ε_2) parts of ε and d of very thin layer using measurements of the T, R and the back-side reflectance (R') of the layer. The approach is based on the limited expansion on the Abelès characteristic matrix elements and is especially designed for characterization of very thin layers [9, 10]. It possesses some important advantages: simplicity of use, no need of initial information about ε and d, the number of the multiple solutions is significantly reduced, etc. In the present work, we investigate the uncertainties in the ε_1, ε_2 and d evaluated of semiconductor thin layers. As a model layer we choose a thin film with ε_1 and ε_2 close to those of Ge because of their relatively high values in the visible and near infrared spectral ranges, and the positive values of ε_1 (a difficult case for evaluation of d and ε).

2. Main equations used for the determination of ε and d

The optical behavior of an optically homogeneous isotropic layer bounded by parallel surfaces is described by its 2 × 2 characteristic matrix [11]. The nano-thickness of the investigated layers allows us to make a limited expansion of the matrix elements as a function of $(\varepsilon)^{1/2}d/\lambda$. Retaining the terms to the 4-th power we obtain approximate expressions for T, R and R'. The use of the relations $(1 + R)/T$, $(1 - R)/T$ and $(1 - R')/T$ leads to a set of three equations, which are simpler than those for T, R, R', and can be easily solved for the unknown quantities (ε_1, ε_2 and d):

$$\frac{1+R}{T}=\frac{(0.5\varepsilon_2^2+\varepsilon_1^2-\varepsilon_2^2\varepsilon_1+0.5\varepsilon_2^2n_s^2-\varepsilon_1 n_s^2+\varepsilon_1^2 n_s^2+\varepsilon_1^3)\omega^4 d^4}{6n_s}$$
$$+\frac{(\varepsilon_2 n_s-n_s\varepsilon_2\varepsilon_1)\omega^3 d^3}{3n_s}+\frac{[(1-\varepsilon_1)(n_s^2-\varepsilon_1)+\varepsilon_2^2]\omega^2 d^2}{2n_s}+\frac{2\omega d\varepsilon_2 n_s+(n_s^2+1)}{2n_s}, \tag{1a}$$

$$\frac{1-R}{T}=\frac{\varepsilon_2^2\omega^4 d^4}{6}+\frac{(\varepsilon_2 n_s^2-\varepsilon_2\varepsilon_1)\omega^3 d^3}{3n_s}+\frac{\omega d\varepsilon_2+n_s}{n_s}, \tag{1b}$$

$$\frac{1-R'}{T}=\frac{\varepsilon_2^2\omega^4 d^4}{6}+\frac{(\varepsilon_2-\varepsilon_2\varepsilon_1)\omega^3 d^3}{3}+\omega d\varepsilon_2+1, \tag{1c}$$

where $\omega = 2\pi/\lambda$ is the wave number, n_s is the substrate refractive index and n_0 is the refractive index of air ($n_0 = 1$).

To calculate ε_1, ε_2 and d, an exact analytical approach is used [12] allowing the determination of the global minima in the parameter space. We applied it to solve the above set of nonlinear equations [9].

3. Estimation of the uncertainties of ε and d: Numerical example

The uncertainties of the solutions obtained are studied by making a realistic numerical simulation of spectrophotometric measurements of a nano-layer deposited on a transparent substrate in the 400 – 800 nm spectral range. A layer with values of ε_1 and ε_2 very close to the semiconductor Ge with d = 15 nm is considered. A substrate with ε_s very close to the values for BK7 glass is assumed.

At the first step, R, R' and T of the layer are calculated using the exact characteristic matrix elements. Subsequently, the reflectance and transmittance of the bare substrate are evaluated, and then the energetic coefficients of a thin layer deposited onto a semi infinite substrate, T_{exp}, R_{exp}, R'_{exp} are obtained [13]. A random noise with normal distribution (zero mean and standard deviation 0.01 of R and R' and 0.005 of T) is added in order to simulate a real measurement of the above quantities. Thus 'experimental' front-side reflectance, back-side reflectance and transmittance are generated.

Using the set of equations (1), we calculate ε_1, ε_2 and d in two cases. To distinguish between the errors due to the approximation of the method and the uncertainty arising from the uncertainties in T,

Journal of Physics: Conference Series **223** (2010) 012026 doi:10.1088/1742-6596/223/1/012026

R and R', ε_1, ε_2 and d are calculated without and with noise added to T, R and R'. Figure 1 shows the spectral dependences of ε_1, ε_2 and d, calculated without noise (black curves) and with noise (ε_1 - green curve, ε_2 - red curve and d - blue curve). A pronounced influence of the uncertainties in R, R' and T on calculated values of ε_1, ε_2 and d at $\lambda > 620$ nm is observed.

As a measure of the accuracy of the solutions we use the relative uncertainties $\Delta d/d$, $\Delta\varepsilon_1/\varepsilon_1$ and $\Delta\varepsilon_2/\varepsilon_2$ between the model complex permittivity parameters and thickness (ε_1, ε_2, d) and the values estimated (ε_1^*, ε_2^*, d^*). The spectral dependences of $\Delta d/d$, $\Delta\varepsilon_1/\varepsilon_1$ and $\Delta\varepsilon_2/\varepsilon_2$ are presented in figure 2.

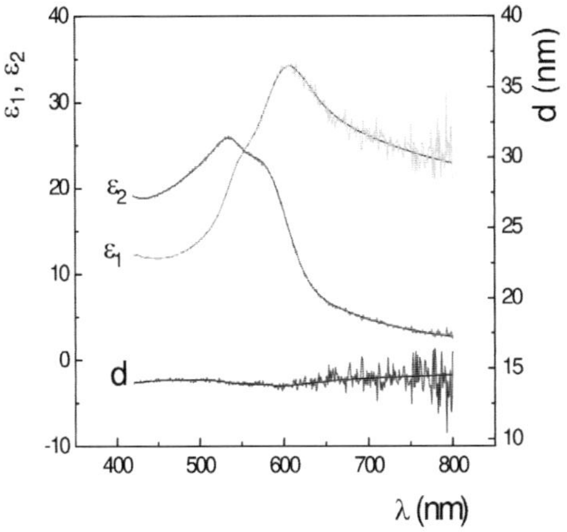

Figure 1. Dispersion of d, ε_1 and ε_2, calculated considering the uncertainties in T, R and R' (colored curves) and without taking into account uncertainties in T, R and R'(black curves).

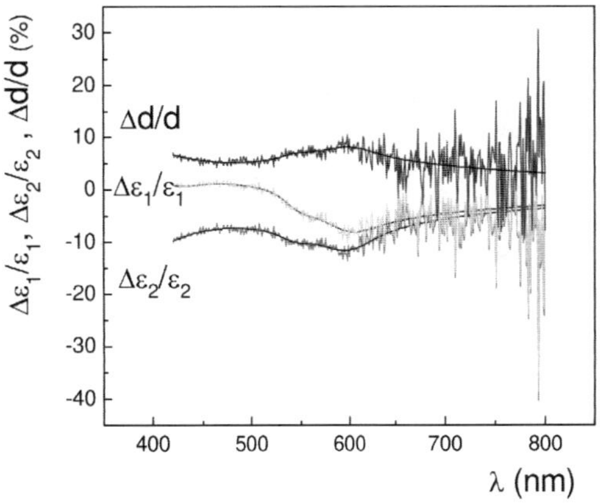

Figure 2. Dispersion of $\Delta d/d$, $\Delta\varepsilon_1/\varepsilon_1$ and $\Delta\varepsilon_2/\varepsilon_2$, calculated considering the uncertainties in T, R and R'(colored curves) and without taking into account uncertainties in T, R and R'(black curves).

The black curves represent the spectral dependence of $\Delta d/d$, $\Delta\varepsilon_1/\varepsilon_1$ and $\Delta\varepsilon_2/\varepsilon_2$, calculated without noise IN R, R' and T. The blue, green, and red curves present the spectral dependence of $\Delta d/d$, $\Delta\varepsilon_1/\varepsilon_1$ and $\Delta\varepsilon_2/\varepsilon_2$, respectively, calculated with noise added to R, R' and T. In the first case, the maximum value of $\Delta\varepsilon_1/\varepsilon_1$ is ~ 8%, of $\Delta\varepsilon_2/\varepsilon_2$ is ~ 11% and of $\Delta d/d$ is ~ 8% at $\lambda = 600$ nm, whereas in the second case, the maximum value of $\Delta\varepsilon_1/\varepsilon_1$ increases to ~ 40%, of $\Delta\varepsilon_2/\varepsilon_2$ to ~ 40% and of $\Delta d/d$ to ~30%.

Our calculations show that the influence of the experimental uncertainties of T, R, and R' and the error due the approximation of the method on the calculated d, ε_1, and ε_2 can be reduced if we select one value of $d(\lambda)$ and recalculate ε_1 and ε_2 values with it. The value of d for which the absolute value of $\Delta d/d$ due to the method approximation is minimal is chosen as the one closest to the model one. In the present example, $\Delta d/d$ has a minimal value 3,3% in the spectral region form 460 nm to 500 nm, where d is 14.5 nm. With this value, which fixes the physical thickness, we recalculate ε_1 and ε_2 using (1a) and (1b) only. The spectral dependences of the recalculated values of ε_1 and ε_2 are shown in figure 3.

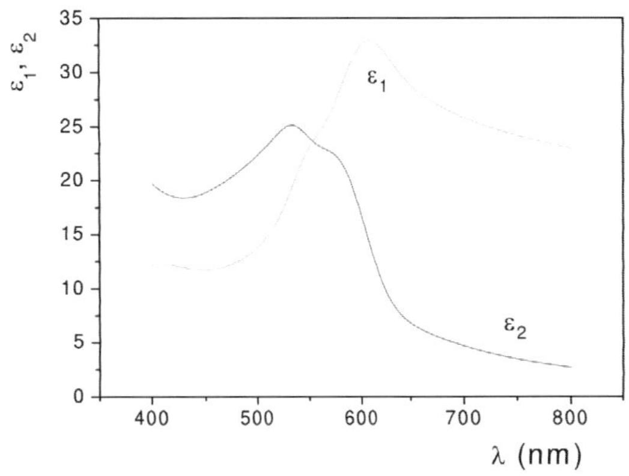

Figure 3. Dispersion of the calculated ε_1 and ε_2, obtained considering the uncertaintiesin T and R.

It is seen that the influence of the experimental uncertainties on them is appreciably reduced. In figure 4, the dispersion of the recalculated values $\Delta\varepsilon_1/\varepsilon_1$ and $\Delta\varepsilon_2/\varepsilon_2$ are plotted. The black curves present the spectral dependence of $\Delta\varepsilon_1/\varepsilon_1$ and $\Delta\varepsilon_2/\varepsilon_2$ calculated without experimental noise added to R and T. The colored curves represent the spectral dependence of $\Delta\varepsilon_1/\varepsilon_1$ (green) and $\Delta\varepsilon_2/\varepsilon_2$ (red), calculated with experimental noise. The significant decrease of the values of $\Delta\varepsilon_1/\varepsilon_1$ and $\Delta\varepsilon_2/\varepsilon_2$ due to the uncertainties in the measured R, R' and T, compared to those in figure 2 is obvious. The methodological error is also slightly reduced. For example, at $\lambda = 600$ nm, the relative error (due to the approximation of the method) of $\Delta\varepsilon_1/\varepsilon_1$ decreases from ~ 8% to ~ 3.8%, and of $\Delta\varepsilon_2/\varepsilon_2$ – from ~ 11.5% to ~ 7.6%.

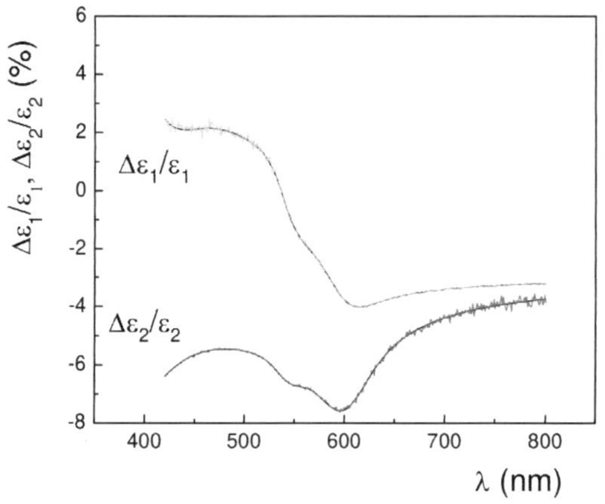

Figure 4. Dispersion of $\Delta\varepsilon_1/\varepsilon_1$ and $\Delta\varepsilon_2/\varepsilon_2$, calculated considering the uncertainties in T and R (colored curves) and without taking into account uncertainties in T, R and R' (black curves).

4. Conclusion

A straightforward approach for the determination of ε_1, ε_2 and d of very thin layers is applied to determine ε_1, ε_2 and d of semiconductor layers. The influence of the methodological error and the uncertainties in T, R and R' on the calculated values of ε_1, ε_2 and d are analyzed. The calculations performed show that the effect of the uncertainties in the measurable quantities is significantly higher than that of the methodological error and increases considerably with the decrease of the d/λ ratio. The accuracy of the solution increases by about a factor of ten if the value of d is estimated first and then the values of ε_1 and ε_2 are recalculated in the whole spectral region under investigation.

References

[1] Ying X, Luo J, Wang P, Cui M, Zhao Y, Li G and Zhu P 2003 *Diamond and Related Materials* **12** 719
[2] Siemroth P and Schülke T 2000 *Surf. Coat. Technol.* **133-134** 106
[3] Fang Y H, Kuo P C, Chen P W, HsuWei-Chih, Chou C Y and Wu T H 2006 *Appl. Phys. Lett.* **88** 261917
[4] Chambouleyron I, Martínez J, Morettin A and Mulato M 1997 *Appl. Opt.* **36** 8238
[5] Pekker D and Pekker L 2003 *Thin Solid Films* **425** 203
[6] Sun X, Hong R, Hou H, Fan Z and Shao J 2006 *Chin. Opt. Lett.* **4** 366
[7] McGahan W, Johs B and Woollam J 1993 *Thin Solid Films* **234** 443
[8] Bondar E, Kulyupin Yu and Popovich N 1978 *Thin Solid Films* **55** 201
[9] Hristov B, Gushterova P and Sharlandjiev P 2007 *J. Opt. Adv. Mat.* **9**/1 217
[10] Gushterova P, Sharlandjiev P and Hristov B 2008 *Appl. Opt.* **47** 5117
[11] Born M and Wolf E 1983 *Principles of Optics* (Oxford: Pergamon) p 55
[12] Korn G and Korn T 1961 *Mathematical handbook for scientists and engineers* (McGraw-Hill, New York)
[13] Vriens L and Rippens W 1983 *Appl. Opt.* **22** 4105

Transport and magnetic properties of epitaxial LSMO thin films grown on MgO single-crystal substrates

V Štrbík[1,3], M Španková[1], M Reiffers[2], J Kováč[2] and Š Beňačka[1]

[1]Institute of Electrical Engineering,
Slovak Academy of Sciences, 9 Dúbravská cesta, 841 04 Bratislava, Slovak Republic
[2]Institute of Experimental Physics,
Slovak Academy of Sciences, 47 Watsonova, 040 01 Košice, Slovak Republic

E-mail: elekstrb@savba.sk

Abstract. Thin epitaxial $La_{0.67}Sr_{0.33}MnO_3$ (LSMO) films were deposited on single crystal MgO substrates. The electrical transport and magnetic properties of the films in the temperature range 4 - 350 K were investigated and strong correlation between them was registered. Magnetoresistance up to 52 % at temperature T = 256 K and magnetic field B = 5 T was achieved. The results obtained indicate that LSMO films with such properties are suitable for application as "barrier" layers in superconducting-ferromagnetic-superconducting heterostructures, but optimization of LSMO film thickness is needed.

1. Inroduction

The perovskite manganite $La_{0.67}Sr_{0.33}MnO_3$ (LSMO) is an interesting material from several points of view. It is a material for colossal-magnetoresistance applications [1], for bolometric purposes [2] and is also used as a barrier material in the heterostructures superconductor (S) – ferromagnet (F) – superconductor (S) [3]. Structures consisting of oxide high-temperature superconducting (HTS) and ferromagnetic films (thickness < 100 nm) constitute an attractive model for investigating the interplay (proximity effect) between the two fundamental condensed-matter phenomena of superconductivity and ferromagnetism [4]. SFS layered heterostructures are promising for applications in devices manipulated by the spin-state of charge carriers; in particular, the S/F based Josephson technology holds great promise in nanoelectronics, e.g. as a physical realization of the qubit of quantum computation [5]. In this paper we report the results of the preparation and investigation (electrical and magnetic properties) of epitaxial LSMO thin films fabricated on MgO single-crystal substrates.

2. Experimental

The LSMO films were deposited by means of an on-axis dc magnetron sputtering system (Torus 2C, LESKER) onto single-crystal MgO (001) polished substrates at a temperature of 800 °C. The deposition was performed in an Ar (80%) + O_2 (20%) gas mixture at a total pressure of 30 Pa. The thickness of the LSMO films was about 500 nm. In order to increase the oxygen content, the LSMO

[3] To whom any correspondence should be addressed.

films were annealed in-situ in O_2 (10^4 Pa) at 800 °C for an hour. After annealing, the temperature was lowered down to room temperature at a rate of 4 K/min [6].

The crystallographic orientation perpendicular to the film surface was determined by recording X-ray diffraction (XRD) patterns in the θ - 2θ configuration. The degree of the preferred orientation, perpendicular to the substrate, was derived from the full-width at half maximum (FWHM) of the rocking curve. The microstructure of the LSMO film was investigated by transmission electron microscopy (TEM) (JEOL 1200EX).

The electrical transport properties were measured by the 4-point dc method in a cryostat in temperature range 4 - 350 K. The magneto-resistive and magnetic properties of the films were measured on PPMS (physical properties measurement system) and MPMS (magnetic properties measurement system) equipment (Quantum Design) in temperature range 150 - 350 K and magnetic field up to 9 T.

3. Results

The XRD analyses showed (00l) oriented growth of the LSMO films with FWHM value of about 1 degree measured on the (001) LSMO peak due to the 8% mismatch of the lattice parameters of the LSMO film and the MgO substrate. The TEM investigations confirmed the epitaxial growth of the LSMO films and a 20 nm thin film surface of a polycrystalline character (only in the case of annealed samples).

The electrical resistivity ρ of the LSMO thin films was investigated in the temperature range 4 - 330 K. For the high-quality films at a temperature of $T_p \gtrsim 300$ K a resistivity peak (transition metal-insulator) was observed. For temperatures higher than T_p (insulating character) the resistivity vs. temperature dependence ($\rho(T)$) can be described by the variable range hopping (VRH) model (figure 1). The rapid decrease of the resistance for $T < T_p$ (metal character) can be interpreted mainly by the electron-electron scattering, electron-magnon scattering and impurity (defects) grain-boundary scattering.

The non-monotonic dependence of the resistivity on the temperature can be characterized by a temperature coefficient of resistivity $(TCR) = [(1/\rho)\times d\rho/dT]\times 100\%$ and its maximal value TCR_{max} indicates the most suitable temperature region for bolometric applications. Our best LSMO films reached $TCR_{max} \approx 3\%\ K^{-1}$ at 290 K.

The magnetoresistivity (MR) of the LSMO films is shown in figure 2. The magnetoresistivity is defined by the relation $MR = [(\rho_B-\rho_0)/\rho_0]\times 100\%$, where ρ_0 and ρ_B are the resistivities at zero and B magnetic field applied, respectively. The MR maximum was registered at the temperature T = 256 K; the MR depends nearly linearly on the parallel magnetic field applied. MR

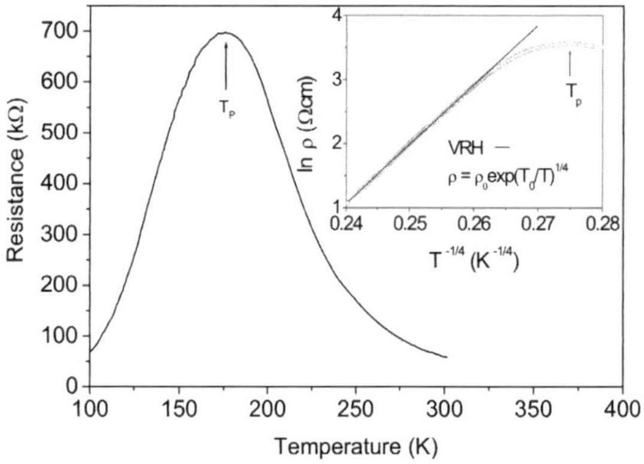

Figure 1. Resistivity vs. temperature dependence of LSMO films. The inset represents the fit of the variable range hopping (VRH) model to the experimental data for T > T_P (paramagnetic state).

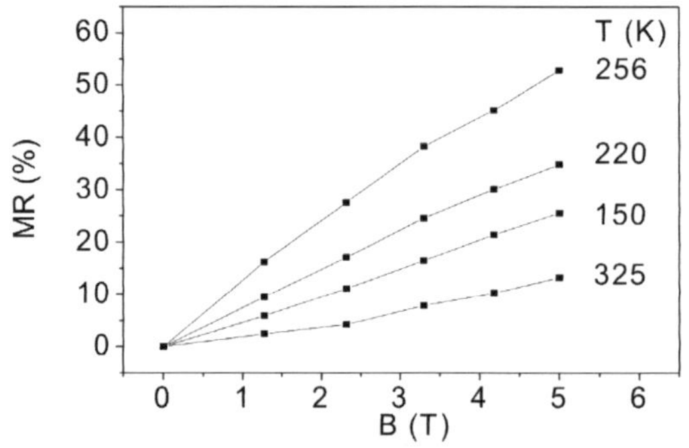

Figure 2. Magnetoresistivity of LSMO thin film. The MR maximum was achieved at B = 5 T and T = 256 K. The resisitivity was changed by 52% compared to value at 0 T.

dependences taken at other temperatures are also shown in figure 2.

The further study showed the strong correlation between the magnetic and transport properties of the LSMO films. The MR maximum was achieved at T = 256 K, the minimum of $d\chi'/dT$ (figure 3) was obtained at T = 250 K (χ' being the real part of the susceptibility) and the TCR_{max} was found at T = 248 K. The correlation bentween T_p and T_{Curie} was mentioned above. The width of the ferromagnetic transition is about 100 K, as is indicated by $d\chi'/dT$ (T) dependence in figure 3.

Thus, the LSMO films are ferromagnetic below 200 K, so that it is possible to use such films as barrier layers in superconductor-ferromagnetic-superconductor heterostructures operating at temperatures 77 K or lower.

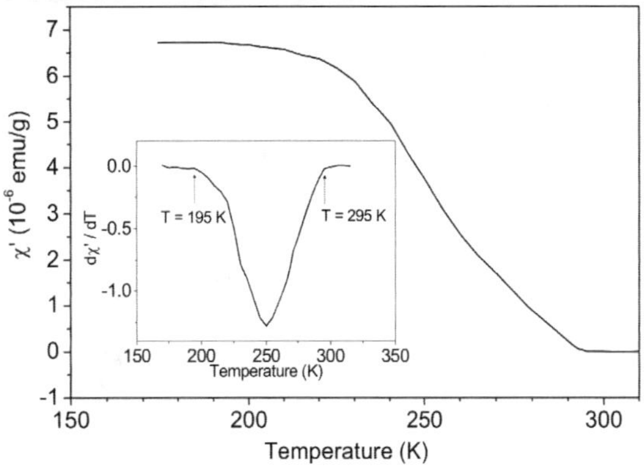

Figure 3. Plot of susceptibility vs. temperature in all measured temperature region. Inset represents dχ'/dT dependence with the important points of ferromagnetic transition of LSMO thin film. Beginning at 295 K and end at 195 K.

4. Conclusion

Relatively thick (about 500 nm) epitaxial LSMO films were grown on single crystal MgO substrates. The electrical (resistivity peak at $T_p \approx$ 290 K, TCR) properties indicate good results on such type of substrate. The magnetic measurements yielded maximal magnetoresitivity at a temperature T = 256 K and magnetic field B = 5 T, ferromagnetic transition in the temperature range 200 - 300 K, Curie temperature T_{Curie} = 293 K. For temperatures below T_p (or T_{Curie}) the LSMO films behave as a metal (from electric point of view) and as a ferromagnet (from magnetic point of view). We believe that such properties allow us to use the LSMO thin films as "barrier" layers in HTS SFS structures in a regime of long-range proximity effects with the LSMO films thickness being in the range of 5 - 20 nm.

Acknowledgments

This work was supported by the VEGA Slovak Grant Agency under projects 2/7125/27 2/0007/09; the Slovak Research and Development Agency under contract No. VVCE-0058-07; the the Centre of Excellence of the Slovak Academy of Sciences and P.J. Šafárik University. The liquid nitrogen for the experiment was sponsored by U.S. Steel Košice, s.r.o.

References

[1] Jin S, Tiefel T H, McCormack M, Fastnacht R A, Ramesh R and Chen L H 1994 *Science* **264** 413
[2] Goyal A, Rajeswari M, Shreekala R, Lofland S E, Bhagat S M, Boettcher T, Kwon C, Ramesh R and Venkatesan T 1997 *Appl. Phys. Lett.* **71** 2535
[3] Alvarez G A, Wang X L, Peleckis G, Shi D Q and Dou S X 2007 *Physica* C **460-462** 438
[4] Buzdin A I 2005 *Rev. Mod. Phys.* **77** 935
[5] Ortlepp T, Ariando, Mielke O, Verwijs C J M, Foo K F K, Rogalla H, Uhlmann F H and Hilgenkamp H 2006 *Science* **312** 1495
[6] Špankovà M, Chromik Š, Vávra I, Sedláčková K, Lobotka P, Lucas S and Stanček S 2007 *Appl. Surf. Sci.* **253** 7599

Journal of Physics: Conference Series **223** (2010) 012028 doi:10.1088/1742-6596/223/1/012028

Magnetic field reversion by SC/FM bilayers in the critical temperature vicinity

E Mateev[1], B Blagoev and T Nurgaliev

Emil Djakov Institute of Electronics,
Bulgarian Academy of Sciences, 72 Tzarigradsko Chaussee, 1784 Sofia, Bulgaria

E-mail: emateev@abv.bg

Abstract. The electrical properties of bilayer structures consisting of a superconducting (SC) and a ferromagnetic (FM) layer deposited on a LAO substrate were studied by a contactless technique in the temperature range from 77 K in the vicinity of the critical temperature. Two series of bilayers were investigated: - SC (YBCO) layers with thickness varying from 35 to 80 nm on a FM (LSMO) layer with a fixed thickness of 40 nm, and - a fixed-thickness YBCO layer (70 nm) on LSMO layers with thickness varying from 10 to 40 nm; YBCO layers deposited under the same conditions as reference. An interesting effect was observed, namely, when the temperature is reduced below the critical value, the bilayer inverts the magnetic field passing through it - the signal picked up by the receiving coil inverts its phase. The measurements of the critical current and temperature showed that when the YBCO layer is thicker than the LSMO layer, the latter acts as a buffer layer improving the SC performance. These is in contrast with the results published earlier by other authors on proximity effect and spin injection for bilayers and super lattices of SC/FM layers that are an order of magnitude thinner. The SC performance worsens only in the case when the YBCO and LSMO layers have a thickness of approximately 40 nm.

1. Introduction

Cuprate bilayers of SC over FM deposited on a monocrystal with a matching crystal lattice are interesting from a fundamental [1, 2, 3] as well as experimental [4, 5] points of view; it is also believed that they have a potential for spintronics applications [6]. Most of the experimental work has been performed for superlattices with layer thickness of several nm (lattice cells). They show worsening of the electrical characteristics of SC (compared to the single SC layers), namely a decrease of the critical temperature up to a complete loss of superconductivity [4]. These results are useful for theoretical considerations; however, structures consisting of thicker SC and FM layers are also interesting from a practical point of view.

The aim of the present work is to investigate the electrical properties – critical temperature and critical current density at a given temperature (77 K) - of the bilayer structures YBCO on LSMO on LAO with various thicknesses of the SC and FM layers.

[1] To whom any correspondence should be addressed.

2. Experiment

The LSMO layers were deposited in an off-axis single RF magnetron sputtering system at a substrate temperature of 780 °C. Immediately after deposition, they were annealed (in the same system) for 30 minutes in a 600 Torr O_2 ambient at 540 °C. The working gas mixture was 50:50 Ar + O_2 at a pressure of 30 mTorr. The YBCO layers were deposited over the LSMO layers in an off-axis dual DC magnetron sputtering system at a substrate temperature of 780 °C. Immediately after deposition they were annealed (in the same system) for 60 minues in 600 Torr O_2 ambient at 540 °C. The working gas mixture was 70:30 percent Ar + O_2 at a pressure of 300 mTorr.

The critical temperature T_{cr} and critical current density J_{cr} were measured by a contactless technique using the dual coil system shown in figure 1.

The T_{cr} measurements were carried out at the fundamental frequency, while to measure J_{cr}, the third harmonic technique [7] was used.

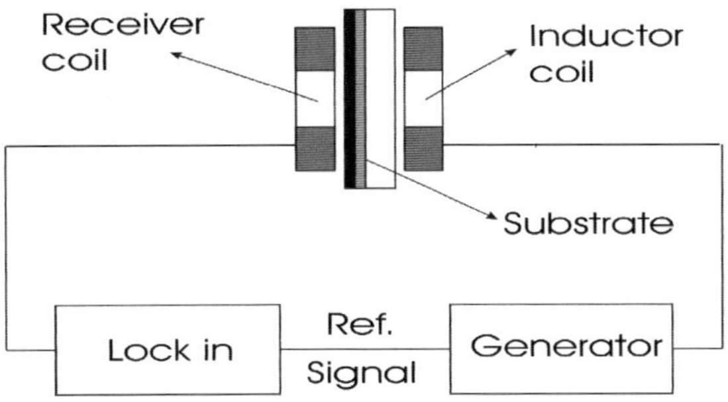

Figure 1. Measuring contactless technique with dual coil system.

3. Results

Two groups of samples were prepared and measured. In the first one, the LSMO thickness was varied from 10 to 40 nm, while the YBCO thickness was fixed at 70 nm. Two samples with a YBCO layer only were also prepared and measured and the results were added to those of the first group. In the second group, the LSMO thickness was fixed at 40 nm and the YBCO thickness was varied. The critical temperatures and critical current densities (at 77 K) of the first group are shown in figures 2 and 3.

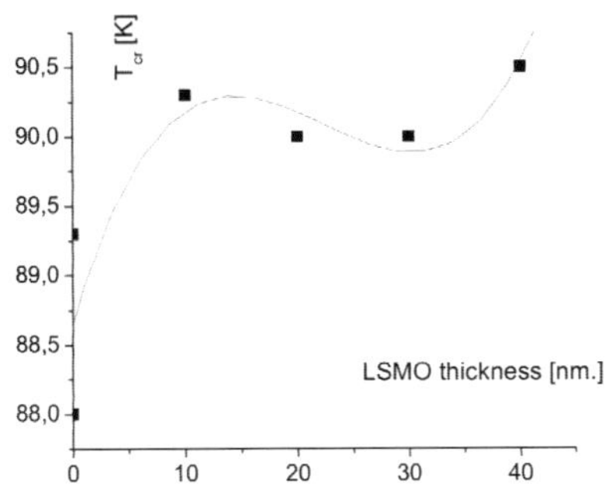

Figure 2. Critical temperatures T_{cr} for the 70 nm YBCO group of samples.

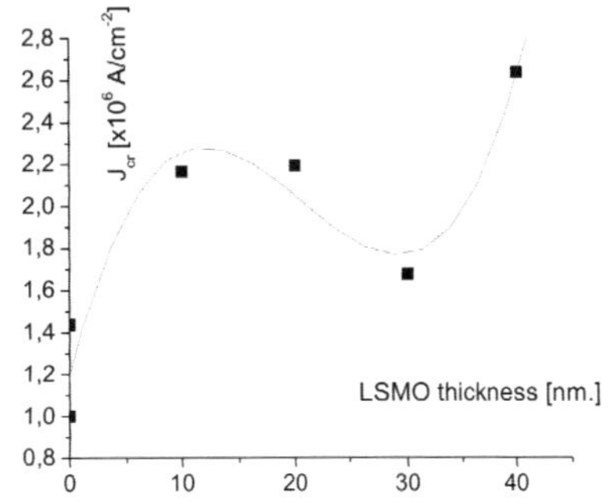

Figure 3. Critical current densities (at 77 K) for the 70 nm YBCO group of samples.

The critical temperature of the single YBCO layer is somewhat lower than in the case of YBCO over LSMO. The theory predicts in some cases a positive ΔT_{cr} due to the contact with the FM [3]. On the other hand, no substantial dependence of the T_{cr} on the LSMO thickness is observed (within the reproducibility of the experiment). This suggests an alternative explanation – the LSMO acts as a buffer layer and improves the YBCO crystal structure. The critical current density dependence on the FM thickness shows a similar behavior.

The critical temperatures and critical current densities (at 77 K) of the second group are shown in figure 4 and figure 5 respectively.

The SC performance of the bilayers shows a sharp drop at YBCO thickness of 40 nm – the same as of the LSMO underlayer. At higher YBCO thicknesses, the SC electrical characteristics are improved as the SC thickness is raised.

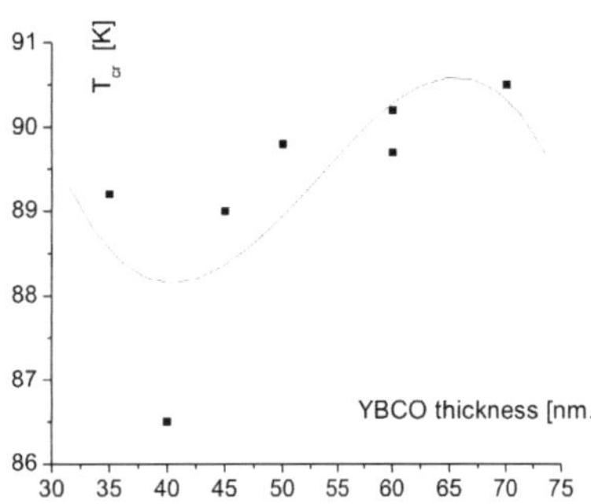

Figure 4. Critical temperatures T_{cr} for the 40 nm LSMO group of samples.

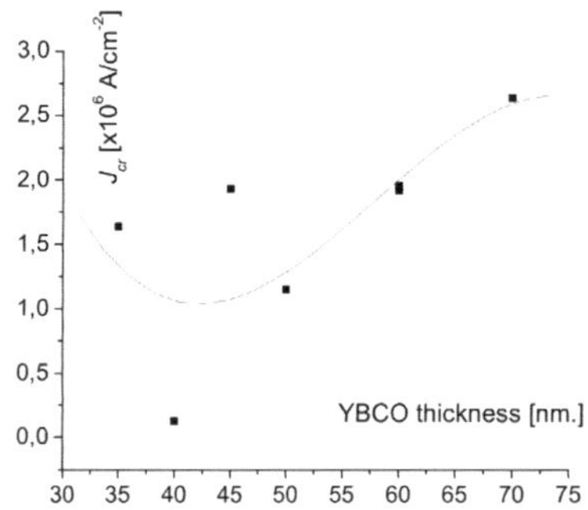

Figure 5. Critical current densities (at 77 K) for the 40 nm LSMO group of samples.

A very interesting effect is observed in the plots of the signals picked up during the critical temperature measurements using the dual coil system shown in figure 1. The plots are shown in figure 6 for FM thicknesses 35, 40 and 45 nm.

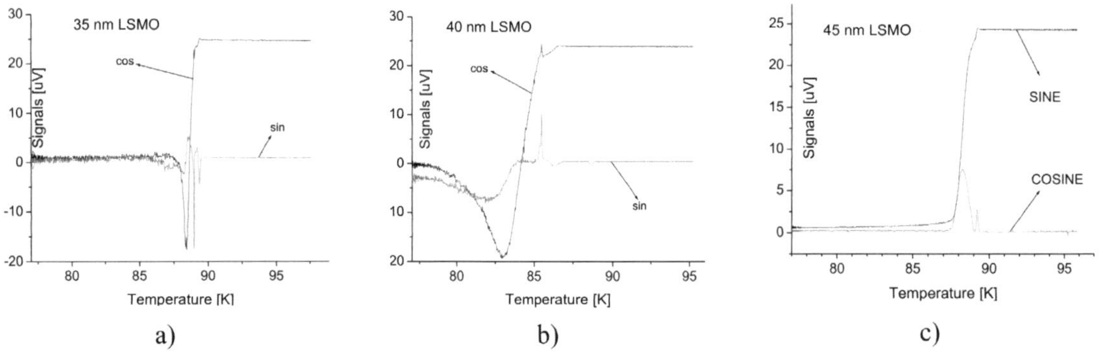

Figure 6. Signals picked up during the critical temperature measurements for FM thicknesses: a) 35 nm, b) 40 nm, c) 45 nm.

The phase of the reference signal is chosen so as to give zero sine signal at room temperature; thus, the cosine signal is the main signal. The critical temperature is located at the onset of the diamagnetic screening.

For a thickness of 45 nm, the cosine signal received has the same sign at all temperatures with only its value changing. Similar curves were obtained for thicker bilayers and single YBCO layers, the only difference being a shift in the critical temperature – the temperatures of SC screening onset are shown in figures 2 and 4. In the cases of layers with thickness 35 nm and 40 nm, a temperature interval appears where the received signal changes its sign.

The amount of FM material in the layers studied is too small to produce a measurable magnetic field, as checked at room temperature.

4. Discussion

In 1964, Ginsburg [8] pointed out that the surface of a SC can exhibit properties different from those of the bulk and specific surface SC can exist. A surface related phenomenon was also observed in high-T superconductors [9]. The authors measured the critical temperature using a single coil contactless setup and obtained a higher T_{cr} value when the probe coil was placed on the SC side covered with Ag than when the probe coil was on the opposite side.

The phase sensitive measurements of YBCO layers with thicknesses 35 and 40 nm show that the received signal changes its sign within a certain temperature interval. This means that the averaged magnetic field on the plane of the receiver coil has a direction opposite to the usual case. We assume that this is an effect of surface superconductivity. The latter amplifies the SC response to the external magnetic field; it is overscreened within a temperature interval below but close to the critical temperature when the induced current reaches the critical value.

5. Conclusion

For YBCO thicknesses exceeding that of the LSMO underlayer, the latter acts as a buffer and improves to some extent the electrical SC characteristics.

An unexpected effect is observed. For specific combinations of SC an FM layer thicknesses, the magnetic field changes its direction when penetrating the bilayer structure. This effect arises at temperatures when critical current densities are reached by the diamagnetic currents induced. A sine component appears in the received signal (together with higher odd harmonics [7]). This effect is the result of currents induced in the SC, rather than magnetization of the FM.

References

[1] Izyumov Yu A, Proshin Yu N and Khusianov M G 2002 *Physics-Uspekhi* **172** 113
[2] Sa de Melo C A R 2003 *Physica* C **387** 17
[3] Pokrovsky V L and Wei H 2004 *Phys. Rev.* B **69** 104530
[4] Löhneysen H V, Beckmann D, Goll G, Willard F P, Stalzer H, Stokmaier M and Sürgers C 2007 Proximity effect between superconductors and ferromagnets *Physica* C **460-462**/1 322-6
[5] Przyslupski P 2005 Magnetism and superconductivity in oxide ferromagnet/superconductor heterostructures *Physica Status Solidi* C **2**/5 1625 – 32
[6] Volf S A, Awschalom D D, Buhrman R A, Daughton J M, von Molnar S, Roukes M L, Chtchelkanova A Y and Treger D M 2001 *Science* **294** 1488
[7] Nurgaliev T, Mateev E, Blagoev B, Miteva S and Neshkov L 2005 Investigation of contact-less harmonic generation in HTS nano-films *Proc. 5th Workshop Nanoscience & Nanotechnology* ed E Balabanova and I Dragieva (Heron Press Sofia Bulgaria) pp 101-3
[8] Ginsburg V L 1964 *Sov. Phys. JETP* **20** 1549
[9] Timchev S S *http://www.arxiv.org:0911.2107*

Optimization of carbon steel electron-beam hardening

P Petrov

Emil Djakov Institute of Electronics,
Bulgarian Academy of Sciences, 72 Tzarigradsko Chaussee, 1784 Sofia, Bulgaria

E-mail: pitiv@ie.bas.bg

Abstract. A summary is presented of the investigations on the key conditions ensuring electron-beam surface hardening of carbon steels. A numerical model is developed and used to calculate the temperature field caused by high frequency electron beam scanning. The numerical calculation results show that the heating and cooling rate depend weakly on the electron-beam power and are strongly influenced by the sample speed of motion, namely, they increase as this speed is increased. The efficiency of the process of electron-beam hardening of metals increases with the speed of motion of the samples treated.

1. Introduction

The use of high-energy fluxes (HEF), such as electron and laser beams, for surface treatment of metals and alloys is a promising trend with growing application in the industrial production. This treatment ensures enhancement of the physical and mechanical properties and the quality of components made of steels, cast-iron, aluminum, aluminum alloys, copper and titanium alloys, etc. Surface treatment of materials through electron beam and laser energy fluxes is characterized by a number of specific advantages over the conventional methods, such as very small size of the zones treated, precise adjustment and high density of the input energy, high heating rates, good reproducibility and a possibility for a higher degree of process automation. The action of an HEF on a metallic surface induces heat propagation from the surface to the depth of the metallic target during which the rate of heating and cooling may reach up to 10^{10} K/s, and, in some cases, even greater [1-5]. In the case of swept-line electron-beam hardening, the cooling and heating rate are in the range $10^3 \div 10^4$ K/s [3] and reach $10^9 \div 10^{10}$ K/s in the case of pulsed electron-beam techniques [4]. In general, the possibility for hardening is determined by the chemical composition of the material. It must be emphasized that the study of the thermal processes (heating and cooling), taking place during treatment of materials by HEF are of considerable theoretical and practical importance with the results finding applications in the optimal control of technological processes.

The aim of this study was to develop a model describing adequately the thermal processes during steel hardening by high-frequency electron-beam scanning (the swept-line electron-beam hardening technique) and identify the important conditions affecting the process.

2. The process of electron-beam surface hardening

The experiment on applying the swept-line electron-beam hardening technique is shown schematically in figure 1. An electron beam focused onto the surface of the sample under treatment is deflected perpendicularly to the direction of its motion. The experiments were performed on

Journal of Physics: Conference Series **223** (2010) 012029 doi:10.1088/1742-6596/223/1/012029

Leybold Hereaus EWS300-15/60 equipment with electron beam power of Q = 0.5÷3 kW, scanning frequency f=1÷10 kHz, width of the zone treated L = 14 mm, and speed of sample motion V of 0.5 to 5 cm/s.

The steel used in this study for experiments and for theoretical calculations was a steel with nominal composition (wt.%) of 0.42 %C, 0.96% Cr, 0.6%Mn, 0.37% Si, balance Fe. The samples were machined to the dimensions 60 x 40 x 10 mm and were conditioned as follows: oil hardening at 840÷850 °C and subsequent tempering at 560÷580 °C.

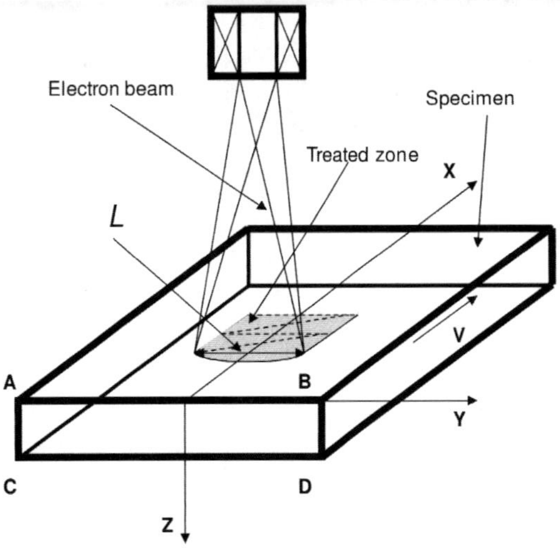

Figure 1. Schematic representation of the process of electron beam hardening.

3. Heat model

The processes of heat propagation during electron beam hardening are studied by using the heat conduction equation:

$$C(T)\rho(T)\frac{\partial T}{\partial t}=\frac{\partial}{\partial x}\left[\lambda(T)\frac{\partial T}{\partial x}\right]+\frac{\partial}{\partial y}\left[\lambda(T)\frac{\partial T}{\partial y}\right]+\frac{\partial}{\partial z}\left[\lambda(T)\frac{\partial T}{\partial z}\right]+Q, \tag{1}$$

under initial conditions (t = 0)

$$T(x, y, z, 0) = T_0, \tag{2a}$$

or

$$T(x, y, z, 0) = T(x, y, z), \tag{2b}$$

and boundary conditions

$$T(x,y,z)\big|_{S_1}=T=const, \tag{3a}$$

$$T(x,y,z)\big|_{S_2}=T(t), \tag{3b}$$

$$\lambda\frac{\partial T}{\partial n}+\alpha(T-T_\infty)=0\big|_{S_3}; \tag{3c}$$

in the case of an impenetrable boundary α = 0, the following condition is obtained

$$\lambda\frac{\partial T}{\partial n}=0, \tag{3d}$$

where $\rho(T)$ is the density [kg/m^3], $C(T)$ is the specific heat [J/kg K], $\lambda(T)$ is the thermal conductivity coefficient [W/mm K], $k=\lambda/c\rho$ is the thermal diffusivity [mm^2/s], α is the heat transfer coefficient [W/mm^2 K], $T(x, y, z, t)$ is the temperature [°C], with T_0 being the initial temperature and T_∞, the surrounding medium temperature, S_1, S_2, S_3 are the boundary surfaces, n is the normal to the surface, Q is the energy generated per unit volume [W/mm^3], t is the time [s].

Equation (1) does not take into account the processes of convective heat transfer within the molten pool. Conditions (2) and (3) allow one to account for the initial temperature field non-uniformity in the different zones of the hardened zone, and for the heat exchange conditions at the boundaries. Using Q one can model the heat absorption or release during phase transformations. Thus, the set of equations (1-3) describes the main processes of heat propagation during electron-beam hardening.

In [6,7], an algorithm is presented for solving set (1-3) by applying the finite elements method. The main points in solving the equations in the case of a two-dimensional problem are adopted in accordance with [6] and have to do with finding the minimum of the functional:

$$\chi = \int_V \frac{1}{2}\left[\lambda(T)\left(\frac{\partial T}{\partial x}\right)^2 + \lambda(T)\left(\frac{\partial T}{\partial y}\right)^2 + \left(2C(T)\rho(T)\frac{\partial T}{\partial t} - 2Q\right)T\right]dV + \int_S \left[\frac{1}{2}\alpha(T - T_\infty)^2\right]dS. \tag{4}$$

The minimization of functional (4) is carried out with respect to the elements of discretization of the zone considered (triangles with straight sides and three nodes, *i*, *j*, *k*). After presenting (4) in a matrix form, minimizing its values at the elements nodes, integrating and substituting the time derivatives by finite differences, the problem is reduced to solving the matrix equation:

$$\{[C] + \theta\Delta t[K]\}\{T\}_t = \{[C] - (1-\theta)\Delta t[K]\}\{T\}_{t-\Delta t} + \Delta t\{F\}, \tag{5}$$

where $\{T\}_t$, $\{T\}_{t-\Delta t}$ are the values of the temperature in the nodes at the current t and previous $t - \Delta t$ moments of time, θ is a coefficient characterizing the different solution methods ($\theta = 0$, Euler-Cauchy; $\theta = 1/2$, Crank-Nicholson; $\theta = 2/3$, Galerkin; $\theta = 1$, reverse differentiation).

To solve (5) we used the Gauss method – Zenkeevich algorithm [6]. The temperature is calculated at the center of gravity of the elements; the rates of cooling and the momentary structural state of the metal on the seam and in the near-seam zone are determined. The algorithm used to construct the computing process [5,6] is as follows:

1. The cross-section considered is discretized into a finite number of triangular elements. The main principle followed is grid densification in the zones with large temperature gradients.
2. The nodes lying on the boundary are determined for the cases when a boundary condition with heat exchange with the surrounding medium is adopted.
3. The time step Δt is determined depending on the current time increase. A step-wise (Δt, $2\Delta t$, $5\Delta t$, $10\Delta t$) increase of the time-step is envisaged that is inversely proportional to the temperature gradient.
4. A scheme is adopted for introducing the heat source, and for the heat distribution on the elements nodes.
5. The matrices of the elements are constructed and introduced in global matrices.
6. The equation set (8) is solved following the Gauss method – Zenkeevich algorithm [6].
7. The temperature and the rates of cooling are calculated at the center of gravity of the elements. The momentary structural state is determined.
8. The temperature and the structural fields are constructed.

The finite-differences scheme used to solve the problem in the time space is built on the basis of the central differences scheme of Crank-Nicholson. This scheme is unconditionally stable, but deviations can still arise of the numerical values from the real values sought [6]. The magnitude of the solution oscillation depends mainly on the material's properties, the elements size and the time-step size. The material's properties are usually known and are set specifically for each separate case. The variables, then, that can be varied in order to eliminate the numerical results oscillations are the size of the elements and of the time step. Their simultaneous reduction would reduce the numerical solution oscillation. It is erroneous to change one of the variables while keeping the other constant. For example, in the case of a rough grid and a small time step, one can obtain results that have no physical sense for the particular case considered.

The electron-beam energy is introduced in the elements nodes of an arbitrary element and is defined as:

$$Q_{net} = \frac{\eta_e Q_{source} / dL}{V t_s A}, \quad t_s = \frac{d}{V}, \tag{6}$$

where $Q_{source} = UI$ is the energy of the source; U is the accelerating voltage [kV]; I is the electron-beam current [mA]; η_e is the process efficiency; d is the electron-beam diameter [mm]; L is the width of the zone treated; V is the sample's speed of motion [mm/s]; t_s is the time of action of the source [s]; and A is unit area [mm^2].

The time t of the heat source action is calculated as the ratio of the electron beam diameter d in the treated plane to the sample's speed of motion V.

4. Results and discussion

The microhardness distribution measured within the depth of the layer is shown on figure 2. It is seen that the hardness reaches values of 550-650 HV_{01} to a depth of 400 μm. As the depth is increased, the hardness decreases rapidly down to the matrix hardness ~300HV_{01}. The total depth of the hardened layer reaches 550 μm. Figure 3 shows the maximum temperature (defined on the basis of the thermal cycles calculated within the depth of the hardened zone) as a function of the sample's speed of motion. It is seen that the maximum temperature decreases with the increase of the sample's speed of motion. The layers within the depth reach a temperature maximum later than the layers that are closer to the surface, the rates of heating and cooling of surface layers being higher than these in depth. The temperature field is nonstationary and highly inhomogeneous. Very high temperature gradients are observed within the depth of the hardening layer in the processes of both heating and cooling.

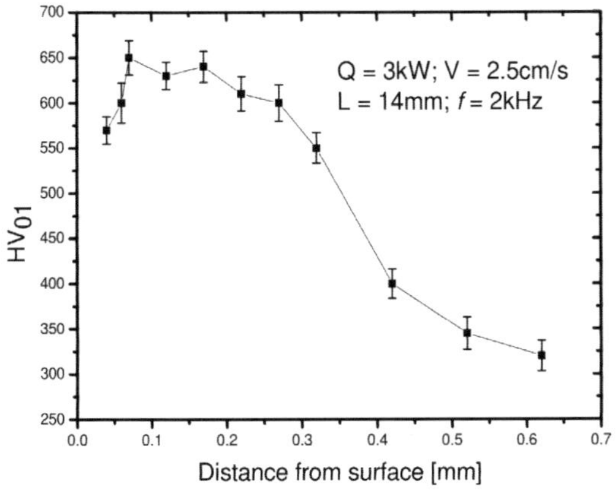

Figure 2. Microhardness distribution along the hardened layer depth.

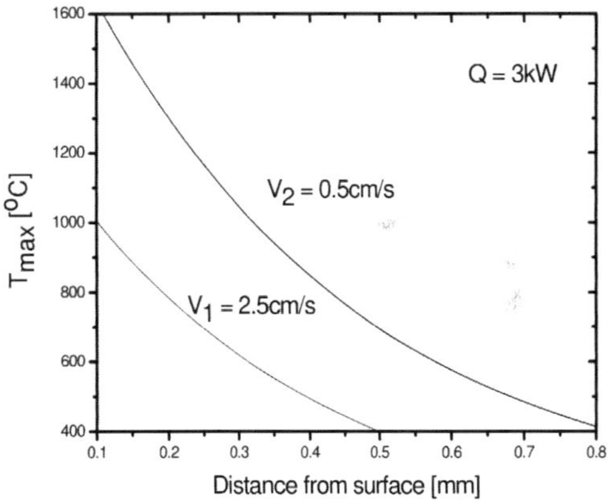

Figure 3. Maximum temperature calculated vs. distance from surface.

The principal electron beam hardening process variables are the beam power and speed of motion of the sample treated. The heating and cooling rates are investigated as functions of the electron beam power and the sample's speed of motion. The average rates of heating and cooling within the temperature interval of 800 to 500°C are defined on the basis of the thermal cycles calculated within the depth of the hardened zone (with power of the electron beam of 0.5 to 3 kW, speed of motion of the sample treated of V = 0.5 ÷ 5cm/s, width of the zone treated L = 14 mm and electron beam diameter d = 0.1mm). The data for the temperature dependence of the thermo-physical parameters of the steel were taken from [8].

Figure 4 and figure 5 show the heating rate calculated depending on the sample's speed of motion and on the electron-beam power. As one can see in figure 4, the heating rate increases with the speed of motion (for a constant electron-beam power of Q = 3 kW). For rates higher than 3 cm/s, the curve is reaching saturation, which means that a further increase of the speed of motion of the sample under treatment influences weakly the rate of heating. Further, the heating rate increases linearly with the electron-beam power (figure 5). This is mainly connected with the very fast processes of electron-beam energy transfer from the surface layers of the sample under treatment.

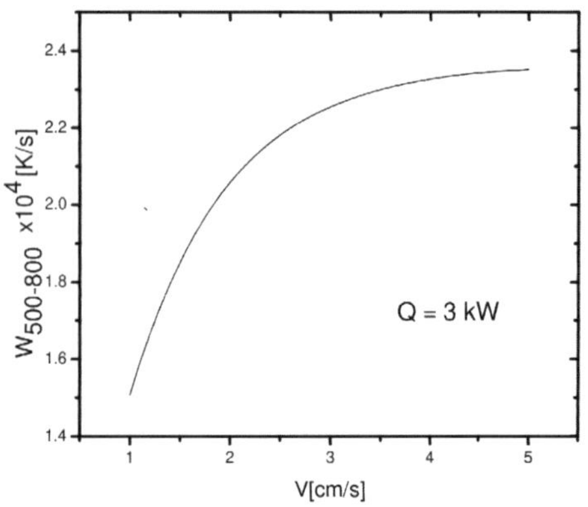

Figure 4. Calculated heating rate depending on sample's speed of motion.

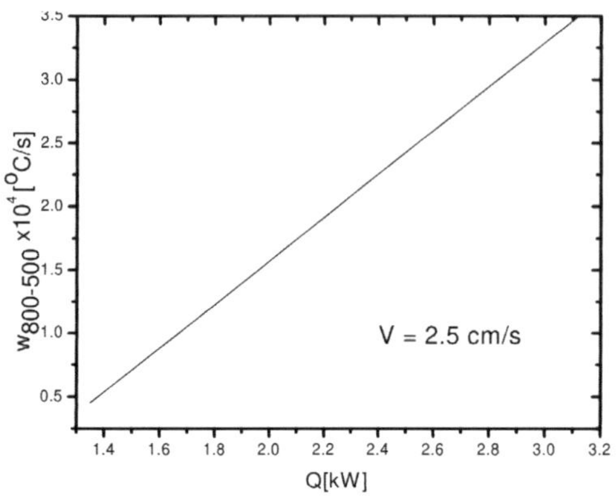

Figure 5. Calculated heating rate depending on electron beam power.

Figure 6 and figure 7 show the cooling rate as a function of the sample's speed of motion and the electron-beam power. The cooling rate depends weakly on the electron beam-power (figure 4), but is strongly influenced by the speed of motion of the sample processed (figure 5). The cooling rate increases with the sample's speed of motion. This effect may be used for electron-beam treatment, since the hardening effect is mainly determined by the rate of cooling. Therefore, the process of electron beam hardening of metals should be performed at higher speeds of motions of the samples to be hardened. The results of the experimental measurements of the microhardness of the hardened zones and the calculated hardening temperature T_h and rate of cooling $W_{800\text{-}500}$ allowed us to construct curves that can be used to determine the optimal conditions of carbon steel samples hardening by swept-line electron-beam hardening. Figure 8 shows the dependence of the experimentally measured microhardness on the temperature of hardening at a cooling rate of 3 x 10^4 K/s. An increase of the microhardness is observed up to the temperature ~900 °C, after which the microhardness practically does not increase. One can observe two characteristic values of the temperature, namely, T_{hs} and T_{hm}. The former temperature, T_{hs}, at which hardening of the layer treated is observed, is determined by the carbon content (% C) in the initial material. The latter temperature, T_{hm}, is the temperature after which no rise is observed in the experimentally measured microhardness of the hardened layer. T_{hm} is higher than the temperature of A_{C3} transformation and decreases from ~1200 °C (at C = 0.2%) down to ~900 °C (at C= 0.4%).

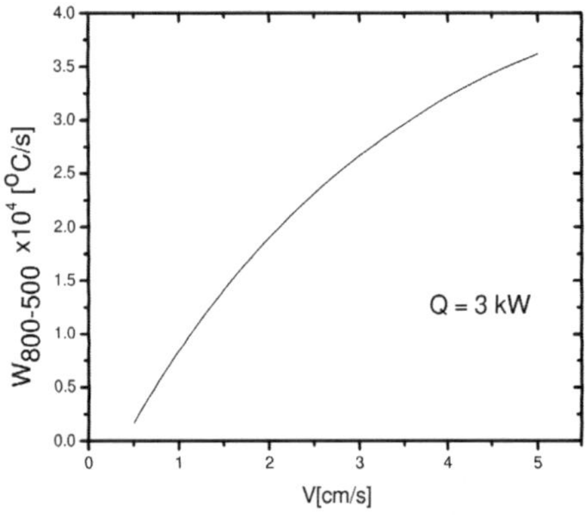

Figure 6. Calculated cooling rate depending on sample's speed of motion.

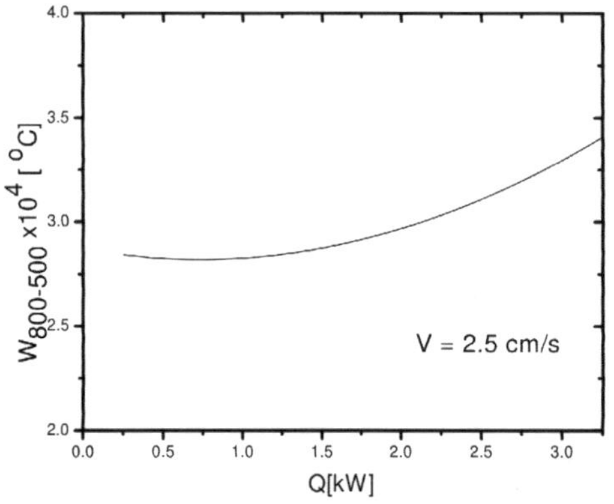

Figure 7. Calculated cooling rate depending on electron beam power.

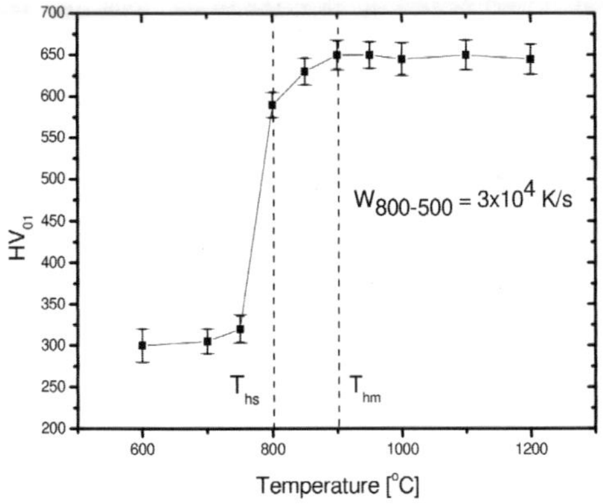

Figure 8. Measured hardness vs. calculated heated temperature.

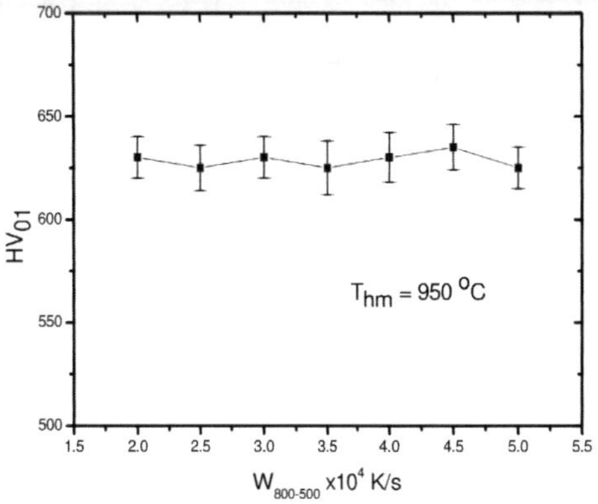

Figure 9. Maximum hardness measured vs. calculated cooling rate.

Figure 9 presents the dependence of the experimentally measured microhardness on the cooling rate in the range between $2x10^4$ °K/s and $5x10^4$ °K/s. As one can see, the microhardness does not change as the rate of cooling is raised. A change in the microhardness could be expected at considerably higher rates of cooling, up to $10^7 \div 10^9$ K/s, which are typical for the pulsed electron-beam hardening. The dependences HV = $f(T_h)$ and HV = $f(W_{800\text{-}500})$, allow one to determine the optimal conditions for hardening of carbon steel samples by swept-line electron-beam hardening.

5. Conclusions

A numerical thermal model is developed in view of studying the processes of heating and cooling during swept-line electron-beam hardening. The numerical calculations demonstrate that the heating and cooling rate increase with the speed of motion (constant electron-beam power) and electron beam power (constant specimen speed of motion). Efficiency of the process of swept-line electron-beam hardening of carbon steels increases with the increase of the sample's speed of motion.

References

[1] Schiller S and Pancer S, 1988 *Annual Review of Materials Science* **18** 121
[2] Arata Y 1987 *Trans. JWRI* **13** 121
[3] Zenker R and Müller M 1987 *Neue Hütte* **32** 127
[4] Proskurovsky D, Rotshtein V, Ozur G, Markov A, Nazarov D, Shulov V, Ivanov Yu and Buchheit R 1998 *J. Vac. Sci. Technol.* A **16** 2480
[5] Petrov P and Dimitrov D 1998 A computer modeling of surface electron beam transformation hardening *Proc. 6th Int. Conf. Welding and Melting by Electron and Laser Beams* (Toulon France June 1998) pp 101-8
[6] Segerlind L 1976 Applied Finite Element Analysis (John Wiley, New York) p 388
[7] Petrov P 1992 *Int. J. Joining of Materials* **4** 110
[8] Vargraftik N 1956 *Thermophysical properties of the materials* (Energoizdat, Leningrad, USSR) p 317

Journal of Physics: Conference Series **223** (2010) 012030 doi:10.1088/1742-6596/223/1/012030

Biocompatibility of dielectric Ta_2O_5 coatings in *in vitro* tests

N Donkov[1], E Mateev[1], A Zykova[2,5], V Safonov[3] and V Luk'yanchenko[4]

[1]Emil Djakov Institute of Electronics,
Bulgarian Academy of Sciences, 72 Tzarigradsko Chaussee, 1784 Sofia, Bulgaria
[2]Institute of Surface Engineering, 4 Zalutinskaya Str., Kharkov, Ukraine
[3]National Science Center "Kharkov Institute of Physics and Technology",
1 Academicheskaya Str., Kharkov, Ukraine
[4]Inmasters Ltd, 80 Pushkinskaya Str., 61024 Kharkov, Ukraine

E-mail: zykov@bi.com.ua

Abstract. A study was performed on the impact of the structure and properties of e-beam evaporated Ta_2O_5 films on the cell/material response. The surface properties and structure the films were investigated by means of XPS and XRD. The cyto toxicity and cyto compatibility were estimated by *in vitro* tests. Other parameters, such as the surface free energy (SFE) and fractional polarity, were determined by means of the Wu, and Owens-Wendt-Rabel-Kaelble methods. The dependence was followed of the cells adhesive behavior on the coatings surface composition and parameters. The best biological response parameters (total cell number, detached cells percentage, proliferation ratio) were obtained in the case of annealed Ta_2O_5 coatings with stoichiometric composition and the largest values of the SFE polar part and the of the fractional polarity.

1. Introduction

The interest in electret materials and coatings applications has considerably increased in various areas of science and technology in recent years. The implants applied now for surgical treatment with dielectric coatings in an electret state create a normal biopotential in the osteosynthesis area that prevents atrophy and necrosis formation, bone tissue deformation and surface strains of large joints, thus reducing the treatment period and minimizing the postoperative complications. Electret coating deposition necessitates ensuring high purity and a definite stoichiometric composition of the dielectric coatings in the electret state. Thus, the major factors are the optimum regime of fabrication and precise control of the technological process.

Tantalum and tantalum-based compounds are promising for biomedical applications. Ta-based implants exhibit high fracture toughness, corrosion and wear resistance, and chemical stability. The results of test on implantation of Ta in both soft and hard tissue of rats showed the good biocompatibility and osteogenesis of this metal [1]. TaC and TaN materials posses relatively high hardness due to the covalent nature of their bond [2] and demonstrate high thermal stability and superior corrosion resistance [3]. The blood compatibility of TaN films was shown to be better than that of Ta [4].

[5] To whom any correspondence should be addressed.

Journal of Physics: Conference Series **223** (2010) 012030 doi:10.1088/1742-6596/223/1/012030

In vitro tests of cell adhesion on material and coating surfaces are the basic tools to determine the material surface/tissue response on a cellular level [5, 6]. The effects of materials composition, surface chemistry and surface topography on cell adhesion and proliferation have been studied extensively [7, 8]. The surface energy is another fundamental material property that can influence the cell behavior [9]. In this study, Ta pentoxide ceramic coatings were found to be promising materials for various biomedical applications.

2. Materials and methods

The samples were formed on glass substrates. The main parameters of the process were described in our previous study [10]. The evaporation process was carried out at initial vacuum of 9.3×10^{-4} Pa, operational vacuum of 4×10^{-3} Pa, anode current of 50 mA and calculated evaporation power of 350 W [11]. The deposition rate under these conditions was 28 nm/min. The layer thickness and the deposition rate were controlled by a digital thin-film deposition monitor MSV-1843/H MIKI-EEV operating at 6 MHz. The surface properties and structure of the e-beam evaporated Ta_2O_5 films were investigated by means of XPS and XRD. The X-ray photoelectron spectroscopy measurements were carried out using ESCALAB MkII (VG Scientific) electron spectrometer at a base pressure in the analysis chamber of 5×10^{-10} mbar (during the measurement 1×10^{-8} mbar), using an AlKα X-ray source (excitation energy h_n = 1486.6 eV). The instrumental resolution measured as the full width at half maximum (FWHM) of the Ag3d5/2 photoelectron peak was 1 eV. The energy scale was corrected to the C1s - peak maximum at 285 eV for electrostatic charging.

Analysis was carried out of surface parameters, such as the surface free energy and the fractional polarity. The contact angles were measured by means of a tensiometric technique. Prior to contact angle measurements, the samples were ultrasonically cleaned in acetone and deionized water and dried. The advancing contact angle was measured by means of the Wilhelm's technique (Kruss K12) at temperature 20 °C [10].Standard liquids with well-known values of surface tension, component of dispersion and polar interaction were used. The surface free energy (SFE) and its polar and dispersion components were determined by means of the Wu [12] and the Owens-Wendt-Rabel-Kaelble [13] methods.

The *in vitro* experiments on cytotoxicity and cytocompatibility were carried out in a culture of fibroblasts. During cell cultivation on coated samples, the cell cytology, morphology and proliferation activity were determined by means of optical microscopy after 24 h and 3 and 5 days cultivation. Rat hypodermic cellular tissue was extracted to obtain the initial fibroblast culture. The suspension of extracted cells was centrifuged at 750 rev/min for 15 min. The sowing cell area was 3×10^5 cell/ml density of cultural medium. The fibroblast cultivation in a 3 ml of Dulbecco Modified Eagle's Medium (DMEM, Sigma) was made by using a monolayer culture in a thermostat (temperature 37 °C for 5 days). After cultivation, fixation in an acetic acid and methyl alcohol (1:3) solution and azure-eosin coloration, the cellular proliferation on the cover glasses was determined by optical microscopy (Micros). The experiments were triplicated.

3. Results and discussion

XPS survey spectra of the e-beam evaporated Ta_2O_5 films (initial, non-annealed sample; sample annealed at 450°C in O_2) were obtained. All spectra consist of well-defined XPS lines of Ta 4f, 4d, 4p and 4s; O 1s; C 1s. All binding energies of the high-resolution spectra were calibrated with the C 1s binding energy of 285.0 eV. Figure 1 shows the high-resolution Ta 4f and 4 O1s XPS spectra of the structures investigated. Ta 4f doublets are typical for e-beam evaporated Ta_2O_5 and have two peaks: Ta 4f7/2 at ~ 26.3-26.6 eV and Ta 4f5/2 whose binding energy higher by 1.9 eV. The Ta 4f lines of the films deposited agree well with the Ta 4f doublet representative of the Ta-0 bond in Ta_2O_5. The Ta4f7/2 peak is located at 26.5 and 26.6 eV, and the Ta 4f5/2 one is at 28.4 and 28.5 eV for the annealed and non-annealed films, respectively. These results clearly demonstrate that annealing improves the stoichiometry of the Ta_2O_5 films. The O1s spectra further support this assumption. The O 1s peaks of the deposited layers are centered at binding energies of 530.9 and 530.8 eV for the annealed and non-annealed films, respectively, which is consistent with data reported for Ta_20_5. The

FWHM of both peaks is 1.9 eV. The O/Ta ratio estimated from the spectra is ~ 3 for all samples. The annealing process changes the Ta_20_5 coatings structure from amorphous one (for as-deposited films) to a orthorhombic phase, the surface topography, from smooth to nano-crystalline (up to 20 nm) and improves the crystallinity of the Ta_20_5 films. At the same time, the dielectric constant of the films increases from 17 to 28 with the increase of the annealing temperature due to the improvement in the film crystallinity and packing density [14].

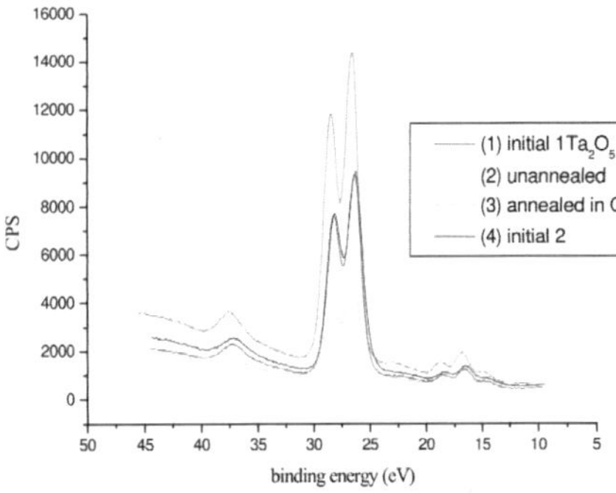

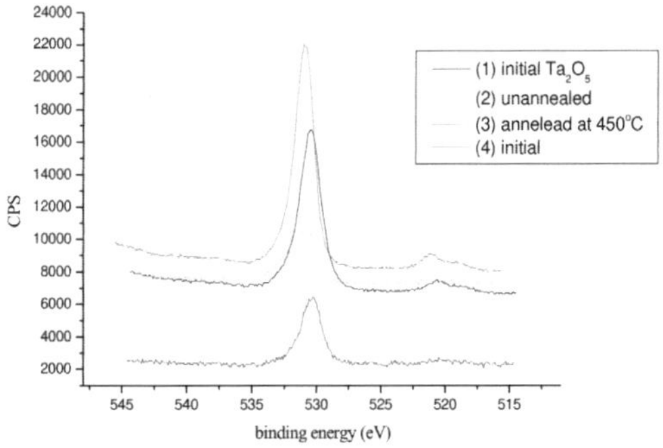

Figure 1. High-resolution XPS spectra of Ta 4f and 4 O 1s spectra of e-beam evaporated Ta_2O_5 films.

The calculation of the surface free energy of solids using contact-angle measurements is based on the surface energy balance condition: The values of the surface free energy and its polar and dispersion components, as calculated by the Wu method for two liquids and the Owens-Wendt-Rabel-Kaeble method for the liquid system α-bromonaphthalene- formamide-ethylene glycol-diidomethane-glycerol-water were determined from contact angle measurements at 20°C (table 1).

Table 1. Values of the total surface free energy and its dispersion and polar components, and the fractional polarity (by Owens-Wendt-Rabel-Kaelble method).

Type of coatings	Component of surface free energy [mN/m]			
	Dispersion part γ^d	Polar part γ^p	Total γ	Fractional polarity $\gamma^p/(\gamma^d+\gamma^p)$
Glass substrate	30.9	29.8	61.7	0.48
Ta_2O_5 non-annealed)	26.8	8.1	34.9	0.23
Ta_2O_5 (annealed)	39.7	14.8	54.5	0.27

Breaking the surface bonds with oxygen by either thermal or plasma treatment results in shear in the more chemically stable hydrophilic surface region.

After a-three-day stay in the culture, the fibroblast cells were well spread on both the control and coated surfaces. The cell morphology was typical for cells on a coated surface. The cell structural organization corresponded to the initial fibroblast. After five days of cultivation, the cell density increased for all samples. Most of the cells were ripe fibroblasts with strongly marked phenotype. The effect of the surface properties on cell proliferation was investigated by counting the cells attached to the glass and coating surfaces after incubation (table 2).

Previous studies have examined the effect of surface energy on cell functions such as adhesion, proliferation and differentiation. In some cases, the cell functions are enhanced on hydrophilic surfaces, in other cases, on hydrophobic. In our study, the SFE values were in the range 35 - 55 mN/m. The annealing process results in shear in the chemically stable hydrophilic surface region.

The more detailed study of the effect of the surface free energy on cell spreading and proliferation requires that one should account for the dispersive and polar components of surface free energy and the fractional polarity [15]. The polar part of SFE changes from 8.1 for non-annealed to 14.8 for annealed films, while the fractional polarity changes from 0.23 to 0.27, respectively (table 2). Fibroblast cells were well spread both on the glass (control) and the coated surfaces. Differences in the cell attachment and spreading on e-beam evaporated Ta_2O_5 coating surfaces was observed (table 2).

Table 2. Total cell number and detached cell percentage on the surface of Ta_2O_5 (non-annealed and annealed) coatings after 3 and 5 days of cultivation.

Type of coatings	3 days		5 days	
	Total number of cells	Number of detached cells and percentage	Total number of cells	Number of detached cells and percentage
Glass (control)	$2,34\cdot10^4\underline{+}2,35\cdot10^3$	$1,17\cdot10^3\underline{+}1,77\cdot10^2$ 5%	$3,77\cdot10^4\underline{+}1,97\cdot10^3$	$3,06\cdot10^3\underline{+}2,15\cdot10^2$ 8%
Ta_2O_5 (non-annealed)	$2,18\cdot10^4\underline{+}2,32\cdot10^3$	$1,57x10^3\underline{+}1,53\cdot10^2$ 7%	$3,48x10^4\underline{+}1,72x10^3$	$3,94x10^3\underline{+}3,36\cdot10^2$ 11%
Ta_2O_5 (annealed)	$2,62\cdot10^4\underline{+}2,67\cdot10^3$	$1,78\cdot10^3\underline{+}2,05\cdot10^2$ 6%	$3,95\cdot10^4\underline{+}2,02\cdot10^3$	$2,92\cdot10^3\underline{+}2,49\cdot10^2$ 7%

4. Conclusion

The biocompatibility of Ta_2O_5 films was estimated by in vitro tests. The results demonstrated the good cyto compatibility of e-beam evaporated Ta_2O_5 coatings especially in the case of annealed films with good stoichiometric Ta_2O_5 composition (table 2). The best biological response parameters (cell number, proliferation function, morphology) were obtained in the case of materials with the largest values of the SFE polar part component and of the fractional polarity. The results show that the surface properties are strongly influenced by the preliminary treatment. By varying the deposition and treatment conditions, one can control the surface parameters of the e-beam evaporated Ta_2O_5 films and the cell response.

Acknowledgment
The work was supported under a program for collaboration between the Bulgarian Academy of Science and the National Academy of Science of Ukraine.

References

[1] Matsuno H, Yokoyama A, Watari F, Uo M and Kawasaki T 2001 *Biomaterials* **22** 1253-8
[2] Miyazaki T, Kim H M, Kokubo T, Ohtsuki C and Nakamura T 2002 *Biomaterials* **23** 827-34
[3] Georgiev G, Fescheschiev N, Popov D and Uzunov Z 1989 *Vacuum* **36** 595-99
[4] Leng X Y, Sun H, Yang P, Chen J Y, Wang J, Wan G J, Huang N, Tian X B, Wang L P and Chu P K 2001 *Thin Solid Films* **398/399** 471-81
[5] Meyer U, Bushter A, Wiesmann H P, Joos U and Jones D B 2005 *J. Eur. Cells Mater.* **9** 39-49
[6] Anselme K 2000 *Biomaterials* **21** 667-81
[7] Hunter A, Archer C W, Walker P S and Blunn G W 1995 *Biomaterials* **16** 287-95
[8] Dhayal M, Cho S I, Moon J Y and Zykova A 2008 *J. Appl. Surf. Sci.* **254** 3331-8
[9] Van der Valk P, Van Pelt A W J and Busscher H J 1983 *J. Biomed. Mater. Res.* **17** 807-17
[10] Donkov N, Zykova A, Safonov V and Mateev E 2009 *Problems of Atomic Sci. & Technol. Series: Plasma Physics* **15** 153-5
[11] Todorova Z, Donkov N, Ristic Z, Bundaleski N, Petrovic S and Petkov M 2006 *J. Plasma Process and Polymers* **3**/2 174-8
[12] Wu S 1970 *J. Phys. Chem.* **74** 632
[13] Owens D K, Wendtn R C 1969 *J. Appl. Polym. Sci.* **13** 1741
[14] Uthanna S, Jagadeesh C S V, Sreedhara R P and Mohan R G 2008 *J. Phys.: Conf. Series* **114** 01203
[15] Hallab N J, Bundy K J, O'Connor K and Moses R L 2001 *J. Tissue Engin.* **7** 55-71

Si nanoclusters generated in Si-SiO$_2$ structures implanted with different doses of Si ions

S Kaschieva[1,5], A Gushterov[2], P Gushterova[3] and S N Dmitriev[4]

[1]Georgi Nadjakov Institute of Solid State Physics,
Bulgarian Academy of Sciences, 72, Tzarigradsko Chaussee, 1784 Sofia, Bulgaria
[2]Institute of Nanostructure Technologies and Analytics,
University of Kassel, 40 Heinrih-Plet Str., 34109 Kassel, Germany
[3]Central Laboratory of Optical Storage and Processing of Information,
Bulgarian Academy of Sciences, Acad. G. Bonchev Str. Bl. 101, 1113 Sofia, Bulgaria
[4]Flerov Laboratory of Nuclear Reactions, Joint Institute of Nuclear Research,
141980 Dubna, Moscow region, Russia

E-mail: kaschiev@issp.bas.bg

Abstract. Topography changes of a SiO$_2$ surface and formation of Si nanoclusters, induced by high-energy electron irradiation of ion implanted Si-SiO$_2$ structures, are studied by atomic force microscopy (AFM) measurements. Si ions with energy of 15 KeV and doses of 10^{12} or 10^{16} cm^{-2} are implanted in the Si-SiO$_2$ structures with an oxide thickness of 20 nm. The ion energy is chosen to produce maximum ion damage at the Si-SiO$_2$ interface. Some of the implanted samples are simultaneously irradiated by 20 MeV electrons with a flux of about $1\text{x}10^{15}$ cm^{-2}. AFM measurements indicated that the SiO$_2$ topography does not vary significantly only due to the ion-implantation. Si nanoclusters are observed in a SiO$_2$ layer after MeV electron irradiation. The measurements show that the size and shape of the Si nanoclusters depend significantly on the dose of the previously implanted ions.

1. Introduction

Silicon nanocrystals generated in a SiO$_2$ layer have attracted great interest due to their excellent optical and electronic properties [1-2]. The most important property of these nanocrystals is their intense photoluminescence in the visible and red/near-infrared region of the optical spectrum [2-3]. This is why Si nanocrystals are potential candidates for Si-based optoelectronic devices. Klimenkov et al. investigated the influence of the electron irradiation energy on Si-SiO$_2$ structures implanted with different ion doses. They showed that high-energy electron irradiation causes movement of clusters or atoms on the surface [4] and, alternatively to the thermal treatment, cluster formation can be stimulated by high-energy electron irradiation [5].

In this paper we demonstrate the effects of the high-energy electron irradiation on the topography of SiO$_2$ samples implanted with a dose of 1.5×10^{12} and 1.0×10^{16} cm^{-2} Si ions (Si$^+$). It is shown that Si accumulates and Si nanoclusters are formed on the SiO$_2$ surface after high-energy electron irradiation.

[5] To whom any correspondence should be addressed.

We assume that free oxygen is generated during MeV electron irradiation as a result of Si-O bond breaking and radiation-stimulated movement through defects takes place.

In addition, we followed the effect of the Si^+ dose on the shape, height and density of Si nanoclusters.

2. Experimental details

The native oxide and contaminations from the surface of n-type CZ Si <100> wafers were removed by HF cleaning and the wafers were then oxidized in dry O_2 at 950 °C. The SiO_2thickness was determined to be 20 nm by ellipsometry.

After oxidation, the SiO_2-Si samples were implanted through the oxide by 15 KeV Si^+ with a dose of 1.5×10^{12} and 1.0×10^{16} cm^{-2}. The ion energy is chosen so as to produce maximum ion damage at the Si-SiO_2 interface.

The samples were irradiated with electrons in a Microtron MT-25 system (Flerov Laboratory of Nuclear Reactions, Joint Institute of Nuclear Research - Dubna, Russia). Some of the implanted samples were simultaneously exposed to 20 MeV electrons perpendicular to the SiO_2 surface in a vacuum chamber under a pressure of about 1×10^2 Pa. Because of the high energy of the electrons, they penetrated through the whole sample (SiO_2 and Si wafer), creating radiation defects. The flux of the electrons was 1.2×10^{15} el/cm^2. The sample temperature was controlled during the irradiation; it remained near room temperature. The distance between the Microtron window and the samples was 150 mm.

The surface topography of the implanted and irradiated Si-SiO_2 structures was studied by an atomic force microscope (AFM) (CP II, Veeco Instruments) in tapping mode. The AFM images were evaluated using Gwyddion software [6]. The size of the surface area investigated was 1 μm × 1 μm for Si^+ implantation with a dose of 1.5×10^{12} cm^{-2} and 5 μm × 5 μm for a dose of 1×10^{16} cm^{-2}. The SiO_2-Si structures were measured before and after MeV electron irradiation.

3. Results and discussion

Figure 1 presents the AFM images of Si^+ samples implanted with a dose of 1.5×10^{12} cm^{-2} before (a) and after 20 MeV electron irradiation (b). It can be seen that after Si^+ implantation the SiO_2 surface is rough, but unambiguous nanostructures cannot be observed (figure 1a). It is possible for implanted Si atoms to migrate and form small precursor precipitations, because implanted ions and implantation-induced damages are in some activated state. The following 20 MeV electron irradiation (with a flux of 1.2×10^{15} el/cm^2) generates silicon nanoclusters, which are observed by AFM on the SiO_2 surface (figure 1b). The mean lateral size (l_{mean}) of these nanostructures is in order of 50 nm and the mean height (h_{mean}), between 2.3 to 4.2 nm (table 1). We presume that MeV electron irradiation break Si-O bonds and the free oxygen moves through radiation defects so that Si nanostructures are generated in the SiO_2. This is in good agreement with our earlier results [8-9], where it was shown that oxygen accumulates at the SiO_2-Si interface after MeV electron irradiation of SiO_2-Si structures. In our opinion, the nanoclusters created by MeV electron irradiation have the same structure as the nanoclusters generated by thermal annealing [10] and, therefore, have crystal structure. One of the advantages of MeV electron irradiation compared to thermal annealing is that Si nanoclusters can be observed at the SiO_2 surface without additional temperature treatment.

The topography of a high-energy electron irradiated SiO_2-Si sample implanted with Si^+ with a dose in the order of 10^{16} cm^{-2} is shown in figure 2. In this case l_{mean} of the Si nanoclusters is 35 nm, which is lower than l_{mean} of the Si nanoclusters in the sample implanted with a lower dose of Si. On the contrary, h_{mean} of the Si nanoclusters increases in comparison to those implanted with a lower dose of Si and has values in the range of 9 to 16 nm. The comparison between the AFM images in figure 1 b) and figure 2 show that the nanocluster size depends on the dose of the Si ions implanted in SiO_2-Si structures. The variation of the nanocluster size can be related to the reduction of the oxygen concentration on the oxide surface. The height of the nanoclusters increases, but the lateral size and the surface density decrease with increasing the dose of implanted Si atoms, as can be seen in figure 2. The ion implantation breaks many of the Si-O bonds and also may knock out atoms from their sites. However, most of the atoms, silicon or oxygen, bond again shortly after being dislodged. One can thus explain the presence of nanoclusters with larger sizes observed after electron irradiation in silicon implanted samples with dose 10^{16} cm^{-2} in figure 2 and in table 1.

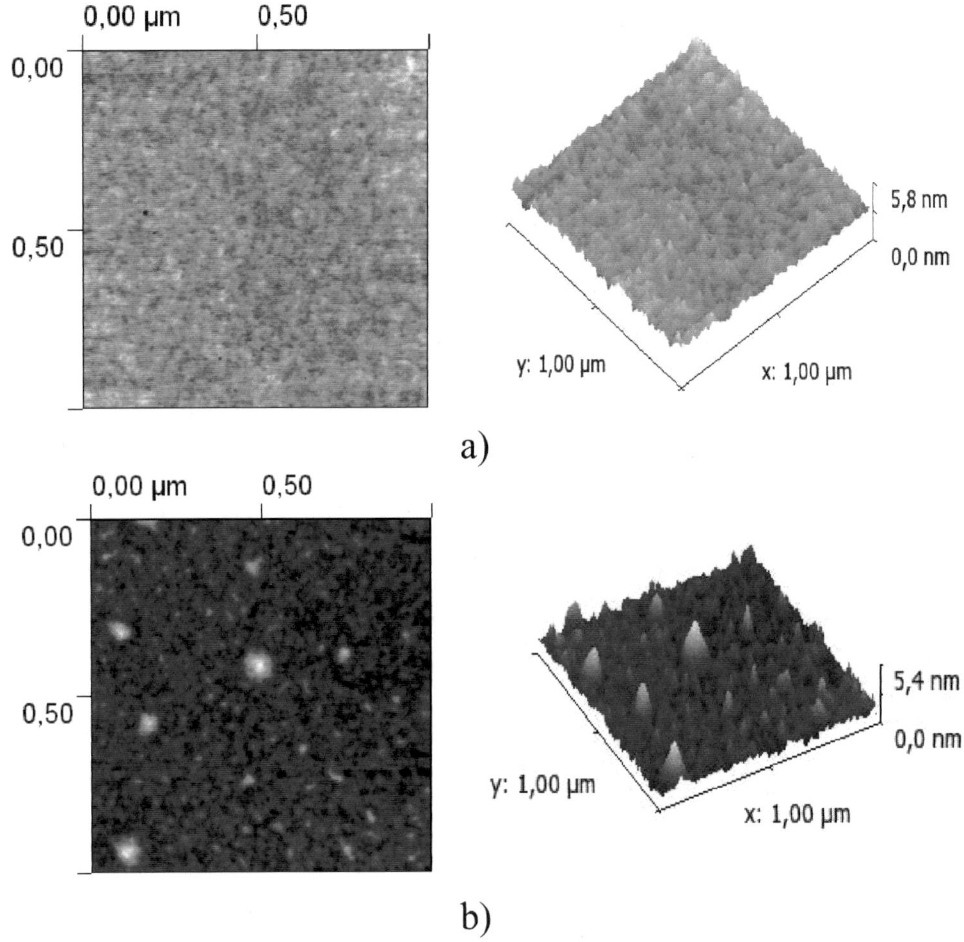

Figure 1. AFM images of surface topography of Si^+ implanted Si-SiO_2 sample (a dose of 1.5×cm^{-2}): (a) before and (b) after electron irradiation with acceleration voltage of 20 MeV.

Table 1. Mean height and mean length of the Si nanoclusters, created by MeV electrons.

Sample	l_{mean} [nm]	h_{mean} [nm]
Si^+10^{12}+20 MeV	50	2.3 to 4.2
Si^+10^{16}+20 MeV	35	9 to 16

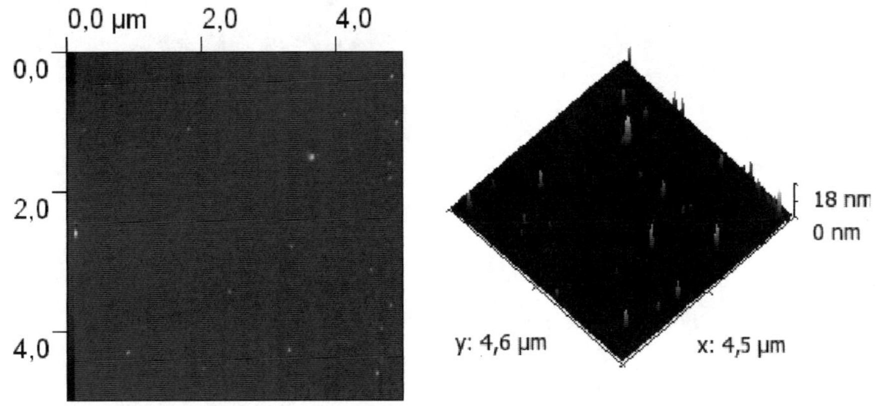

Figure 2. AFM images of Si^+ implanted Si-SiO_2 sample with a dose of 10^{16} cm^{-2} and 20 MeV electron irradiation.

4. Conclusion

The effect of MeV electron irradiation on the surface topography of Si-SiO_2 structures implanted with silicon ions with a dose of 1.5×10^{12} and 1.0×10^{16} cm^{-2}, is characterized by AFM measurements. The SiO_2 surface topography demonstrates that high-voltage electron irradiation (at a flux of 1×10^{15} cm^{-2}) creates Si nanoclusters not only in the bulk of SiO_2 layers, but also on its surface in both groups of silicon implanted samples. The shape and the density of the Si nanoclusters depend on the dose of implanted ions. The lateral size and the surface density of the nanoclusters are smaller and their height, larger for the samples implanted at a higher dose of Si ions.

References

[1] Chen X Y, Lu Y F, Wu Y H, Cho B J, Liu H, Dai D Y and Song W D 2003 *J. Appl. Phys.* **93** 6311
[2] Sias U S, Behar M, Boudinov H and Moreira E C 2007 *Nucl. Instr. & Meth.* B **6** 257
[3] Pavesi L, Mazzoleni C, Tredicucci A and Pellegrini V 1995 *Appl. Phys. Lett.* **67** 3280
[4] Klimenkov M, Matz M and Borany J 2000 *Nucl. Instr. & Meth.* B **168** 367
[5] Nepijko S, Klimenkov M, Kuhlenbeck H and Freund H 1999 *Vac. Sci. Technol.* A **17** 577
[6] Martinez C C M, Madrid R E, and Felice C J. 2008 *Sensors and Actuators B* **133** 682-6
[7] Nordlund K J, Arvi T T, Meinander K and Samela J 2008 *Appl. Phys.* A **91** 561–6
[8] Kaschieva S 1999 *Generation and annealing of radiation defects in MOS structures* (DSci Thesis Dubna Russia) (in Russian)
[9] Kurmaev E, Shamin S, Galakhov V, Kirilova M, Kurennykh T, Viykhodets V and Kaschieva S 1997 *J. Phys. Cond. Matter* **9** 6969
[10] Dunaevskii M S, Grob J J, Zabrodskii A G, Laiho R and Titkov A N 2004 *Semiconductors* **38** 1294

Optical properties of Si^+ implanted PMMA

S Balabanov[1], T Tsvetkova[1,5], E Borisova[2], L Avramov[2], L Bischoff[3] and J Zuk[4]

[1]Georgi Nadjakov Institute of Solid State Physics,
Bulgarian Academy of Sciences, 72 Tzarigradsko Chaussee, 1784 Sofia, Bulgaria
[2]Emil Djakov Institute of Electronics,
Bulgarian Academy of Sciences, 72 Tzarigradsko Chaussee, 1784 Sofia, Bulgaria
[3]Forschungszentrum Dresden-Rossendorf e.V.,
P.O.Box 510119, D-01314 Dresden, Germany
[4]Maria Curie-Sklodowska University, 20-031 Lublin, Poland

E-mail: tania_tsvetkova@yahoo.co.uk

Abstract. In the present work, low energy ion beam irradiation was used for surface modification of polymethyl-methacrylate (PMMA) using silicon (Si^+) as the ion species. After high doses ion implantation of Si+ in the polymer material, a characterization of the optical properties was performed using optical transmission measurements in the visible and near infra-red (IR) wavelength range. The optical absorption increase observed with the ion dose was attributed to ion beam induced structural changes in the modified material.

1. Introduction

In recent years there has been growing interest to the modification of surfaces and thin layers of polymers. The ion beam technique is widely used, as it is a flexible and powerful tool for surface engineering of different materials, including polymers. The increased attention to the ion implantation of polymer materials is due to the possibility to control precisely the technological parameters for fundamental research purposes, and also in view of some possible applications to device fabrication. Thus, optical, mechanical, electrical and chemical properties can be selectively modified using ion bombardment [1]. The ion irradiation of polymers is followed by various processes, such as: macro-molecular destruction, cross-linking, free radicals formation, carbonization and oxidation [2-4]. Therefore, understanding the influence of certain structural re-arrangements on the suitable properties of polymers opens a way to design devices with required parameters, involving such applications as optical filters, absorbers, reflectors, etc. [5].

The net result of ion beam treatment of polymer materials is that it leads to alteration in their physical and chemical properties, i.e. modification of the electrical and optical properties, changes in the surface wettability and adhesive bonding, as well as improvement of surface hardness and wear resistance. Some of the beam-induced chemical effects involve: reduction of crystallinity, partial graphitization of the surface layer, formation of unsaturated C=C bonds, formation of methyl groups due to chain scission, etc. [6,7].

In the present work, low energy ion beam irradiation was used for surface modification of polymethyl-methacrylate (PMMA) using silicon (Si^+) as the ion species. The ion bombardment effects

[5] To whom any correspondence should be addressed.

on the optical properties of the polymer materials were investigated using optical transmission measurements in the visible and near infra-red (IR) wavelength range.

2. Experimental

Polymethyl-methacrylate (PMMA) samples, available commercially as bulk samples with a thickness of 2 mm, were studied. All samples were shaped as squares (side of 10 mm) or circles (diameter of 10 mm).

Danfysik 1090 ion implanter was used to carry out the Si^+ ion implantation at room temperature (RT) and relatively low energies (E = 30÷50 keV), with ion doses ranging from D = 1×10^{13} cm^{-2} to 1.2×10^{17} cm^{-2}. The ion beam current was kept under 2 $\mu A/cm^2$ during the implantation process, so as to maintain the target temperature below 80°C, at which the polymer material starts to decompose. The pressure in the target chamber was kept under 10^{-6} Pa during implantation.

The optical transmission and reflection spectra of the PMMA samples were measured by double-beam spectrophotometers Specord UV-VIS (400 ÷ 800 nm) and Cary 5E (330 ÷ 3300 nm).

3. Results and discussion

The optical transmission spectra of Si^+ implanted (E = 50 keV) PMMA samples are presented in figure 1(a). Wide range of ion doses were used: D = 10^{14} ÷10^{17} cm^{-2}. There is no significant change in the optical transmission for the lower doses – up to $D_3 = 1\times10^{15}cm^{-2}$ – while a well expressed increase in the absorption is observed with the further dose increase, reaching quickly saturation at $D_4 = 3{,}2\times10^{15}cm^{-2}$ and higher doses. Similar behavior in the IR absorption is also observed in figure 1(b).

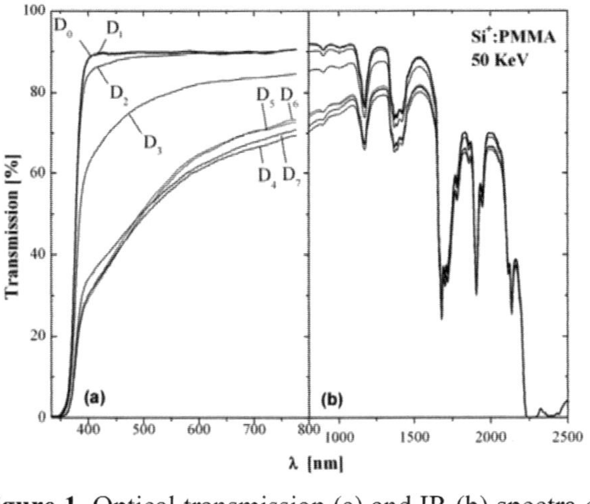

Figure 1. Optical transmission (a) and IR (b) spectra of Si^+ implanted PMMA at ion beam energy E=50 keV and different ion doses: D_1-1×10^{14}; D_2=3.2×10^{14}; D_3=1×10^{15}; D_4=3.2×10^{15}; D_5=1×10^{16}; D_6=3.2×10^{16}; D_7=$1\times10^{17}cm^{-2}$; D_0 - unimplanted PMMA.

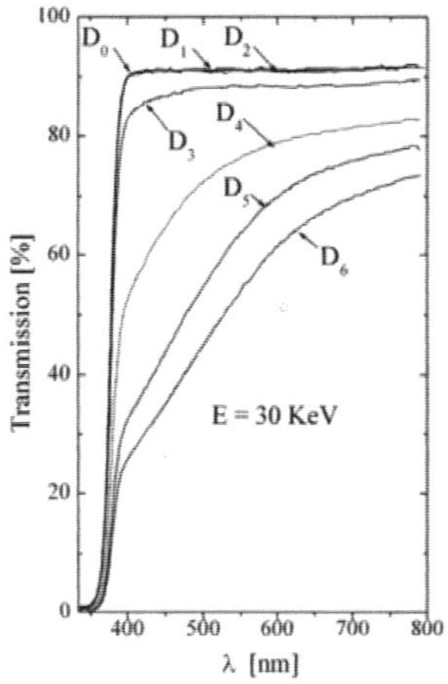

Figure 2. Optical transmission spectra of Si^+ implanted PMMA at ion beam energy E=30 keV and different ion doses: D_1=1×10^{13}; D2=5×10^{13}; D_3=2.5×10^{14}; D_4=1×10^{15}; D_5=5×10^{15}; D_6=$10^{16}cm^{-2}$; D_0 - unimplanted PMMA.

A more gradual increase in the absorption with the dose is registered for Si^+ implanted PMMA samples at lower energy (E = 30 keV) and relatively low doses (D = 1×10^{13} ÷ 2.5×10^{16} cm^{-2}) – figure 2. In this case, noticeable absorption increase is observed at a similar low dose (D_4 = 1×10^{15} cm^{-2}), but then continues to increase gradually with the dose, without reaching saturation.

Figure 3 presents the transmission spectra in the visible range for two PMMA samples implanted at two different energies with the same ion dose, D = 1×10^{15} cm^{-2}. Slightly greater absorption is observed for the sample implanted with E = 30 keV, as compared to the one for E = 50 keV, probably due to the slightly larger peak concentration in the Si^+ ions distribution for the lower energy and the same ion dose [8].

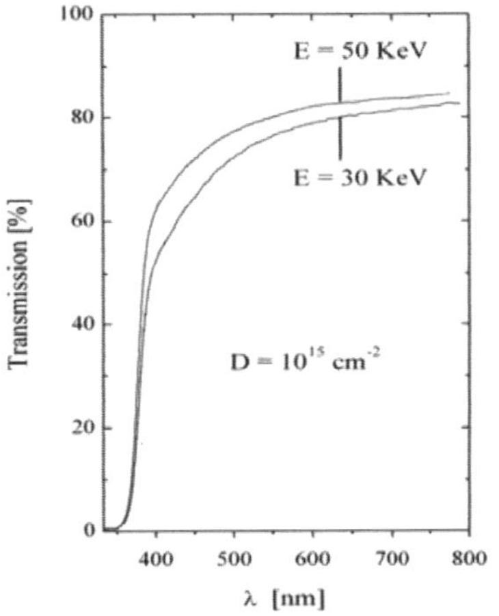

Figure 3. Optical transmission spectra of Si^+ implanted PMMA samples with ion dose $D=1\times10^{15}cm^{-2}$ at two different ion energies: E_1=50keV and E_2=30keV.

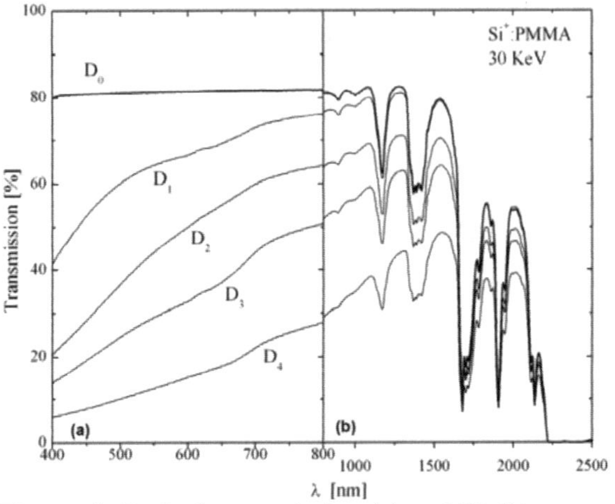

Figure 4. Optical transmission (a) and IR (b) spectra of Si^+ implanted PMMA at ion beam energy E=30 keV and different ion doses: $D_1=1\times10^{15}$; $D_2=5\times10^{15}$; $D_3=2.5\times10^{16}$; $D_4=1.2\times10^{17}cm^{-2}$; D_0 - unimplanted PMMA.

A more pronounced effect of absorption increase with the dose, again without reaching saturation, is shown in figure 4 for still higher ion doses (D = $1\times10^{15} \div 1.2\times10^{17}$ cm^{-2}) at a lower energy (E = 30 keV), i.e. an even higher peak concentration in the Si^+ ions distribution. This is demonstrated both in the visible (figure 4(a)) and IR (figure 4(b)) range.

Using the transmission data one can derive the dependence T(D), i.e. the transmittance v/s the ion implantation dose (figure 5) for one wavelength in the visible range – 500 nm, and for one in the IR – 1300 nm. T(D) for the other wavelengths in respective ranges are essentially the same. As seen in figures 5 (a,b), T(D) decreases in a non-monotonic way, in contrast to the behavior observed in other implanted polymers at comparable doses [9-11].

This declination type of behavior seems as a 'break in transmission' and occurs at a certain 'breaking dose', D_B. In fact, at a slightly lower dose the photoluminescence (PL) of Si^+ implanted PMMA is enhanced, and at $D \geq D_B$ it is continuously quenched [12]. Presumably, the increased defect introduction, above some critical concentration, results in an increased degree of structural disorder.

It appears that D_B is related to the energy of implantation. T(D) obtained from transmission spectra of another set of PMMA samples, also implanted with Si^+ but at lower energy E = 30 keV (figures 5(c,d)), do not exhibit any declination behavior.

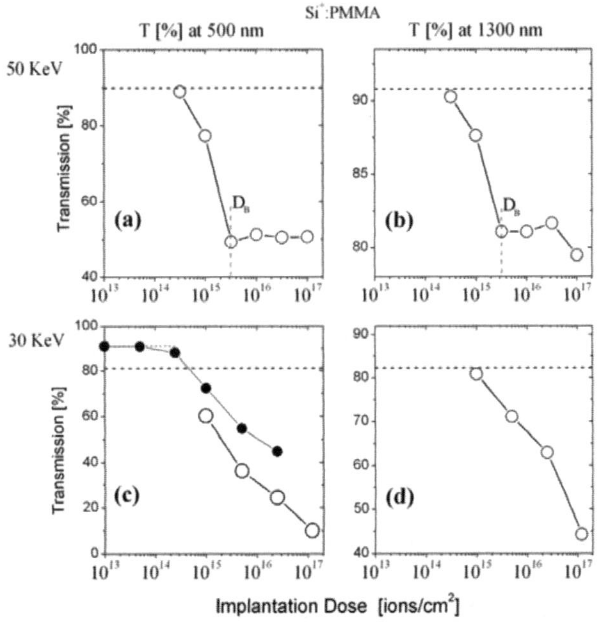

Figure 5. Dose dependence of the optical transmittance (T) for Si^+ implanted PMMA samples with: ion energy E=50keV at λ_1=500 nm (a) and λ_2=1300 nm (b); and ion energy E=30keV at λ_1 (c) and λ_2 (d).

Figure 5(c) includes also the dependence T(D) (solid circles) for an additional set of PMMA samples coming from another source with probably different impurities content, crystalline to amorphous phase ratio, etc., which is the reason for the lack of complete overlapping between the results of the two sets of samples implanted with the same energy and range of doses. Otherwise, the reproducibility of the above given results is within a relatively small error (±1,5%, hence no error bar has been introduced in the figures.) As is seen, T(D) decreases gradually. In our view, the lack of declination means that the polymer surface in this case is not so hardly structurally modified as for the Si^+ implantation at E = 50 keV. Probably, for the higher energy case, some critical level of energy deposition into the surface layer is reached, so that greater destruction occurs via a mechanism resulting in a significant change of the defect states density.

4. Conclusion

The results obtained in this work demonstrate that the optical absorption of Si^+ implanted PMMA could be significantly altered, both in the visible and the IR range, depending on the ion energy and dose. For higher energies, the ion beam induced absorption increase with the dose exhibits declination point type of behavior, which is absent in the lower energy case. This is presumably due to an increased defect introduction above some critical concentration causing an abrupt increase of structural disorder.

Additional experiments with Raman and X-ray photo-electron spectroscopy (XPS) are in run to help identify the ion beam induced structural changes associated with the observed optical properties modification. The correlation between the Si^+ implantation effects on the optical and PL behavior of ion implanted polymers is also of interest in view of further uses in optoelectronics and photonics.

Acknowledgements
This work was supported by an International Atomic Energy Agency project (No:1294/RBF). The support of Forschungszentrum Rossendorf e.V., Germany, is also acknowledged.

References

[1] Marletta G 1995 *Materials and Processes for Surface and Interface Engineering* (Dordrecht: Kluwer Academic Publishers) p 597
[2] Lueck H B 1982 *Nucl. Instr. & Meth.* B **202** 497
[3] Popok V N, Odzaev V B et al 1997 *Nucl. Instr. & Meth.* B **129** 60
[4] Balabanov S and Krezhov K 1999 *J. Physics D: Appl. Phys.* **32** 2573
[5] Vargo T G, Gardella J A et al 1993 *Surface Characterization of Advanced Polymers* ed L Sabbattini and P G Zambonin (Weinheim: VCH Publ Co) p 163
[6] Charlesby A 1960 *Atomic Radiation and Polymers* (Oxford: Pergamon Press)
[7] Baleva M, Zlateva G, Tsvetkova T, Balabanov S and Bischoff L 2009 *J. Optoelectr. Adv. Mat.* **11** 1420
[8] Ziegler J F, Biersack J P and Littmark U 1985 *The Stopping and Range of Ions in Solids* (Oxford: Pergamon)
[9] Stepanov A L, Popok V N, Khaibulin I B and Kreibig U 2002 *Nucl. Instr. & Meth.* B **191** 473
[10] Stepanov A L and Khaibulin R I 2004 *Rev. Adv. Mater. Sci.* **7** 108
[11] Stepanov A L 2005 *Metal-Polymer Nanocomposites*, ed L Nicolais and G Carotenuso (London, UK: John Wiley & Sons, Ltd.) p 241
[12] Tsvetkova T, Balabanov S, Avramov L, Borisova E, Angelov I, Sinning S and Bischoff L 2009 *Vacuum* **83** 252

Journal of Physics: Conference Series **223** (2010) 012033 doi:10.1088/1742-6596/223/1/012033

Photoluminescence of Si^+ and C^+ implanted polymers

T Tsvetkova[1,5], S Balabanov[1], L Avramov[2], E Borisova[2], I Angelov[3] and L Bischoff[4]

[1]Georgi Nadjakov Institute of Solid State Physics,
Bulgarian Academy of Sciences, 72 Tzarigradsko Chaussee, 1784 Sofia, Bulgaria
[2]Emil Djakov Institute of Electronics,
Bulgarian Academy of Sciences, 72 Tzarigradsko Chaussee, 1784 Sofia, Bulgaria
[3]Institute of Organic Chemistry,
Bulgarian Academy of Sciences, Acad. G. Bonchev Str., Bl. 9, 1113 Sofia, Bulgaria
[4]Forschungszentrum Rossendorf e.V., P.O.Box 510119, D-01314 Dresden, Germany

E-mail: tania_tsvetkova@yahoo.co.uk

Abstract. Visible photoluminescence (PL) of ion implanted polymers was studied. Different polymer materials were used for the purpose: polypropylene (PP), poly-tetrafluor-ethylene (Teflon), ultra-high-molecular-weight-polyethylene (UHMWPE) and UHMWPE+Bi. Ion implantation with Si^+ and C^+ was performed at energies of 30 keV with doses in the range $10^{13} - 10^{17}$ cm^{-2}. The results show that a PL enhancement (PLE) effect may occur for some polymer materials if proper implantation energy and doses are employed, the effect in the case of some polymer materials implanted with Si^+ and C^+ being considerable. While the effect is observed for all doses of C^+ implanted in UHMWPE, PLE is only observed for the lowest dose of Si^+ ($D=1\times10^{15}$ cm^{-2}) implanted in Teflon, the further dose increase resulting in PL quenching only, presumably due to processes of structural degradation The appearance of ultra-violet (UV) range PL in the case of Si^+ implanted UHMWPE could be originating from the formation of Si-related new defect sites, but more data are needed to explore this effect further into the deeper UV range ($\lambda<350$ nm).

1. Introduction

The ion implantation technique is an attractive tool for material properties modification of polymers due to the possibility for precise control of the technological parameters both for fundamental research purposes and for various possible applications. The optical, electrical, mechanical and chemical properties can be selectively modified using ion bombardment [1]. Important advantages of the ion implantation method, compared to the other doping methods, are the possibility for precise dopants concentration control, the ability to resolve their spatial distribution – in-depth dopant profiling by designing the implantation parameters, and surface localization by using masks or focused ion beams.

The understanding of the structural re-arrangements effect on the relevant optical properties of ion-bombarded polymer surfaces reveals new approaches to the design of devices with desired parameters, e.g. for applications of high-performance polymers for optical filters, absorbers, reflectors, luminescent devices, etc. [2]. This knowledge is of considerable importance for the application of the ion beam synthesis (IBS) techniques to produce new compounds and advanced materials in the

[5] To whom any correspondence should be addressed.

irradiated polymer surfaces by implanting ions of various chemical elements [3,4]. The optical properties of ion implanted polymers, being of particular interest for modern optoelectronics and photonics, have been extensively explored [5,6].

The aim of the present work was to apply the ion beam technique to produce polymer materials yielding interesting photoluminescence (PL) properties. IN particular, it was of interest to study the PL dependence on the ion implantation conditions and implanted ion species.

2. Experimental

Various polymers, commercially available as bulk samples with thickness of 2 mm, were studied: ultra-high-molecular-weight-polyethylene (UHMWPE), polypropylene (PP), and poly-tetrafluor-ethylene (Teflon). The samples were shaped as squares (side of 10 mm) or circles (diameter of 10 mm).

Ion implantation with Si^+ was performed at room temperature (RT) and relatively low ion beam energy (E = 30 keV), with ion doses ranging from $D=1.10^{13}$ to $D=1.2.10^{17}$ cm^{-2}, using a Danfysik 1090 implanter. In order to keep the target temperature lower than 80° C, at which the polymer can be decomposed, the beam current was kept under $2\mu A/cm^2$. The pressure in the target chamber was kept under 10^{-6} Pa during implantation. Additionally, implantation by carbon ions was performed with the aim to distinguish the Si^+ implants induced effects from those introduced by self-ion implantation.

The photoluminescence (PL) measurements were performed at RT using two different excitation sources – Ar laser (λ_{ex} = 488 nm) and N_2 laser (λ_{ex} = 337 nm). The N_2 laser was used to obtain PL spectra of some of the UHMWPE samples. The laser power used was 14 μJ/pulse, with 10 Hz repetition rate. The detector was a Sony ILX511 2048-element linear silicon CCD array, with estimated sensitivity 86 photons/count, signal-to-noise ratio 250:1, and effective range 200-1100 nm.

The PL spectra excited by the Ar^+-laser (Spectra Physics 171) were registered by an optical multi-channel analyser (2048 channels) coupled with a Ocean Optics S2000 spectrometer. The entrance slit was set at 100 μm. The samples were illuminated by a laser beam with diameter of about 1 mm and power 10 mW and 30 mW @ TEM_{00}. During the integration cycle of 5 seconds we did not observe any nonreversible temperature changes, degradation or damages to the polymer materials.

3. Results and discussion

Figure1(a) shows the PL spectra of Si^+ implanted UHMWPE samples (E=30 keV, $D=1\times10^{15}$–2.5×10^{16} cm^{-2}). For PL excitation, an N_2 laser (λ_{ex}=337 nm) was used and the PL spectra were measured in the range 350 – 650 nm. The Si^+ ion implantation results in PL quenching in the whole visible range of the spectrum, while emergence of an UV range (λ~350 nm) PL signal is observed, increasing with the dose. No such UV range PL effect is found for the case of C^+ implantation in the same polymer material (UHMWPE), as shown in figure1(b). Instead, only visible range PL is registered, with C^+ implantation resulting just in its modification in the same range. The PL spectral changes observed in the case of Si^+ implanted UHMWPE, and the UV range emission emergence in particular, could be originating from Si^+–related defect complexes [7].

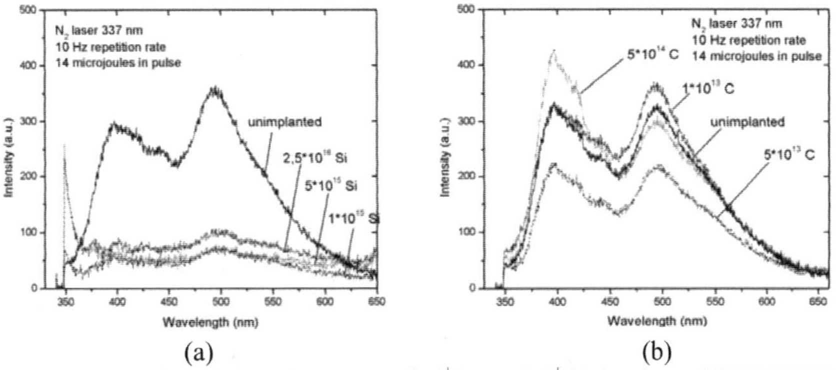

Figure 1. PL (λ_{ex}=337 nm) spectra of Si^+(a) and C^+(b) implanted UHMWPE samples at ion beam energy E=30 keV and different doses. PL of reference samples is also given.

When PL measurements are performed with the same material (UHMWPE) employing an Ar^+ laser as the excitation source (λ_{ex}= 488 nm), the visible range PL modification effects for the two kinds of ions, Si^+ and C^+, are again very different (figure 2). While there is a considerable photoluminescence enhancement (PLE) of the emission for the C^+ implanted UHMWPE samples compared to the reference sample (figure 2(a)), the Si^+ implantation results in significant PL quenching for all implantation doses used (figure 2(b)). This difference could be explained bearing in mind the higher mass of the Si^+ ions, compared to the C^+ ones, hence a greater degree of radiation induced defects is introduced by the Si^+ implants, resulting in increased disorder and related processes of degradation of the polymer structure [8].

Bombardment with the lighter C^+ ions, accompanied by a considerable energy dissipation in the polymer surface, is more likely to induce increased cross-linking and hence a more ordered structure of the polymer material [9].

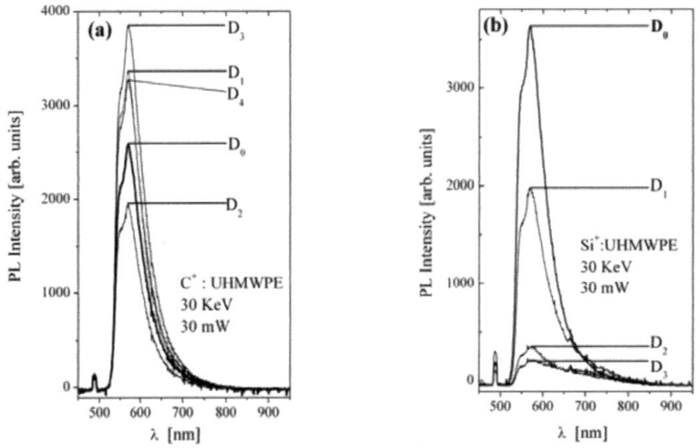

Figure 2. PL (λ_{ex}=488 nm) from UHMWPE samples implanted at E=30 keV by: (a) C^+ at ion doses: $D_1=1\times10^{13}$; $D_2=5\times10^{13}$; $D_3=5\times10^{14}$; $D_4=1\times10^{15}$ cm^{-2}; D_0-unimplanted polymer; and (b) Si^+ at ion doses: $D_1=1\times10^{15}$; $D_2=5\times10^{15}$; $D_3=2.5\times10^{16}$ cm^{-2}; D_0 - unimplanted polymer.

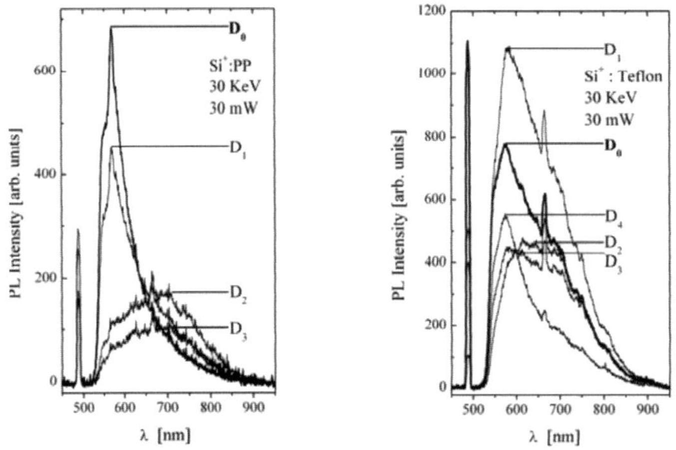

(a) (b)

Figure 3. PL (λ_{ex}=488 nm) from Si^+ implanted (E=30keV) PP (a) and Teflon (b) samples with ion doses: $D_1=1\times10^{15}$; $D_2=5\times10^{15}$; $D_3=2.5\times10^{16}$; $D_4=1.2\times10^{17}$ cm^{-2}; D_0-reference.

For Si^+ implants in PP (figure 3(a)), only PL quenching is registered for all available doses, as already the lowest dose (D= 1×10^{15} cm^{-2}) is obviously inducing degradation processes. Some noticeable PLE is however observed for Si^+ implants in Teflon (figure 3(b)), but it is only present for the lowest dose available (D=1×10^{15} cm^{-2}) while a further dose increase results in PL quenching only, presumably due to processes of structural degradation as discussed above.

The PL dependence on the ion dose is shown in figure 4, where the difference in the PL behavior, resulting from Si^+ implants in Teflon (figure 4(a)) and the one induced by the C^+ implants in UHMWPE (figure 4(b)), is displayed. While considerable PLE is found even for the higher C^+ ion doses available, for the Si^+ implants PLE is only recorded for the lowest dose used (D=1×10^{15} cm^{-2}). The reproducibility of the above given results is within a relatively small error (±2,5%), hence no error bar has been introduced in the figures. Thus, self-ion implantation (C^+ implants) is accompanied by a lower rate of disorder introduced in the ion bombarded polymer matrices, with possibilities to accommodate higher C^+ doses while preserving significant PLE with further implications for possible applications.

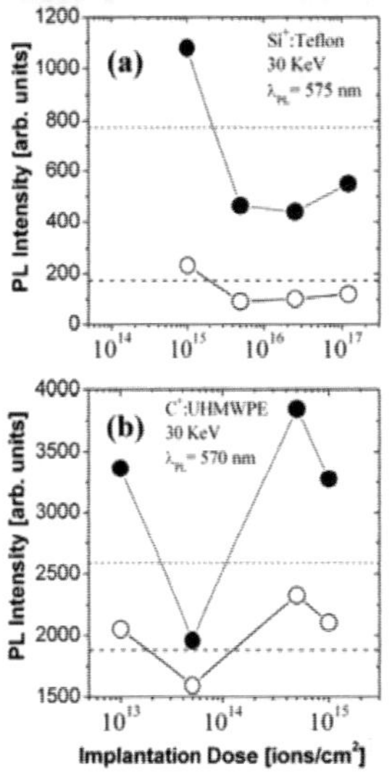

Figure 4. Maximum PL intensity (at λ_{PL}) vs ion doses for ion implanted polymers: Si^+ in Teflon (a), and C^+ in UHMWPE (b). The dashed line is for the PL intensity (at λ_{PL}) of the unimplanted polymer. For laser power 10 mW – open circles, for 30 mW – solid circles.

4. Conclusion

The interesting PLE effect observed in the case of some polymers implanted with Si^+ and C^+ is of considerable magnitude, with some specifics for the different polymer materials. While the effect is observed for all doses of C^+ implanted in UHMWPE, PLE is only observed for the lowest dose (D=1×10^{15} cm^{-2}) of Si^+ implanted in Teflon, a further dose increase resulting only in PL quenching, presumably due to processes of structural degradation. No PLE was observed in the case of Si^+ implants in PP; instead PL quenching is registered for all available doses, as already the lowest doses are obviously introducing significant degradation processes. The emergence of UV range PL observed in the case of Si^+ implanted UHMWPE could be originating from the formation of Si-related new defect sites, but more data are needed to explore the effect further into the deeper UV range ($\lambda<350$ nm). Additional Raman and X-ray photo-electron spectroscopy (XPS) measurements are needed to identify the structural changes associated with the PL effects observed in the implanted polymer materials. Better understanding of the registered ion beam induced PLE in some polymers has wide implications for future optoelectronic and photonic applications due to its relevance to their electroluminescence behavior.

Acknowledgements

This work was supported by an International Atomic Energy Agency project (No:1294/RBF). The support of Forschungszentrum Rossendorf e.V., Germany, is also acknowledged.

References

[1] Marletta G 1995 *Materials and Processes for Surface and Interface Engineering* (Dordrecht: Kluwer Academic Publishers) p 597

[2] Vargo T G, Gardella J A et al. 1993 *Surface Characterization of Advanced Polymers* ed L Sabbattini and P G Zambonin (Weinheim: VCH Publ. Co.) p 163

[3] Balabanov S and Krezhov K 1999 *J. Phys D: App. Phys.* **32** 2573
[4] Evelyn A L, Ila D et al. 1999 *Nucl. Instrum. & Meth.* B **148** 1141
[5] Stepanov A L 2004 *J. Techn. Phys.* **74** 1
[6] Sloof L H, van Blaaderen A, Polman A, Hebbink G A, Klink S I, van Veggel F C J M and Hofstraat J W 2002 *Appl. Phys. Lett.* **91** 3955
[7] Yu Y H, Li L, Lei Y M, Zhao J, Mao D S, Zou S C, Sundaraval B, Luo E Z, Wong S P, Xu J B and Wilson I H 2000 *Ion Implantation Technology* ed H Ryssel et al. (IEEE) p 765
[8] Baleva M, Zlateva G, Tsvetkova T, Balabanov S and Bischoff L 2009 *J. Optoelectr. Adv. Mater.* **11** 1420
[9] Charlesby A 1960 *Atomic Radiation and Polymers* (Oxford: Pergamon Press)

Ion – beam assisted process in the physical deposition of organic thin layers

D Dimov[1,3], A Georgiev[2], E Spassova[1], J Assa[1] and G Danev[1]

[1]Acad. J .Malinowski Central Laboratory of Photoprocesses,
Bulgarian Academy of Sciences, Acad. G. Bonchev Str., Bl.109, 1113 Sofia, Bulgaria
[2]University of Chemical Technology and Metallurgy,
8 Kl. Ohridski Blvd., 1756 Sofia, Bulgaria

E-mail: dean@clf.bas.bg

Abstract. A novel method was developed for physical deposition of thin polyimide layers by applying an argon plasma assisted process. The influence was investigated of the plasma on the combined molecular flux of the two thermally evaporated precursors – oxydianiline and pyromellitic dianhydride. The effects observed on the properties of the deposited films are explained with the increased energy of the precursor molecules resulting from the ion-molecular collisions. As could be expected, molecules with higher energy possess higher mobility and thus determine the modification of the films structure and their electrical properties.

1. Introduction

Polyimides (PI) are polymers with various applications thanks to their inherently good mechanical properties, high chemical resistance, low dielectric constant and high thermal stability. These polymers are, therefore, suitable as materials for insulating or protective layers and optical coatings [1, 2]. Physical evaporation of organic PI molecules is feasible at relatively low temperatures of deposition due to the thermal instability of the organic molecules. As a result, the thin organic layers formed have structural defects or micropores and cracks. To overcome this disadvantage, the use of electron assisted deposition (EAD) of PI precursors - oxydianiline(ODA) and pyromellitic dianhydride (PMDA) - was described by us in a previous paper [3]. This study was related to energy-stimulated processes based on the electron-treatment of the molecular flow of the PI precursors in the gas phase. The results obtained stimulated the development of an experimental technique for deposition of thin PI layers by applying a low-energy argon plasma assisted process. The aim of the present paper is to reveal the effects related to such as in view of the possibilities of applying this new type of process for production of thin PI layers.

[3] To whom any correspondence should be addressed.

2. Experimental

2.1. Sample preparation

The PI layers were formed by vacuum co-deposition of the precursors ODA and PMDA from two independent heated vessel sources (figure 1) on a static substrate. The deposition rate was 0.2 – 0.4 nm/sec. The base pressure was 4×10^{-2} Pa. The evaporation temperatures were 120 - 145° C for PMDA and 100 - 110°C for ODA. The layers were grown using Ar plasma assisted treatment processes: the molecular fluxes of the evaporated precursors cross the Ar plasma. Part of the ionized Ar molecules and electrons collide with the organic molecules of the precursors and act on the growing deposited layers. The plasma source was water-cooled in order to prevent excessive heating of the substrates. The processes parameters were as follows: cathode current 21 A; anode current 1,4 A; anode voltage 100 V. The layers obtained were transformed into PI by a 5-min microwave (MW) treatment (2.45 GHz, 700 W) followed by a 15-min thermal treatment at 300°C in air [4].

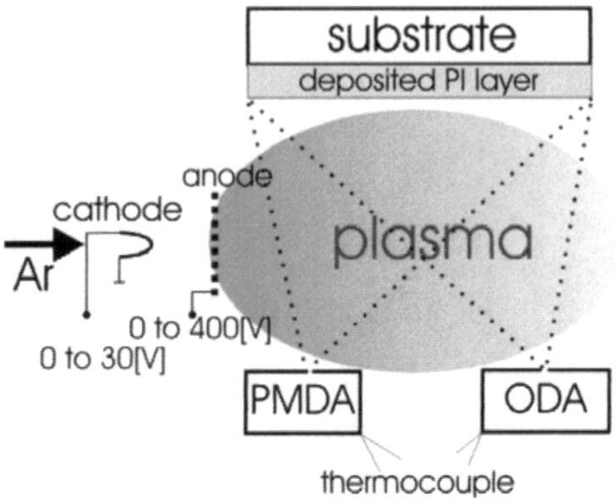

Figure1. A detailed schematic of the system for plasma assisted deposition (PAD).

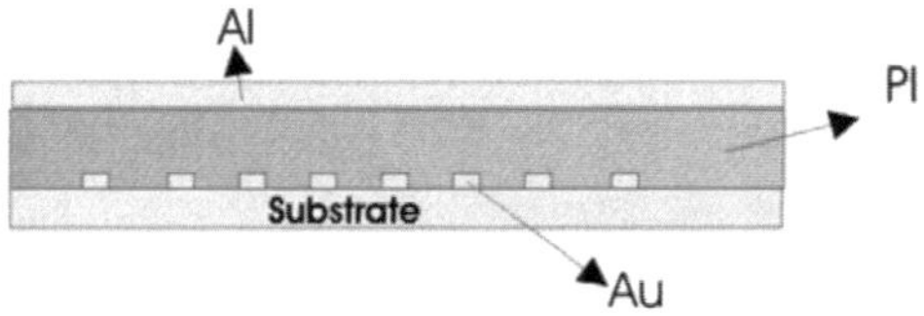

Figure2. Arrangement for current – voltage (I-V) measurement.

2.2. Methods of investigation

The film microhardness (Mhd) was determined by the Knoop prism method known to be sensitive for measurements of thin films. The load value was 1.25 mPa for all samples studied.

The surface morphology of the layers was characterized by scanning electron microscopy (SEM), Philips 515. The FTIR spectra (PI on KBr substrate) were recorded by a Bruker interferometer in the range of 4400 - 450 cm^{-1} with a resolution of 2 cm^{-1}.

2.2.1. Electrical measurements. The layers investigated were deposited on soda-lime glass substrates with previously prepared bottom Au electrodes. Aluminium top electrodes were then evaporated in vacuum perpendicularly to the Au electrodes. The resulting Au / PI / Al sandwich structures (figure 2) were again thermally treated at 200°C for 4 hrs to make the metal / polymer contacts more effective. Current-voltage (I-V) characteristics were measured in the DC (Direct Current) mode at room temperature in a vacuum of 1 Pa by a computer controlled Keithley 617 electrometer. The conductivity (σ) was determined as the slope of the linear regression of the Ohmic part of the I-U curves.

3. Results and discussion

The FTIR spectroscopy investigation results are shown in figures 3 and 4. According to the kinetic theory of collisions, the reaction rate of depends mainly on the energy factor, i.e. the number of collisions between the particles of the reagents. These collisions are called effective and chemical interaction takes place only between the particles taking part in them. The reaction rate depends also on the so-called factor of orientation (a possible space volume factor) between the reagents. It expresses the probability for the appropriate spatial orientation needed for the accomplishment of effective collisions between them. In figure 3 the typical bands of PI are observed prior to and after the thermal treatment of the vacuum deposited layer (the band at 1384 cm^{-1}) and the ratio of the band areas at 1384 cm^{-1} and 1500 cm^{-1} is used for assessing the degree of imidization δ [5]. In this case $\delta = 0.89$ following the thermal treatment.

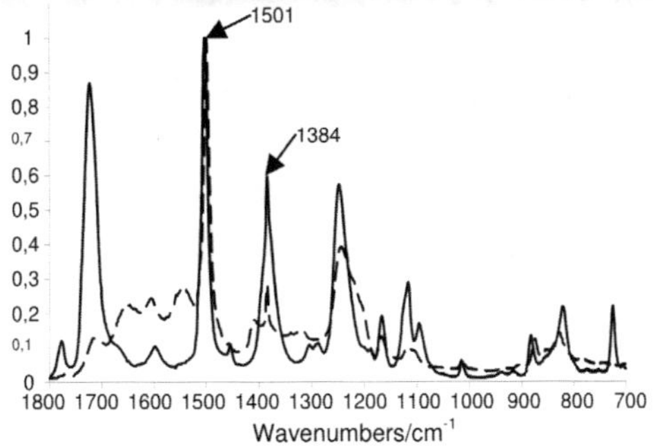

Figure 3. FTIR spectra of 500 nm thick PI films: - - - - as deposited, ——— thermally treated.

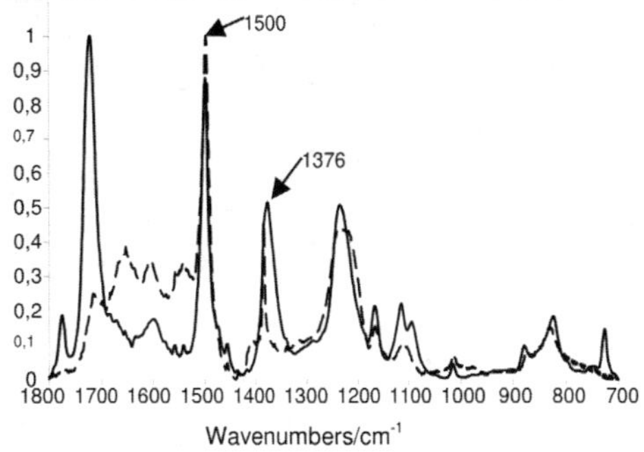

Figure 4. FTIR spectra of 500 nm thick PI films, obtained with Ar plasma: - - - - as deposited, ——— thermally treated.

Applying treatment in Ar plasma with raising the current to I_a=0.60 A leads to δ gradually rising to 0.99. In this case, a process of imidization is observed even in the PI layer that was not thermally treated. Therefore, the application of PAD facilitates the imidization process.

Figure 5 presents the volt-ampere (I/U) characteristics of PI layers obtained without (a) and with (b) assisted treatment in Ar plasma. The deviations from the linear dependency at the beginning of the graphs (a) and (b) are probably due to contact imperfections such as, e.g., contact barriers. In the section where Ohm law is observed, i.e. the ratio U/I is linear, the resistance (R) of the material can be calculated from the equation R = U/I and this result could be interpreted as ohmic conductivity. The dependencies I(U) in the linear scale satisfy the equation of a straight line which in its most common type is expressed by the equation U = A + B.I, where A and B are the respective coefficients of this equation.

The parameter σ can be determined from the data obtained as well. For the σ thus calculated, values of 1/2.911 S for an untreated in Ar plasma PI and 1/8.47 S for treated in Ar plasma PI are obtained.

It is well-known that the electrical characteristics of the PI layers are especially sensitive to the pre-history of the layer formation. Chemical and structural defects, unfinished imidization, inclusions of radicals or parts of molecules obtained in the bond cleavage, cracks, pores and uncontrollable adsorption of gases from the residual atmosphere – all these phenomena exert an influence on the electrical parameters of the layers under study.

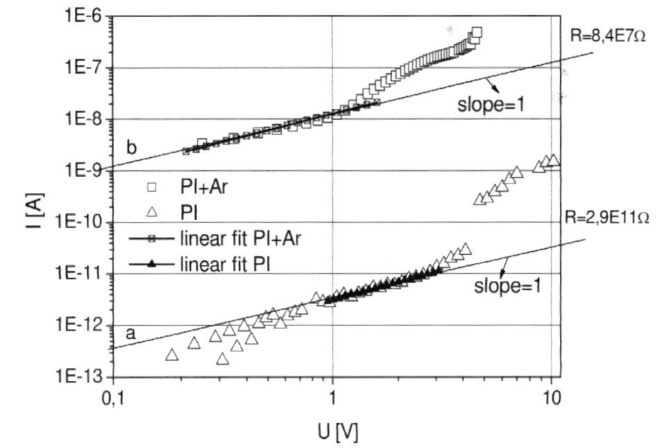

Figure 5. Current-voltage characteristics of a 1200 nm thick PI layer thermally treated after deposition and obtained: a) without treatment in Ar plasma; b) with treatment in Ar plasma.

If the assumption that PAD contributes to the imidization process and the production of layers with less chemical and structural defects is true, then the four-fold increase of σ of the PI layers obtained in the case of PAD application in Ar plasma is probably due to the reduction of the free volumes in the PI matrix, i.e. a thickening of the layers is achieved. This assumption is confirmed also by the SEM micrographs of the plasma treated and plasma untreated PI precursors PMDA (SEM in figure 6, a and в) and ODA (SEM in figure 6, c and d).

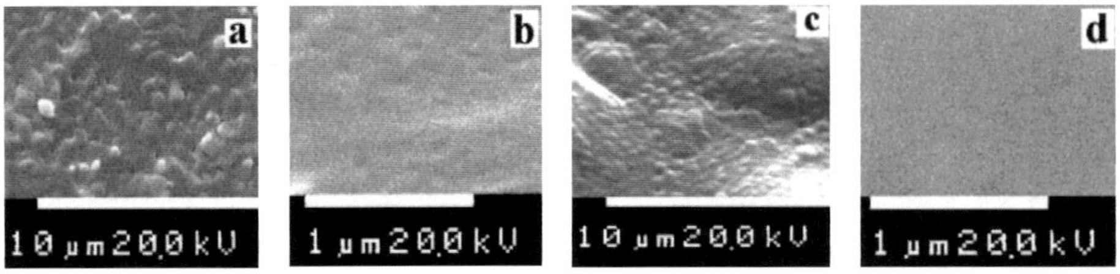

Figure 6. SEM micrographs of PMDA (a, b), ODA (c, d); (a, c) obtained without Ar plasma;(b,d) – obtained with Ar plasma assisted deposition.

It is obvious that the ion treatment accounts for the production of layers of a smoother as well as more even and thicker surface structure [6]. As a consequence of the layer thickening, the value of the measured Mhd of the PI layers increases by more than 40 percent.

4. Conclusion

It can be assumed that the plasma flux crossing the molecular flows of the PI precursors enhances the imidization process by partly activating the precursor molecules in the gas phase. Their increased kinetic energy, as well as their chemical activation, not only assist the formation of defectless layers but also the solid-phase reaction of PI synthesis. Thus, the possibility arises of shortening the duration of the post deposition treatment of the layers obtained. The films exhibit higher micro-hardness and lower electric resistance (by a factor of 4). The results obtained contribute to the controllability of the processes and widen the potential of physical deposition of organic layers by introducing a low-energy action on the molecular organic flux to produce layers of desired properties, such as a lower resistance and increased density, as is our case.

Acknowledgements

This work was supported by the National Science Fund at the Ministry of Education and Science of Bulgaria under contracts VUH – 09/05, B-TH- 205/06 and DO 02-254/ 18.12.2008.

References

[1] Mittal K L 2003 *Polyimides and Other High Temperature Polymers* **2** (V.S.P. Intl Science) p 241
[2] Vora R H, Krishnan P S G, Veeramani S and Goh S H 2003 *Polyimides and Other High Temperature Polymers* **2** (V.S.P. Intl Science) p 3
[3] Dimov D, Spassova E, Karamancheva I, Zhivkov I and Danev G 2004 Electron-assisted deposition of thin organic layers *Vacuum* **76** 223-6
[4] Dimov D, Georgiev A, Spassova E, Karamancheva I Shopov Y and Danev G 2007 Microwave assisted processes for producing thin layer materials in the field of nanotechnology *J. Optoelectr. Adv. Mater.* **9**/2 494-7
[5] Spassova E 2003 Vacuum deposited polyimide thin films *Vacuum* **70** 551-61
[6] Dimov D, Danev G, Assa J, Spassova E 2008 Ion beam assisted process in the physical deposition of polyimide *Proc. 15th Int. School on Condensed Matter Phys.* (Varna, Bulgaria)

Gold nanoparticles as nanoheaters and nanolenses in the processing of different substrate surfaces

N Nedyalkov[1,3], S Imamova[1], P Atanasov[1] and M Obara[2]

[1]Emil Djakov Institute of Electronics,
Bulgarian Academy of Sciences, 72 Tzaridradsko Chaussee, 1784 Sofia, Bulgaria
[2]Department of Electronics and Electrical Engineering,
Faculty of Science and Technology, Keio University,
3-14-1 Hiyoshi, Kohoku-ku, Yokohama 223-8522, Japan

E-mail: nnn_1900@yahoo.com

Abstract. We present results of our recent study on the heating process and near field localization arising when gold nanoparticles are irradiated by ultrashort laser pulses at wavelength of 800 nm. The system under consideration consists of Au nanoparticles with diameter of 40, 80, or 200 nm in vacuum or deposited on different substrates. Substrate materials with different dielectric properties are used in order to sense and visualize the nanoparticle heating and near field electromagnetic distribution. The theoretical analysis is based on the optical properties obtained by the Mie scattering theory. The absorption coefficients calculations are implemented in a two-temperature heat model for estimation of the nanoparticle temperature. The near field distribution in the vicinity of the particles is calculated by the finite difference time domain (FDTD) method. It is found that at even moderate laser fluences the temperature of the particle can reach a value sufficient for bubble formation in biological tissues. The analysis of the near field distribution shows that when the particle is deposited on a substrate surface, the dielectric properties of the substrate define the spatial distribution and the enhancement of the near field intensity. The observed localization and field enhancement may result in a precise modification of the substrate with a resolution defined only by the nanoparticle size. Such modifications are experimentally observed in different substrates.

1. Introduction

Being an object of the fundamental modern physics as materials with "unusual" properties related to size dependent electron energy states, metal nanostructures found a practical application four centuries ago [1,2]. The specific extinction spectra that for gold structures have a strong peak in the visible region give the brilliant ruby color used to decorate glasses in cathedrals in the 17th century. About a hundred years ago, in 1908, Gustav Mie explained the phenomena describing the absorption and scattering of the electromagnetic radiation by spherical particles. It became clear that the decrease of the particle size to dimensions smaller than the illuminating wavelength can contribute to the possibility of a resonant excitation of collective electron oscillations (plasmons) in these structures.

[3] To whom any correspondence should be addressed.

The plasmon excitation influences their far field optical properties, giving sharp resonances in the absorption and scattering spectra [3, 4] which are not present in bulk material. In 1985, B. Messinger et al. [5] applied the Mie theory to describe the scattering properties of metal nanoparticles in the near field zone. They showed that the power scattered in the vicinity of the particle can be orders of magnitude higher than the incident one for a given geometrical cross-section. It can be shown [6, 7] that this enhancement is related to the charge distribution on the metallic surface and the properties of the near field are evanescent; i. e. they are present in the very close vicinity of the particle. Thus, the size of the metallic structure defines the zone of the near field localization and the influence of the incident wavelength as a limiting parameter for the smallest spatial resolution (an effect related to the diffraction limit) vanishes.

The narrow absorption band and the existence of the enhanced near field that are characteristic of the noble nanoparticles - electromagnetic radiation interaction open new trends in many aspects in the modern technologies: in biological applications as biomedical diagnostics [8], targeted drug delivery [9], photo thermal therapy [10], surface-enhanced Raman spectroscopy (SERS) [11], near field scanning optical microscopy [6], tip enhanced Raman spectroscopy [12] with a resolution on a nanometer scale. The enhanced field in the vicinity of the metal nanoparticle can also result in a permanent modification of the surface in the near field zone [7, 13-17].

However, the physics of the plasmon properties and the near field characteristics need further clarification. While for a single isolated particle the properties of the local enhanced field depend on the particle parameters, and on the properties of the incident irradiation and surrounding medium, for a nanoparticle array they can be additionally modified and influenced by the inter-particle coupling [17-20]. These mechanisms are still not fully understood.

In this work we estimate the heating of gold nanoparticles when they interact with ultrashort laser pulses and define some of the main features of the electromagnetic field in the particle vicinity. The effects of particle heating and near field intensity enhancement are demonstrated experimentally as nanosized modifications of different substrates are shown. The incident irradiation is assumed to consist of ultrashort laser pulses at wavelength of 800 nm. The application of ultrashort laser pulses limits the heat affected zone which is crucial for many applications. The use of radiation at wavelength falling in the tissue transparency window (600 - 1300 nm) gives an opportunity of designing techniques for application of nanoparticles in medicine.

2. Theoretical model

In order to describe adequately the nanoparticle heating process, its optical parameters should be obtained. In this study the absorption cross-section is obtained on the basis of Mie scattering theory using the following equations:

$$C_{abs} = C_{ext.} - C_{sca}, \tag{1}$$

where $C_{ext.}$ and C_{sca} are the extinction and scattering cross sections given by [4]:

$$C_{ext} = \frac{4\pi R^2 \varepsilon_m^{3/2}}{\lambda a^3} \sum_l (2l+1)\,\mathbf{Re}(A_l + B_l), \tag{2}$$

$$C_{sca} = \frac{4\pi R^3 \varepsilon_m^{1/2}}{\lambda a^3} \sum_l (2l+1)(|A_l|^2 + |B_l|^2). \tag{3}$$

Here, λ is the wavelength of the incident irradiation, ε_m is the dielectric function of the surrounding medium, $a = (2\pi R \varepsilon_m^{1/2})/\lambda$, R is the particle radius, and A_l and B_l are the electric and magnetic partial oscillation coefficients. The optical parameters for gold used in the calculation are taken from [21].

The temperature evolution of the particle during and after the interaction with the laser pulse is traced by the one-dimensional two-temperature diffusion model:

$$C_e \frac{\partial T_e}{\partial t} = -\gamma(T_e - T_i) + S, \tag{4}$$

$$C_i \frac{\partial T_i}{\partial t} = \gamma(T_e - T_i), \tag{5}$$

$$S = IC_{abs} / V_p. \tag{6}$$

Here, T_e and T_i are the electron and lattice temperatures, S and I are the laser energy source and laser intensity (time dependent), respectively. $V_p = 4/3\pi R^3$ is the particle volume. *Ci*, *Ce*, and γ are the lattice and electron system heat capacities and the electron-lattice coupling constant, respectively. In the present model the spatial characteristics of the particle heating are neglected and the heating is assumed to be homogeneous.

The coupled equations (4)-(6) are solved using a classical finite difference scheme and by taking into account the temperature dependence of the electron heat capacity according to the formula $C_e = A_e T_e$, where $A_e = 70$ J/m^{3}K^2, and the temperature dependence of *Ci* [22]. The electron heat conductivity of gold is assumed to be constant, as the value of 2×10^3 W/mK is used as an average obtained from the expression $k_e = k_{eo}(T_e/T_i)$, for the electron temperature range from 300 K to 9000 K, and $k_{eo} = 318$ W/mK [23]. The lattice heat conductivity of gold k_i is 318 W/mK [23]. The value of the coupling constant, $\gamma = 3\times10^{16}$ W/m^{3}K is taken from [24].

FDTD simulation [25] is applied in order to study the near field properties around the irradiated nanoparticles. The method is based on the numerical solution of Maxwell's equations and can be used to obtain adequate picture of the electromagnetic field properties in the near and far field zone around structures with arbitrary shapes [26-28]. The simulation system consists of a gold nanoparticle or a 2D nanoparticle array placed on different substrates. Absorbing boundary conditions are applied to the boundary of the simulation cell. For the simulations of nanoparticle arrays, a hexagonal arrangement of the particles is assumed because this arrangement is predominantly obtained by the existing chemical methods for deposition. Simulations with circular and linear polarization of the incident irradiation are performed. The electric field strength of the incident wave is set at 1 V/m.

3. Experimental setup

The gold spherical particles used in the experiments have diameter of 40, 80 and 200 nm, with a standard diameter deviation of less than 8% (BBInternational Corp.). The particles are deposited as a colloid on polished silicon and PMMA substrates by spin-coating. Under these conditions all the substrate surfaces are covered by randomly distributed gold particles. The roughness of the substrates (without particle deposition) is in the order of few nanometers. The samples are irradiated by laser pulses delivered by a Ti: sapphire chirped pulse amplification system that produces pulses with energy of 1 mJ at a repetition rate of 1 kHz, and center wavelength of 800 nm. The laser pulse duration used is 100 fs FWHM. The laser radiation is incident normally to the substrate surface and is focused by a lens with a focal length of 200 mm. The pulse energy is adjusted by a variable attenuator. The experiments are done in air, on a single shot basis. Linear and circular polarization of the incident irradiation is used. The irradiated samples are analyzed by SEM and AFM. The substrate surface is not chemically cleaned after the laser irradiation.

4. Results

4.1. Gold nanoparticles as nanoheaters

Figure 1 presents the absorption cross-section spectra of single gold nanoparticles with radius of 20, 40 and 100 nm calculated by (1)-(3). The clearly expressed peaks in the absorption cross sections are related to the plasmon excitation in the nanoparticles. The maximal values are observed in the range of about

525 to 550 nm for particles with radius R = 20 and 100 nm, respectively. The values of the absorption cross-sections at wavelength of 800 nm are about two orders of magnitude lower than the optimal ones for particles with R = 40 and 80 nm. Although the increase of the particle dimensions results in a broadening of the plasmon absorption band and a shift of the maxima to the longer wavelengths, the value of the cross-section for a particle with radius of 100 nm is about 5 times lower than the peak value. Using the calculated absorption cross-sections the temperature of the particles is obtained using the two-temperature diffusion model. The values of the maximal temperature reached in the nanoparticles when they are irradiated by a 100 fs laser pulse as a function of the incident laser fluence are shown in figure 2.

It should be noted that for biophotonics applications, such as cell membrane modification and when irreversible cell modification is based on bubble formation (as in tumor cell killing) the necessary temperature increase should be at least 170 K [29, 30]. When ultrashort laser pulses are used such heating is realized in nearly isochoric conditions which will result in the development of significant tensile stress [31] in the vicinity of the heating point. This will lead to the formation of a cavitation bubble without the need of heat accumulation, i.e. after a single pulse. As is seen in figure 2, a temperature increase of 170 K can be obtained in gold nanoparticles after a single laser pulse at fluences of about 70 mJ/cm^2 for R = 100 nm and at 100 mJ/cm^2 for particles with R = 40 and 80 nm. It should also be mentioned that the experimentally obtained threshold for biological cell modification (the example is for human breast MDA-MB cells [32]) using direct femtosecond laser pulses irradiation (without nanoparticle mediation) is estimated to be about 500 mJ/cm^2, i.e. five times higher.

4.2. Near field localization

Figure 3 presents the calculated distributions of the electric field around a gold particle with radius of 100 nm – (a) in vacuum, and (b) placed on a Si surface. The incident laser radiation has circular polarization in both cases. The electric field distributions are shown in a cross-sectional plane of the system, (a) - through the center of the sphere and parallel to the incident wave propagation and (b) – through the center of the sphere and the point of contact between the particle and the substrate.

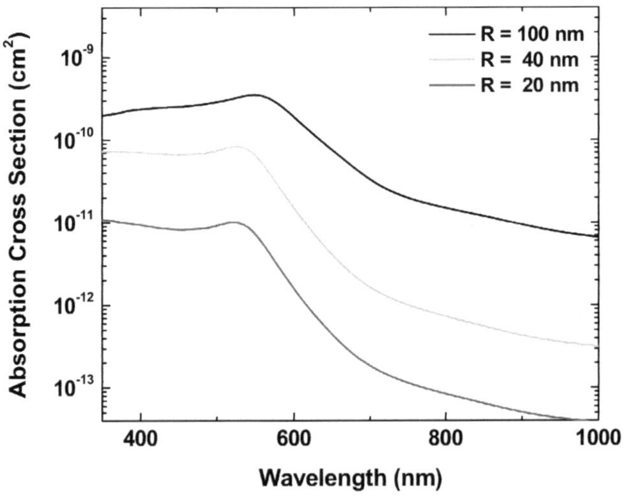

Figure 1. Absorption cross section of gold nanoparticles with different diameters.

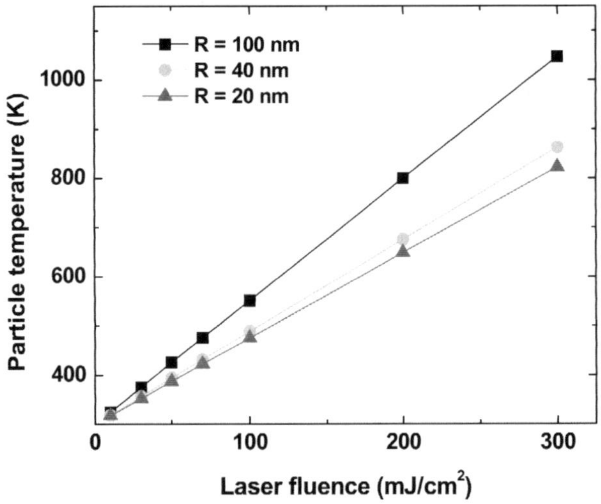

Figure 2. Maximal temperature of gold nanoparticles achieved after interaction with laser pulse with duration of 100 fs at wavelength of 800 nm. The data are calculated using equations (4-6).

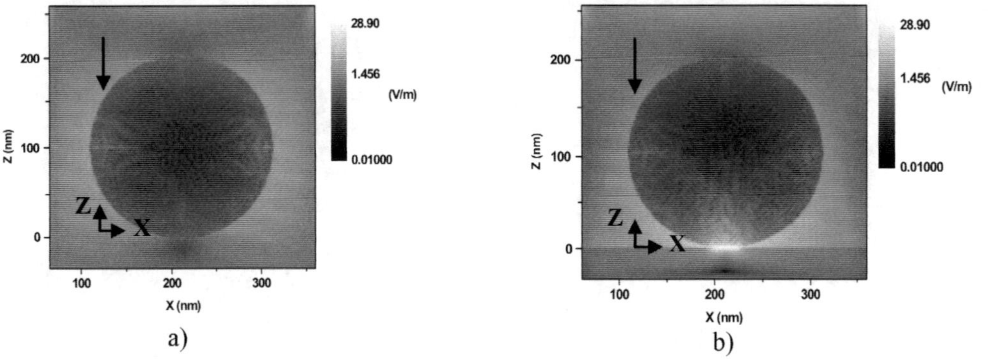

Figure 3. Calculated distribution of the electric field around a gold particle with radius of 100 nm. The incident radiation is at wavelength of 800 nm and the amplitude of the incident electric field is 1 V/m. The direction of propagation of the incident wave is shown by an arrow. (a) a gold particle in vacuum; (b) a gold particle on a Si substrate.

In the case of isotropic vacuum environment and a small particle (when dipole excitation is dominant) the charge oscillation direction follows the direction of the incident electric field. The induced near electric field on the particle surface has a maximal value in the vicinity of the equator and approaches the incident field amplitude on the poles. Although for a particle with the dimensions chosen higher modes can dominate the near field distribution, at the wavelength used of 800 nm only dipole mode can be efficiently excited [7]. When the particle is placed on the substrate surface (figure 3 (b)) the near electric field distribution becomes quite different. As it is seen in figure 3 (b), the field is localized in the vicinity of the point of contact on the Si surface. In the present case, the strong field on the gold particle surface produced by charge accumulation induces image charges in the substrate material. The field distribution both on the gold sphere surface and in the substrate is related to the existence of a component in the direction of the incident electric field and a *z*-component, which appear due to the interaction between the particle and substrate charges. Due to this component, the induced charge density in the bottom of the sphere is significantly higher than that in the upper hemisphere. Since the near electric field amplitude decreases rapidly with the distance, the density of the induced image charges on the substrate surface increases in the direction to the point of contact.

The properties of the near field for a system consisting of a particle deposited on a substrate will depend on the charge density induced on the substrate surface [7]. In its turn, this density depends on the dielectric properties of the substrate material and one should expect different characteristics of the near field for different substrate materials. The simulations made for dielectric substrates with different refractive indices show that the effect of near field localization in the vicinity of the contact point between the particle and the substrate becomes more pronounced (and therefore the value of the field intensity on the substrate surface increases) with the increase of the refractive index of the substrate [7, 17]. A clear localization of the zone with maximal intensity on the substrate surface is observed for substrates with a refractive index higher than 1.7.

In order to validate the theoretical results, experiments on modification of substrates with assistance of nanoparticles are performed. The theoretical results indicate that for a dielectric substrate with a refractive index of 1.488, the near field intensity is not localized in the vicinity of the contact point between the substrate and the particle. A possible modification of the underlying substrate could be related only to the particle heating. Figure 4 shows modification of PMMA and silicon substrates when gold nanoparticles with radius of 100 nm are deposited on them and the systems are irradiated by a single ultrashort laser pulse with fluence of 60 mJ/cm^2 and 200 mJ/cm^2, respectively. The polarization of the incident irradiation is linear (shown by an arrow). In the case of a PMMA substrate, the nanohole shape has radial symmetry and there is no observable influence of the incident irradiation polarization on the hole shape.

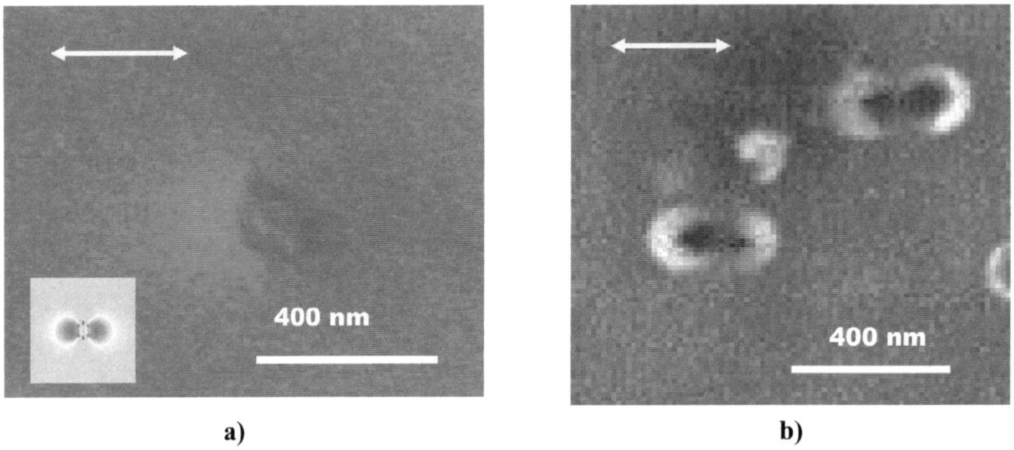

Figure 4. SEM images of nanomodification when gold nanoparticles with radius of 100 nm are deposited on (a) PMMA and (b) Si substrate and irradiated by single laser pulse with fluence of 60 mJ/cm^2 and 200 mJ/cm^2, respectively. The inset in (a) shows the distribution of the near field intensity on the substrate surface predicted by FDTD simulation. The arrow shows the polarization of the incident irradiation.

Such dependence is not seen in the fluence range up to the modification threshold of the native substrate. Furthermore, the rest of the substrate remains unchanged which confirms the idea of local interaction under the particle only. According to the numerical model, the maximal temperature achieved with this fluence is near the melting temperature of the PMMA substrate (440 K). These results indicate that heating has the main contribution to the substrate modification in this case. In the case of a Si substrate, the shape of the modified area corresponds to the field intensity distribution and the nanoholes formed have a shape elongated along the polarization direction. This shows that at these conditions the enhancement of the near field under the gold particle is the governing mechanism for surface modification rather than particle heating. When circular polarization of the incident irradiation is used the hole shape formed shows good radial symmetry [7].

The characteristics of the nanoholes produced depend also on the nanoparticle dimensions [16]. Figure 5 shows SEM images of holes produced when a 0.185 J/cm^2 laser pulse irradiates a Si surface with gold particles with (a) R = 100 nm, (b) R = 40 nm, and (c) R = 20 nm deposited on it. The polarization of the incident radiation is circular. The maximal depths of the holes are 30 nm, 15 nm, and 7 nm, respectively.

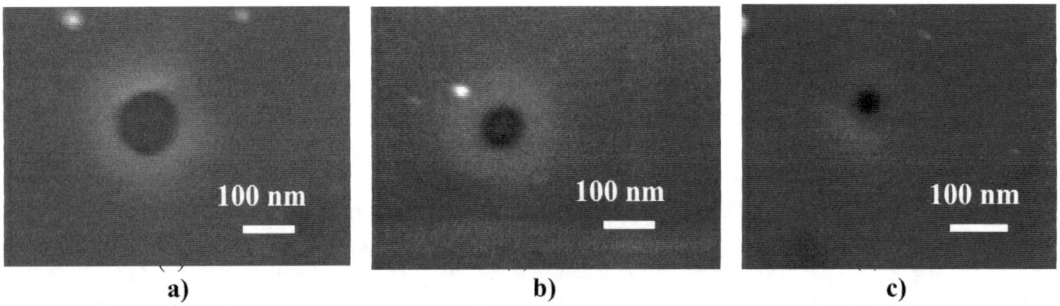

Figure 5. SEM images of holes produced when a 0.185 J/cm^2 laser pulse irradiates a Si surface with gold particles with radius of (a) 100 nm, (b) 40 nm, and (c) 20 nm deposited on it.

4.3. Nanoparticle array

Figure 6 shows experimentally obtained absorption spectrum of gold nanoparticles with radius of 100 nm deposited on glass substrate. The inset shows a SEM image of the deposited particles. The dashed line shows the absorption spectrum of an isolated nanoparticle calculated by Mie theory. It is seen that the absorption of the nanoparticle cluster is efficient in the wavelength range from 550 to about 850 nm in the case of the structure presented. The near field distribution in the vicinity of the nanoparticle array is also different compared to the case of single particle [17, 20]. Figure 7 a) shows the calculated distribution of the electric field intensity on a Si substrate when closely packed nanoparticle array is placed on it.

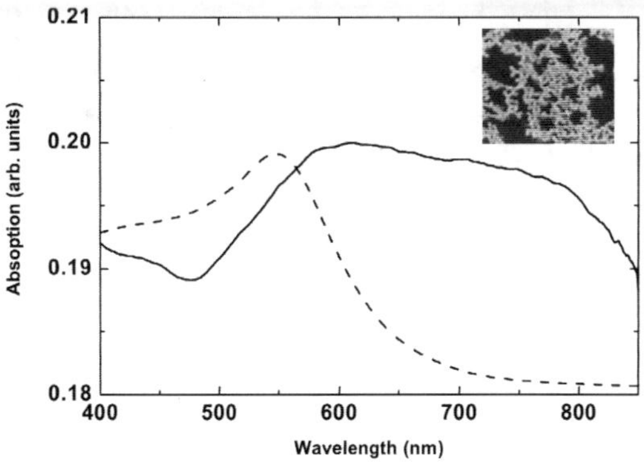

Figure 6. Experimental absorption spectrum of gold nanoparticles with radius of 100 nm deposited on glass substrate (solid line). the dashed line shows the spectrum calculated for a single particle.

A cross sectional view of the field intensity distribution in a nanoparticle chain within the array is shown in figure 7 b). Due to plasmon coupling, the zone with the maximal field intensity in a nanoparticle array is localized between particles, where its value is about two orders of magnitude higher than that on the substrate surface.

Thus, for surface modification applications the cluster heating will be dominant mechanism rather than the direct near field localization and enhancement as it is in the case of single particle.

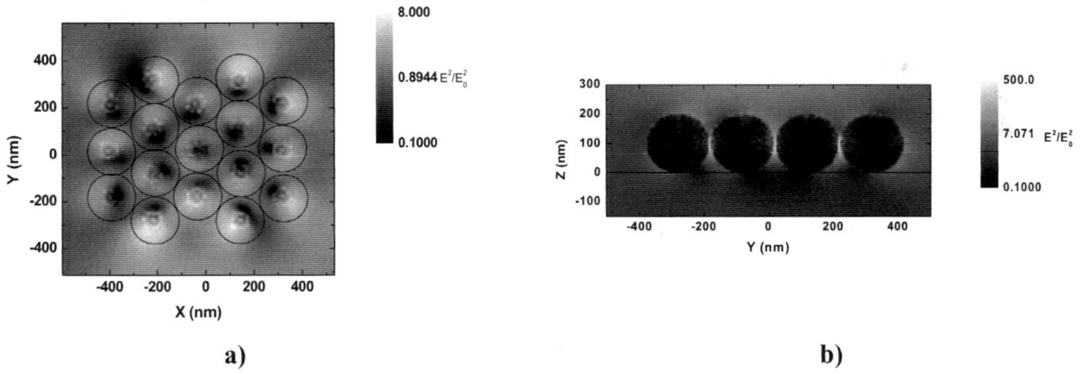

Figure 7. Calculated distribution of the electric field intensity on a Si substrate when a closely packed nanoparticle array is placed on it. b) Cross sectional view of the field intensity distribution in a nanoparticle chain within the array.

5. Conclusions

We present experimental and theoretical results on two effects when laser pulse interacts with gold nanoparticles. On the basis of Mie theory, the particle absorption cross-section is estimated and the maximal nanoparticle temperature is calculated. It is shown that at laser fluences lower than human tissue modification, the nanoparticle can be heated to temperatures that may induce local cell damage. In addition to the particle heating, the properties of the near electromagnetic field are also investigated. It is shown that the field intensity can be strongly enhanced as the enhancement value depends on the properties of the environment. This dependence is used to distinguish between the domination of the two mechanisms studied in the modification of different substrates. In the case of a substrate with a refractive index higher than approximately 1.7, the near field intensity is localized in the vicinity of the contact point between the particle and the substrate. In such a case the near field intensity enhancement plays a dominant role in the substrate modification. When the substrate has a lower refractive index, the heating of the particle has the main contribution. In all cases, the laser irradiation of the samples at fluences lower than the modification threshold of the native surface results in the formation of nanoholes with a size smaller than the gold nanoparticle diameter. When a 2D nanoparticle array is considered, the absorption spectra and near field intensity distribution are found to be different than the case of single particle. For such a structure the interparticle plasmon coupling defines the optical response.

The results obtained can form the basis for the development of techniques for nanomodification of different materials. The heating of nanoparticles can be used in photothermal therapy of biological samples. Furthermore, the present analyses can be used for the appropriate selection of efficient substrates for SERS technique.

Acknowledgments
This work is supported by Ministry of Education and Science of Bulgaria, under contract DO 02-293 with the National Science Found and Grant-in-Aid for the Global COE for High-Level Global Cooperation for Leading-Edge Platform on Access Spaces from MEXT in Japan.

References

[1] Kerker M 1985 *J. Colloid Interface Sci.* **105** 297
[2] El-Sayed M A 2001 *Acc. Chem. Re.* **34** 257
[3] Kreibig U and Vollmer M 1995 *Optical Properties of Metal Clusters* (Springer: Berlin)
[4] Papavassiliou G C 1979 *Prog. Solid St. Chem.* **12** 185
[5] Messinger B J, Ulrich von Raben K, Chang R K and Barber P W 1981 *Phys. Rev.* B **24** 649
[6] Kawata S 2001 *Near-Field Optics and Surface Plasmon Polaritons* (Springer: Berlin)
[7] Nedyalkov N N, Sakai T, Miyanishi T and Obara M 2006 *J. Phys. D: Appl. Phys.* **39** 5037
[8] Kubik T, Bogunia-Kubik K and Sugisaka M 2005 *Curr. Pharm. Biotechnol.* **6** 17
[9] Pissuwan D, Valenzuela S M and Cortie M 2006 *Trends Biotechnol.* **24** 62
[10] Huang X, El-Sayed I H, Qian W and El-Sayed M A 2006 *J. Am. Chem. Soc.* **128** 2115
[11] Moskovits M 2005 *J. Raman Spectrosc.* **36** 485
[12] Wessel J 1985 *J. Opt. Soc. Am.* B **2** 1538
[13] Jersch J and Dickmann K 1996 *Appl. Phys. Lett.* **68** 868
[14] Leiderer P, Bartels C, König-Birk J, Mosbacher M and Boneberg J 2004 *Appl. Phys. Lett.* **85** 5370
[15] Nedyalkov N N, Takada H and Obara M 2006 *Appl. Phys.* A **85** 163
[16] Nedyalkov N, Miyanishi T and Obara M 2007 *Appl. Surf. Sci.* **253** 6558
[17] Nedyalkov N N, Atanasov P. A. and Obara M 2007 *Nanotechnology* **18** 305703
[18] Quinten M 2001 *Appl. Phys.* B **73** 245
[19] Zou S, Janel N and Schatz G C 2004 *J. Chem. Phys.* **120** 10871
[20] Nedyalkov N N, Sakai T, Miyanishi T, and Obara M 2007 *App. Phys. Lett.* **90** 123106
[21] Johnson P B and Christy R W 1972 *Phys. Rev.* B **6** 4370
[22] Chowdhury I H and Xu X 2003 *Numer. Heat Transfer* A **44** 219
[23] Wellershoff S S, Hohlfeld J, Gedde J and Matthias E 1999 *Appl. Phys.* A **69** S99
[24] Hodak J H, Martini I and Hartland G 1998 *J. Chem. Phys.* B **102** 6958
[25] Taflove A and Hagness S C 2000 *Computational Electrodynamics: The Finite-Difference Time-Domain Method* (Artech House: Boston)
[26] Sendur K, Challener W and Peng C 2004 *J. Appl. Phys.* **96** 2743
[27] Krug II J T, Sanchez E J and Xie X S 2002 *J. Chem. Phys.* **116** 10895
[28] Laroche T and Girard Ch 2006 *Appl. Phys. Lett.* **89** 233119
[29] Zharov V P, Letfullin R R and Galitovskaya E N 2005 *J. Phys.* D **38** 2571
[30] Volkov A N, Sevilla C and Zhigilei L V 2007 *Appl. Surf. Sci.* **253** 6394
[31] Vogel A, Noack J, Huttmann G and Paltauf G 2007 *J. Phys.* **59** 249
[32] Cho S, Chang W, Kim J, Whang K, Choi K, and Sohn S 2008 *Appl. Surf. Sci.* **254** 3370

Non-local quantum–kinetic equation for laser-excited semiconductors

T Apostolova

Institute for Nuclear Research and Nuclear Energy,
Bulgarian Academy of Sciences, 72 Tzarigradsko Chaussee, 1784 Sofia, Bulgaria

E-mail tzveta@inrne.bas.bg

Abstract. A generalized non-local Fokker-Planck type kinetic equation obtained on the basis of quantum-mechanical formalism is used to study the conduction electron dynamics in model semiconductors excited by spatially inhomogeneous pulsed laser irradiation. In this approach the electrons are assumed to undergo a drift-diffusion motion in the energy-position manifold.

1. Introduction

Ultra-short laser-induced breakdown studies have revealed the possibility of self-channeling ultrafast light in condensed media with the strong involvement of photon and avalanche ionization capable of delivering high free-electron densities. The full description of the laser-matter interaction process leading to material modification, and the complex spatio-temporal dynamics of the beam propagation in the media from first principles comprises the nonlinear Schrödinger equation for the laser electric field envelope coupled with an equation describing the growth of the free-electron density $n(\vec{r},t)$ [1], [2]. Usually the equation for $n(\vec{r},t)$ employed in the literature is a simple rate equation not accounting for the energy absorption process from the laser field. The slowly varying [1] envelope function $\vec{\mathrm{E}}$ of the electric field of the laser $\vec{E}$ is defined by: $\vec{E}(\vec{r},t)=\sqrt{\frac{\Omega_L\mu_0}{2k_0}}\vec{\mathrm{E}}(\vec{r},t)e^{i(k_0z-\Omega_0t)}+c.c.$ (for a pulsed beam centered on the operating frequency Ω_L, wave number k_0); where the electromagnetic radiation is polarized in the x direction $\vec{\mathrm{E}}=(\mathrm{E},0,0)$. The free electrons motion is governed by the oscillation of the inhomogeneous optical field beating at Ω_L so that an electron current density with a slowly varying envelope induces a variation of the local population of electrons in space and time leading to a space-dependent carrier distribution. Since the time development of the avalanche ionization process depends also very strongly on the values of the field, it follows that at any spatially non-uniform distribution of the electromagnetic radiation density, initial electron density gradients ∇_r could occur that may give rise to more or less strong diffusion currents. Therefore, instead of using a simple rate equation for the electron density $n(\vec{r},t)$ produced by the laser field, we generalize the kinetic equation for the electron energy distribution function to obtain a multivariate Fokker-Planck kinetic equation accounting for the spatial dependence of electron transport. The rates in the equation explicitly include the effect of laser field energy absorption by the free electrons and the joule heating effect.

2. Theoretical picture

A kinetic equation is derived for the conduction-electron distribution function following references [3, 4, 5]. We assume that the frequency of the electromagnetic field satisfies the conditions $1 << \Omega_L \tau_p << c/\bar{v}$ (the electromagnetic field frequency exceeds appreciably the frequency of electron-phonon collisions), $\Omega >> \omega_p$, $\bar{\varepsilon}\tau_p >> 1$ (where τ_p is the momentum relaxation time, ω_p is the electron plasma frequency and $\bar{\varepsilon}$ and $\bar{v}$ are the average energy and velocity of the electron). These conditions are sufficient to guarantee the penetration of the electromagnetic field deeply into the semiconductor and also mean that the wavelength is much larger than the mean-free-path or the De Broglie wavelength of an electron, the wavelength of the phonons which interact with the electrons and allow for the use of dipole approximation of the radiation field. They imply that the dependences of these processes on the electromagnetic field are local. Under the same conditions, the magnetic field contribution to the interaction with the laser field is negligible because it is of the same order of magnitude as the electric quadrupole contribution. We first consider an electron-phonon system in a spatially uniform sinusoidal electric field with a vector potential $A(t) = \mathrm{A}\sin(\Omega_L t)$ and find the solution of the Schrödinger equation

$$i\hbar\frac{\partial\psi(\bar{r},t)}{\partial t} = \frac{1}{2m^*}\left[\bar{p} + \frac{e}{c}\bar{A}(\bar{r},t)\right]^2 \psi(\bar{r},t). \tag{1}$$

Using the standard time-dependent perturbation theory and the calculated electron wave function renormalized by the laser field, we calculate the electron transition rate. Unlike the isotropic transition rate obtained in the absence of the laser field, the presence of linearly polarized electromagnetic radiation results in an anisotropic electronic transition rate characterized by field dependence via a Bessel function term – this effect is known as free-electron absorption of photons. For moderately high intensities of the laser radiation, the Bessel function is expanded to second order. In the diffusive limit $\hbar\omega_q << E^e{}_k$, where $E^e{}_k = \hbar^2k^2/2m^*_e$ and $\hbar\omega_q$ are the free electron kinetic energy in the conduction band with effective mass m^*_e and the phonon energy respectively, the electron distribution function is expanded in Taylor series. Thus the linearized Fokker-Plank type equation is obtained. The classical Joule heating of electrons from the field which is missing in other quantum-kinetic approaches is included by taking into account the local fluctuation of the electron kinetic energies. Source terms such as stimulated interband electron transitions due to a single or multiphoton absorption, impact ionization due to Coulomb interaction between electrons and holes and sink terms such as non-radiative recombination due to a phonon-mediated interaction are included in the equation:

$$\frac{\partial f^e(E,t)}{\partial t} + V(E,t)\frac{\partial f^e(E,t)}{\partial E} - D(E,t)\frac{\partial^2 f^e(E,t)}{\partial\left(E^e_k\right)^2} = A(E,t)f^e(E,t) + S(E,t). \tag{2}$$

The rates in equation (2) are given by the following expressions [3], [4]:

$$V(E,t) = V_T(E) + V_F(E) + \frac{1}{3}\sigma(\Omega_L)\mathrm{E}^2 + A_T(E)\frac{\tau_p}{3}\sigma(\Omega_L)\mathrm{E}^2, \tag{3}$$

$$D(E,t) = D_T(E) + D_F(E) + \frac{2}{3}\sigma(\Omega_L)\mathrm{E}^2 E^e_k + V_T(E)\frac{\tau_p}{3}\sigma(\Omega_L)\mathrm{E}^2, \tag{4}$$

where

$$V_T\left(E_{\vec{k}}\right)=\frac{2\pi}{\hbar}\sum_{\vec{q}\lambda}\left|C_{\vec{q}\lambda}\right|^2\hbar\omega_{\vec{q}\lambda}\times\left[N_{\vec{q}\lambda}^{ph}\delta\left(E_{\vec{k}}-E_{\vec{k}+\vec{q}}-\hbar\omega_{\vec{q}\lambda}\right)-\left(N_{\vec{q}\lambda}^{ph}+1\right)\delta\left(E_{\vec{k}}-E_{\vec{k}+\vec{q}}+\hbar\omega_{\vec{q}\lambda}\right)\right], \tag{5}$$

$$\begin{aligned}V_F=&\frac{2\pi}{\hbar}\sum_{\vec{q}\lambda}\left|C_{\vec{q}\lambda}\right|^2\Im_{\vec{q}}^2\times N_{\vec{q}\lambda}^{ph}\left(E_k-E_{k-q}\right)\left[\delta\left(E_{\vec{k}}-E_{\vec{k}-\vec{q}}-\hbar\omega_{\vec{q}\lambda}+\hbar\Omega_L\right)+\delta\left(E_{\vec{k}}-E_{\vec{k}-\vec{q}}-\hbar\omega_{\vec{q}\lambda}-\hbar\Omega_L\right)\right]\\&+\left(N_{\vec{q}\lambda}^{ph}+1\right)\left(E_k-E_{k-q}\right)\left[\delta\left(E_{\vec{k}}-E_{\vec{k}+\vec{q}}+\hbar\omega_{\vec{q}\lambda}+\hbar\Omega_L\right)+\delta\left(E_{\vec{k}}-E_{\vec{k}+\vec{q}}+\hbar\omega_{\vec{q}\lambda}-\hbar\Omega_L\right)\right],\end{aligned} \tag{6}$$

$$D_T\left(E_{\vec{k}}\right)=\frac{\pi}{\hbar}\sum_{\vec{q}\lambda}\left|C_{\vec{q}\lambda}\right|^2\left(\hbar\omega_{\vec{q}\lambda}\right)^2\times\left[N_{\vec{q}\lambda}^{ph}\delta\left(E_{\vec{k}}-E_{\vec{k}+\vec{q}}-\hbar\omega_{\vec{q}\lambda}\right)-\left(N_{\vec{q}\lambda}^{ph}+1\right)\delta\left(E_{\vec{k}}-E_{\vec{k}+\vec{q}}+\hbar\omega_{\vec{q}\lambda}\right)\right], \tag{7}$$

$$\begin{aligned}D_F=&\frac{\pi}{4\hbar}\sum_{\vec{q}\lambda}\left|C_{\vec{q}\lambda}\right|^2\Im_{\vec{q}}^2\times N_{\vec{q}\lambda}^{ph}\left(E_k-E_{k-q}\right)^2\left[\delta\left(E_{\vec{k}}-E_{\vec{k}-\vec{q}}-\hbar\omega_{\vec{q}\lambda}+\hbar\Omega_L\right)+\delta\left(E_{\vec{k}}-E_{\vec{k}-\vec{q}}-\hbar\omega_{\vec{q}\lambda}-\hbar\Omega_L\right)\right]\\&+\left(N_{\vec{q}\lambda}^{ph}+1\right)\left(E_k-E_{k-q}\right)^2\left[\delta\left(E_{\vec{k}}-E_{\vec{k}+\vec{q}}+\hbar\omega_{\vec{q}\lambda}+\hbar\Omega_L\right)+\delta\left(E_{\vec{k}}-E_{\vec{k}+\vec{q}}+\hbar\omega_{\vec{q}\lambda}-\hbar\Omega_L\right)\right].\end{aligned} \tag{8}$$

The right-hand side of (1) contains the coefficients for thermal spontaneous phonon emission and the one for the field-induced phonon emission.

$$A_T\left(E_{\vec{k}}\right)=\frac{2\pi}{\hbar}\sum_{\vec{q}\lambda}\left|C_{\vec{q}\lambda}\right|^2\times\left[N_{\vec{q}\lambda}^{ph}\delta\left(E_{\vec{k}}-E_{\vec{k}+\vec{q}}-\hbar\omega_{\vec{q}\lambda}\right)-\left(N_{\vec{q}\lambda}^{ph}+1\right)\delta\left(E_{\vec{k}}-E_{\vec{k}+\vec{q}}+\hbar\omega_{\vec{q}\lambda}\right)\right], \tag{9}$$

$$\begin{aligned}A_F=&\frac{\pi}{2\hbar}\sum_{\vec{q}\lambda}\left|C_{\vec{q}\lambda}\right|^2\left(e\left|\vec{q}.\vec{\mathrm{E}}(t)\right|/m^*\Omega_L^2\right)^2\times\left[\delta\left(E_{\vec{k}}-E_{\vec{k}-\vec{q}}-\hbar\omega_{\vec{q}\lambda}+\hbar\Omega_L\right)+\delta\left(E_{\vec{k}}-E_{\vec{k}-\vec{q}}-\hbar\omega_{\vec{q}\lambda}-\hbar\Omega_L\right)\right]\\&+\left[\delta\left(E_{\vec{k}}-E_{\vec{k}+\vec{q}}+\hbar\omega_{\vec{q}\lambda}+\hbar\Omega_L\right)+\delta\left(E_{\vec{k}}-E_{\vec{k}+\vec{q}}+\hbar\omega_{\vec{q}\lambda}-\hbar\Omega_L\right)\right].\end{aligned} \tag{10}$$

In these expressions E denotes the amplitude of the laser field, the indices T and F denote the thermal contribution due to phonon scattering and field induced contribution respectively, $\left|C_{\vec{q}\lambda}\right|$ is the electron- phonon coupling and $\Im_{\vec{q}}=\left(e\left|\vec{q}.\vec{\mathrm{E}}(t)\right|/m^*\Omega_L^2\right)$. The expressions for the field dependent rates A_F, V_F and D_F reveal that in the presence of electromagnetic radiation the electron-phonon scattering can be accompanied by electronic transitions through photon absorption and emission.

3. Generalized Fokker-Planck kinetic equation

By explicitly substituting the spatially dependent envelope of the laser $\vec{\mathrm{E}}(\vec{r},t)$ field in equations (3) - (10) the space and energy dependent rates $V(\vec{r},E,t)$ and $D(\vec{r},E,t)$ are obtained. Equation (2) is different from the standard Fokker-Planck equation since it cannot be written as a conservation law for particle number and current in energy space even in the absence source terms on the right hand side. Defining a current in energy space:

$$J^E(E,t)=V(E,t)f^e(E,t)-\frac{\partial\left[D(E,t)f^e(E,t)\right]}{\partial E},$$

equation (2) can be written as

$$\frac{\partial f^{e}(E,t)}{\partial t}+\frac{\partial J^{E}(E,t)}{\partial E}+2\frac{\partial D(E,t)}{\partial E}\frac{\partial f^{e}(E,t)}{\partial E}=A(E,t)f^{e}(E,t)+\tilde{A}(E,t)f^{e}(E,t)+S(E,t), \quad (11)$$

where $\tilde{A}(E,t)=\frac{\partial V(E,t)}{\partial E}-\frac{\partial^{2}D(E,t)}{\partial E^{2}}$.

For the time being we ignore the third term on the left-hand side of equation (11). The Joule heating and diffusion rates of equation (2) both contain expressions proportional to the electric field envelope and could be thus generalized to include space dependence. The electron distribution function is also generalized to include space dependence. Then a generalized current density term is defined reflecting the electron energy gain or loss and electron energy diffusion:

$$J^{0}(E,\vec{r},t)=V(E,\vec{r},t)f(E,\vec{r},t)-\frac{\partial\left[D(E,\vec{r},t)f(E,\vec{r},t)\right]}{\partial E}.$$

In references [6] and [7], a theory of electron drift and diffusion was proposed allowing for a non-equilibrium energy distribution, such as occurs in high electric fields. The transport equation assumes a form of a Fokker-Planck equation in a four-dimensional (4D) energy-position continuum instead of the 3D position space. In this equation the 4D current density is a sum of a drift and a diffusion terms that depend on the $(E,\vec{r})$ point. The use of this equation assumes that not only the energy but also the position coordinates vary continuously, therefore demanding that the electron energy be exchanged with the lattice in small steps (e.g. phonons of energy $\hbar\omega_q << E$) and that electron velocity be finite.

Following [6] we generalize equation (11) to a continuity equation in the energy-position manifold to describe electron transport in an inhomogeneous laser field by continuous energy and coordinate changes

$$\frac{\partial f^{e}(E,\vec{r},t)}{\partial t}+\frac{\partial J^{0}(E,\vec{r},t)}{\partial E}+\frac{\partial J^{\beta}(E,\vec{r},t)}{\partial x^{\beta}}=\left[A(E,\vec{r},t)+\tilde{A}(E,\vec{r},t)\right]f^{e}(E,\vec{r},t)+S(E,\vec{r},t), \quad (12)$$

where the Greek indices run from 1 to 3.

Due to our choice of laser field polarization in the x direction, we get drift-diffusion equations in $(E,\vec{r})$ space:

$$J^{x}(E,\vec{r},t)=v_{d}(E,\vec{r},t)f^{e}(E,\vec{r},t)-\frac{\partial\left[e\mathrm{E}(\vec{r},t)D_{x}(E,\vec{r},t)f^{e}(E,\vec{r},t)\right]}{\partial E}-\frac{\partial\left[D_{x}(E,\vec{r},t)f^{e}(E,\vec{r},t)\right]}{\partial x}, \quad (13)$$

$$J^{y}(E,\vec{r},t)=-\frac{\partial\left[e\mathrm{E}(\vec{r},t)D_{x}(E,\vec{r},t)f^{e}(E,\vec{r},t)\right]}{\partial E}-\frac{\partial\left[D_{x}(E,\vec{r},t)f^{e}(E,\vec{r},t)\right]}{\partial y}, \quad (14)$$

$$J^{z}(E,\vec{r},t)=-\frac{\partial\left[e\mathrm{E}(\vec{r},t)D_{x}(E,\vec{r},t)f^{e}(E,\vec{r},t)\right]}{\partial E}-\frac{\partial\left[D_{x}(E,\vec{r},t)f^{e}(E,\vec{r},t)\right]}{\partial z}. \quad (15)$$

In equations (13)-(15), $\mathrm{E}(\vec{r},t)$ is the laser field envelope. We explicitly substitute the spatially dependent envelope of the laser field in the expression for single photon excitation across the bandgap:

$$S_{abs} \propto \frac{2\pi}{\hbar}|F_k|^2\left[\frac{2F_k/\pi}{\left(\hbar\Omega_L - E_k^e - E_k^h - E_G\right)^2 + 4|F_k|^2}\right], \text{ where} \qquad (16)$$

$$|F_k|^2 \approx \frac{e^2 \mathrm{E}_{0L}^2(x,y,z,t)}{m_0\Omega_L}\left[\left(\frac{m_0}{m_e^*} - 1\right)\frac{E_G(E_G + \Delta_0)}{2(E_G + 2\Delta_0/3)}\right].$$

The other two terms included in $S(E,\vec{r},t)$ - the impact ionization and the Auger recombination - are not affected directly by the laser field in our derivation.

4. Relation between the rates in energy and coordinate space

Following closely references [8, 9, 10], we establish the connection between the real space drift velocity and the quantum mechanically derived field dependent velocity in energy space. An electron with momentum $\vec{p}_0$ and energy $E(\vec{p}_0)$ at time $t = 0$ is considered. Defining $\partial_F = e\vec{\mathrm{E}}\cdot\nabla_p$, where $\vec{\mathrm{E}}$ denotes the laser electric field, the energy change before any collision occurs is: $E(t) - E_0 = (\partial_F E)_{p_0} t + 1/2(\partial_F^2 E)_{p_0} t^2 + \ldots$ The probability for the electron starting from $\vec{p}_0$ not to experience collisions between 0 and t is $c(t) = \exp\left\{-\int_0^t dt'/[\tau(\vec{p})]\right\}$, where account is taken of the change in τ as the electron momentum changes in time as $d\vec{p}/dt = e\vec{E}$. Upon first order expansion of $\tau(\vec{p}_0 + e\vec{E}t)$, we obtain $c(t) = \exp(-t/\tau_0)[1 + 1/2(t/\tau_0)^2(\partial_F\tau)_{p_0}] + O(E^2)$, where $\tau_0 = \tau(\vec{p}_0)$.

The exponential gives the non-scattering probability if τ is constant while the electron is accelerated by the field, and the second factor is the lowest-order correction due to the change in τ. We obtain the averages over a collision-free flight

$$\langle t\rangle = \tau_0\left[1 + (\partial_F\tau)_{p_0}\right] \quad \text{and} \quad \langle t^2\rangle = 2\tau_0^2\left[1 + 3(\partial_F\tau)_{p_0}\right].$$

Then the average energy gained over a collisionless flight (of order E^2) is:

$$\langle E(t) - E_0\rangle = \left\{(\partial_F E)_{p_0} + \partial_F(\tau\partial_F E)_{p_0}\right\}\tau_0 \text{ so } \langle E(t) - E_0\rangle = e\vec{E}\cdot\left\{\vec{v}_g(\vec{p}_0) + \left[\partial_F(\vec{v}_g\tau)_{\vec{p}_0}\right]\right\}\tau_0, \text{ and}$$

$$V_F(E_0) = \left(\frac{dE}{dt}\right)_F = e\vec{E}\cdot\left\{\vec{v}_g(\vec{p}_0) + \left[\partial_F(\vec{v}_g\tau)_{\vec{p}_0}\right]\right\}, \text{ with } \vec{v}_d(E_0) = \frac{\iint_{E_0}\partial_F(\vec{v}_g\tau(\vec{p}))dS_p/|\vec{v}_g|}{\iint_{E_0} dS_p/|\vec{v}_g|},$$

where $dS_p/|\vec{v}_g|$ are the number of states in a surface element dS_p in momentum space. From the above expressions the relation between the field dependent drift velocity and the drift velocity in coordinate space is obtained: $V_F(E_0) = e\vec{E}\cdot\vec{v}_d(E_0)$ and for the choice of laser field polarization

$$v_{d_x}(E,\vec{r},t) = \frac{V_F(E,\vec{r},t)}{(e\mathrm{E}_{0L})}.$$

The position-space diffusion coefficient is related to the field dependent diffusion coefficient in energy space by $D_x(E,\vec{r},t) = \dfrac{D_F(E,\vec{r},t)}{(e\mathrm{E})^2}$ [8].

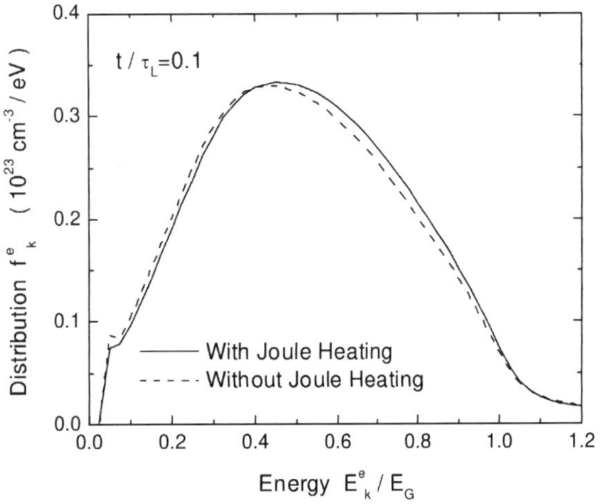

Figure 1. Comparison of the energy spectra of $f^e{}_k$ at $t/\tau_L = -0.5$ including and excluding Joule heating.

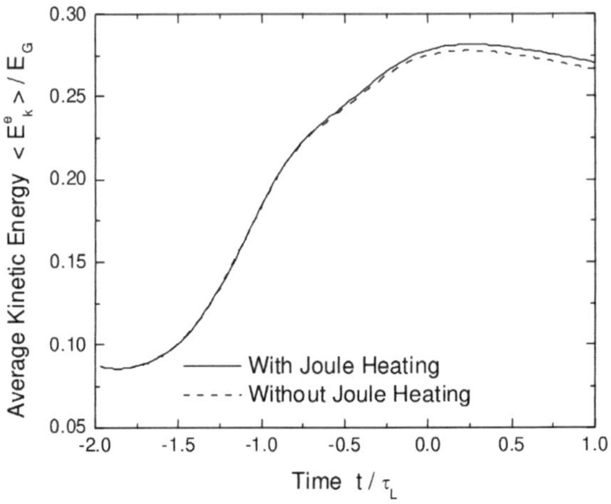

Figure 2. Average electron energy at scaled time of the laser pulse including and excluding Joule heating.

Joule heating is the main energy gain mechanism, so the electrons from low-energy states in the conduction band are transferred to high-energy states by absorbing energy from the laser field. This is illustrated in figure 1 where the peak of the electron energy distribution function exhibits increased amplitude and is shifted to the right (higher electron energy) in the presence of the Joule heating term. Figure 2 shows the dependence of the electron average energy on the Joule heating term. It should be noted that the Joule heating term is only one contribution to the rate $V(E,t)$ (eq.3) which has the physical meaning of energy dependent drift.

5. Conclusion

In our theoretical consideration, we have assumed that the wavelength of the laser radiation is longer than the other characteristic lengths so in the quantum-mechanical derivation the spatial dependence of the electromagnetic field is ignored and an electron-phonon system in a spatially uniform sinusoidal electric field of frequency Ω_L (dipole approximation) is considered. After the derivation of the kinetic equation for the electron energy distribution function, the spatial dependence of the laser field is taken into account by explicitly including in the rates of the equation the spatially dependent envelope function that comprises the amplitude of the laser radiation electric field. The left hand side of the initially obtained kinetic equation is represented as a conservation law for particle number and current density in energy space. Consequently the energy dependent current density is generalized to be energy and space dependent. The kinetic equation is generalized to include additional spatially dependent current densities on the left-hand side of the equation, thus yielding a multivariate Fokker-Planck type equation in the energy-position manifold.

From elementary arguments, without invoking expressions for the electron vector mean-free-path, related to momentum relaxation and mobility, the connection between the field dependent drift velocity in energy space and the drift velocity in real space is obtained.

Acknowledgments

The work was sponsored by the Air Force Office of Scientific Research, Air Force Material Command, USAF, under grant number FA8655-08-1-3080.

Journal of Physics: Conference Series **223** (2010) 012036 doi:10.1088/1742-6596/223/1/012036

References

[1] Skupin S and Berge L 2006 *Physica* D **220** 14
[2] Surdie L, Couarion A, Franco M, Lamouroux B, Prade B, Tzortzakis S and Mysyrowicz A 2002 *Phys. Rev. Lett.* **89** 186601
[3] Apostolova T, Huang D, Alsing P M, McIver J and Cardimona D A 2002 *Phys. Rev.* B **66** 075208
[4] Huang D, Apostolova T, Alsing P M and Cardimona D A 2004 *Phys. Rev.* B **69** 75214
[5] Xu W 1998 *Phys. Rev.* B **57** 12939
[6] Bringuier E 1997 *Phys. Rev.* B **56** 5328
[7] Bringuier E 1998 *Phys. Rev.* B **57** 2280
[8] Bringuier E 1995 *Phys. Rev.* B **52** 8092
[9] Bringuier E 1996 *Phys. Rev.* B **54** 8092
[10] Kurosawa T 1965 *J. Phys. Soc. Jpn.* **20** 937

Study of the charge transport mechanism in AlN:Cr films synthesized by pulsed laser deposition

S Simeonov[1], I Minkov[1,3,4], A Szekeres[1], C Ristoscu[2], G Sogol[2], G Dorcioman[2] and I N Mihailescu[2]

[1]Georgi Nadjakov Institute of Solid State Physics,
Bulgarian Academy of Sciences, 72 Tzarigradsko Chausse, 1784 Sofia, Bulgaria
[2]National Institute for Laser, Plasma and Radiation Physics,
PO Box MG-54, 77125 Bucharest - Magurele, Romania
[3]Technical University of Sofia, 8 Kl. Ohridski Blvd, Sofia 1756, Bulgaria

E-mail: ipminkov@issp.bas.bg

Abstract. Thin films of AlN doped with Cr were formed on p-Si(100) substrates by pulsed laser deposition. Al-AlN:Cr-Si MIS structures were formed and their current–voltage and 1 MHz admittance characteristics were measured. The considerably smaller increase of the conductance at 300 K in comparison to that measured at 77 K is an evidence for tunneling type of conductance in these MIS structures. The current is limited by space charge at deep levels (traps) in these AlN:Cr films which determines the mechanism of the charge transport through the MIS structures. The density of the filled electron traps is in the order of 10^{16} cm^{-3}.

1. Introduction

The possibility for applying diluted magnetic semiconductors (DMS) in spintronics devices and structures has led to the preparation and investigation of different DMS thin films. DMS is formed when metal ions are replaced by magnetic ions of transition metals such as Co, Cr or Mn. Ferromagnetism with Curie temperatures as high as 900°C is observed in Cr-doped AlN (denoted further as AlN:Cr) films [1]. For preparation of these films, the incorporation of Cr ions is accomplished usually by Cr co-evaporation during AlN MBE deposition [2] or by Cr ion implantation into already deposited AlN film [3], or by reactive dc co-sputtering of Al and Cr in a nitrogen gas ambient [4]. The pulsed laser deposition (PLD) technique, due to its capability of reproducing the target stoichiometry in the deposited film, is very suitable for preparation of AlN:Cr films. Prospective AlN:Cr devices and structures formed on Si substrates will be compatible with devices and integrated circuits in contemporary Si planar technology. For spintronics applications, such as resonant tunneling diodes and tunneling magnetoresistance structures, knowledge of the charge transport mechanism in these films is of great importance.

Our research interest was recently focused on AlN thin films doped with Cr, as we have used pulsed laser deposition for film preparation. Some studies of the structural and optical properties of PLD AlN:Cr films have already performed [5]. In this paper we present results of the study of the charge transport mechanism in the PLD AlN:Cr films. For this purpose, MIS structures with PLD AlN

[4] To whom any correspondence should be addressed.

films were formed and their DC current-voltage (I-V) and admittance (capacitance and conductance) characteristics were measured. Through an analysis of the I-V characteristics, the capacitance–voltage (C-V) and conductance–voltage (G-V) characteristics obtained, the charge transport mechanism was established.

2. Experimental details

Thin AlN films doped with Cr were synthesized on single crystalline Si(100) p-type substrates by pulsed laser deposition. The PLD targets were prepared by homogeneously mixing AlN and Cr (10 at.%) powders followed by pressing and sintering. The films were grown using a KrF* ($\lambda = 248$ nm) excimer laser generating pulses of $\tau_{FWHM} \leq 25$ ns at a repetition rate of 10 Hz. During ablation, the target was rotated and translated along 2 orthogonal axes to avoid piercing and improve the morphology of the deposited films. The laser beam was incident at 45°, while the ablation plasma flume was evolving normally to the target surface. The target - substrate distance was 5 cm. After loading the substrate and target, the deposition chamber was pumped down to a residual pressure of 5.0×10^{-4} Pa. Before deposition, the substrate was heated up to 800°C for 1 hour for the purpose of decomposing native SiO_2 on the Si surface. The substrate temperature was kept at 800 °C during deposition, providing conditions for crystalline growth of AlN. The laser fluence on target was 10 J/cm^2. Deposition of one film necessitated 15 000 consecutive laser pulses yielding a ~300 nm thick film. The film synthesis was undertaken in vacuum at a residual gas pressure of 10^{-3} Pa.

To perform electrical measurements, metal–AlN:Cr–silicon capacitors were formed by vacuum evaporation of Al dots with 0.5 mm diameter on the film surface and a continuous Al film on the Si wafer backside. Capacitance–voltage (C-V) and conductance–voltage (G-V) measurements were carried out by using a digital LCR meter E7-12 at room temperature and at 77 K at test voltage frequency of 1 MHz. The DC current–voltage (I-V) characteristics were measured only at room temperature.

3. Results and discussion

Applying the bias voltage to the AlN:Cr-Si structure during the I-V measurements changed the current by several percents. Because of this, we increased the bias voltage from zero to its maximal value at a constant step for a total time of ~30 min. The maximal voltage was kept constant for ~15 min before starting the decrease of the voltage at the same step. The common feature of these characteristics was their hysteresis. The current value measured depended strongly on the measurement sequence and sweep direction. Figure 1 presents the I-V curves recorded with a measurement sequence starting from 0 V toward negative or positive voltages applied to the contact on the AlN:Cr film surface followed by a decrease of the absolute value of the voltage applied toward zero bias.

It has been established that the electron mobility exceeds the hole mobility in AlN by several factors, for example, at 300 K the electron mobility in pure AlN is 800 cm^{2}V^{-1}sec^{-1} [6], while the hole mobility in Mg doped AlN is (10÷26) cm^{2}V^{-1}sec^{-1} [7]. Since in the "forward" direction, i.e. under accumulation, the current density is only about 30-50% larger than that measured in the "reverse" direction, i.e. in inversion (figure 1), it can be concluded that in our PLD AlN:Cr films in both directions the DC current is carried out by electrons. As it is seen, in both sweep directions at a given voltage the current measured when the electric field is increased is larger than that measured with decreasing electric field. Such a current hysteresis is an indisputable evidence that during electron passage through the AlN:Cr film some of electrons are captured by empty deep levels (traps) in these films. These

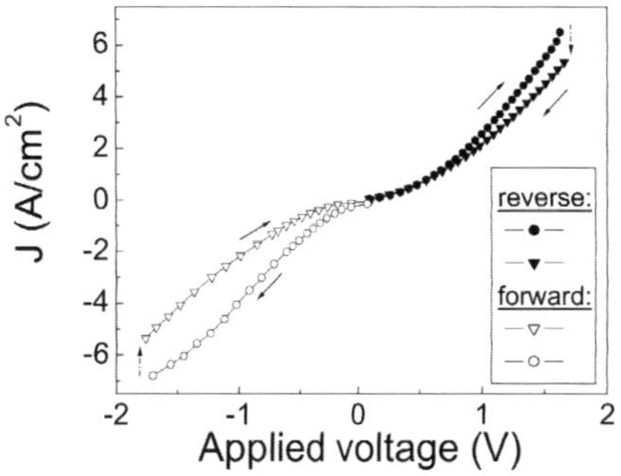

Figure 1. Room temperature I-V characteristics of Al-AlN:Cr-Si capacitors.

captured electrons oppose the current flow leading to lower current density in the directions of decreasing electrical fields. In the case of uniform trap concentration in the film, the density of these filled electron traps, N_t, can be estimated from the electrical field change ($\Delta V/d_{AlN} = qN_t d_{AlN}/\varepsilon_{AlN}$, where d_{AlN} is the thickness of the film and ε_{AlN} is its permittivity). To restore the same current density, as is evident from figure 1, the electrical field applied should be changed by $\Delta V_j/d_{AlN}$, where ΔV_j is the voltage difference between the up and down plots at a given current density. With $\Delta V_j = 0.39$ V at $J = 3$ A/cm^2 (forward direction, figure 1) and $\varepsilon_{AlN} = 30\varepsilon_0$ (determined from the C-V characteristics), the density estimated of filled electron traps is $N_t = 7.2\times10^{15}$ cm^{-3}. Since the permittivity of AlN is frequency dependent [8], ε_{AlN} value at low frequencies is greater than that measured at 1 MHz and, therefore, the real value of N_t is greater than its estimated value. A trap density of $\sim10^{16}$ cm^{-3} is reasonable for an insulator film which, together with the hysteresis observed in the I-V curves, indicate that the current in the AlN: Cr films is limited by space charge. As it is shown in [9], in that case the current is determined by the density of the charge trapped in the film deposited and by the effective velocity of the charge carriers. As both density and velocity are proportional to the applied voltage, the experimentally measured current is proportional to V^{ξ}, where ξ is equal to 2 for traps with single energy level or close to it for traps with normal energy distribution around a fixed energy position. In [9] it has been also established that the trap space-charge limited current at lower electrical fields should be ohmic. This is in good agreement with our observations that the I–V dependence (figure 1) is linear at low electrical fields applied (below 10^3 Vcm^{-1}) and obeys the $J \approx V^{\xi}$ dependence at higher electrical fields as the current increases rapidly and becomes proportional to V^2.

The dielectric properties of an insulating film are usually characterized by its resistivity value ($R=\rho d_{AlN}/A$), where ρ is the specific resistivity of the material and A is the contact area. In general, ρ has a constant value. However, as it is evident from the I–V characteristics in figure 1, in our case the specific resistivity ρ depends on the electric field applied due to the trapped-charge influence on the electrical conduction in the PLD AlN:Cr films. The $\rho(V)$ dependence obtained is given in figure 2.

The C-V and G-V characteristics of the MIS structures are shown in figures 3 and 4, respectively. Keeping the same voltage sweep directions, the plots remain identical during several records. The same situation occurs even when the voltage sweep rate is decreased by a factor of ten. At both temperatures, the accumulation, depletion and strong inversion regions are clearly discernable. The permittivity ε_{AlN} of the AlN film, calculated from the capacitance value in accumulation, is equal to $30.6\varepsilon_0$ at 300 K and to $25\varepsilon_0$ at 77 K. These values are higher than the value $\varepsilon_{AlN} \approx 10\varepsilon_0$ estimated from the optical measurement, as reported in [10]. Such high values of ε_{AlN} have been related to the nitrogen vacancies contribution to capacitance of the AlN [11]. Another evidence for defects in these AlN-Si structures is the extended depletion region between the accumulation and deep inversion regions.

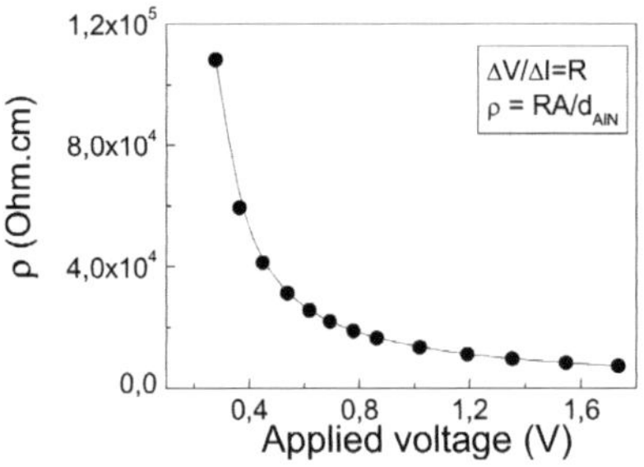

Figure 2 Specific resistivity as a function of applied voltage for Al-AlN:Cr-Si capacitors.

In the C-V plots (figure 3), a maximum near zero bias voltage is observed at both temperatures. Such a capacitance feature is observed in many semiconductor structures and is connected with the back-to-back Schottky model of depletion regions on both sides of the grain boundary [12]. The change of the widths of these layers when voltage is applied causes the capacitance maximum observed near zero bias voltage. The capacitance saturation in accumulation indicates that the trapped charge concentration at defects in the AlN:Cr films does not depend on the bias voltage applied. This is evidence of the fact that all traps are filled with charge carriers during the measurements.

Similarly to the capacitance, the conductance (figure 4) also saturates in accumulation and deep inversion regions due to the complete filling of traps in the AlN:Cr films at these bias voltages. In the regions where the capacitance and conductance are constant, i.e. below –8 and above 8 V, all deep levels lying below the charge carrier quasi–Fermi level in the AlN:Cr film are filled up. At 300 K, the value of the conductance saturation in accumulation is approximately twice as high as that at 77 K, while its value in deep inversion is 50 % higher than that at 77 K. These observations are an indication for tunneling-type conductance in the AlN:Cr-Si MIS structures investigated.

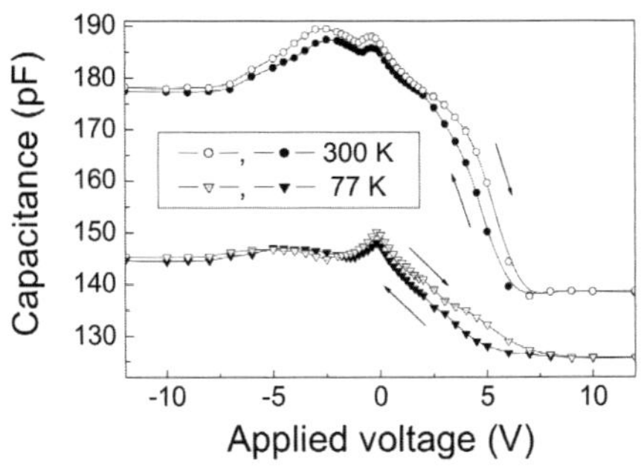

Figure 3. C-V characteristics of Al-AlN:Cr-Si capacitors measured at 77 and 300 K.

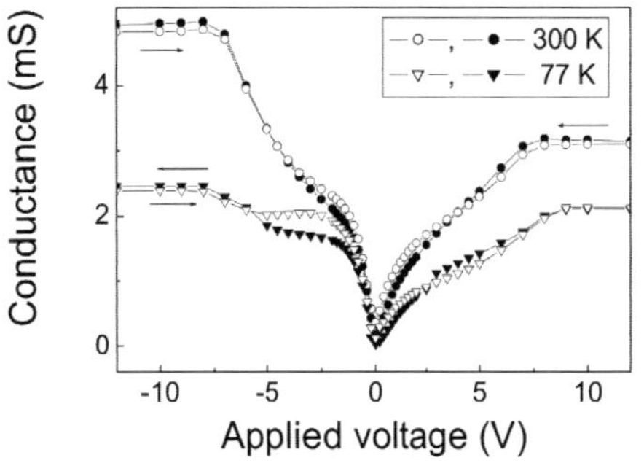

Figure 4. G-V characteristics of Al-AlN:Cr-Si capacitors measured at 77 and 300 K.

4. Conclusions

It was established that deep levels in the PLD AlN:Cr films contribute to the capacitance and conductance of MIS structures formed with AlN:Cr films. At electrical fields higher than 10^3 Vcm^{-1} the conductance is almost constant, which is caused by a complete filling of deep levels below the quasi-Fermi level in the AlN:Cr film. The trapped charge carriers at deep levels influence the current density which leads to a hysteresis in the C-V and I-V characteristics and to a dependence of the specific resistivity on the electrical field applied. Due to the high concentration of deep levels ($N_t \sim 10^{16}$cm^{-3}) in the PLD AlN:Cr films, the charge transport mechanism is trap-space-charge-limited current.

Acknowledgements

The authors acknowledge the support of this work under the Collaboration Agreement between the Bulgarian and Romanian Academies of Sciences for 2007–2009.

References

[1] Liu H X, Wu S Y, Singh R K, Lin G,. Smith D J, Newman N, Dilley N R, Montes L and Simmonds M B 2004 *Appl. Phys. Lett.* **85** 4076

[2] Frazier R M, Thaler G T, Leifer J Y, Hite J K, Gila B P, Abernathy C R and Pearton S J 2005 *Appl. Phys. Lett.* **86** 052101

[3] Pearton C J, Abernathy C R and Fan Ren 2006 *Gallium nitride processing for electronics, sensors and spintronics* (Berlin: Springer) p 291

[4] Kumar D, Antifakos J, Blamire M G and Barber Z H 2004 *Appl. Phys. Lett.* **84** 5004

[5] Szekeres A, Bakalova S, Grigorescu S, Cziraki A, Socol G, Ristoscu C and Mihailescu I N 2009 *Appl. Surf. Sci.* **255** 52271

[6] Yarar Z, Ozdemir B and Ozdemir M 2007 *J. Electronic Mater.* **36** 1303

[7] Wongchotigul K, Chen N, Zhang D P, Tang X and Spencer M G 1996 *Mater. Lett.* **24** 223

[8] Simeonov S, Bakalova S, Szekeres A, Kafedjiiska E, Grigorescu S, Grigorescu S, Socol G and Mihailescu I N 2008 *J. Phys.: Conf. Series* **113** 012050

[9] Lampert M A and Mark P 1970 *Current injection in solids* (Academic Press: New York)
[10] Mohammad S N and Morkoc H 1996 *Progress Quantum Electr.* **20** 361
[11] Bi Z X, Zheng Y D, Zhang R, Gu S L, Xiu X Q, Zhou L L, Shen B, D.J. Chen and Shi Y 2004 *J. Materials Sci.:Materials in Electronics* **15** 317
[12] Bardhadi A, Amziel H, Muller J C and Siffert P 1989 *Springer Proc. Physics, Polycryst. Semicond.* **35** 158

Investigation of thin ZnO layers in view of laser desorption-ionization

A A Grechnikov[1], V B Georgieva[2,5], S S Alimpiev[3], A S Borodkov[1], S M Nikiforov[3], Ya O Simanovsky[3], D Dimova-Malinovska[4] and O I Angelov[4]

[1]Vernadsky Institute of Geochemistry and Analytical Chemistry,
Russian Academy of Sciences, 19 Kosygin Str., 119991 Moscow, Russia
[2]Georgi Nadjakov Institute of Solid State Physics,
Bulgarian Academy of Sciences, 72 Tzarigradsko Chaussee, 1784 Sofia, Bulgaria
[3]General Physics Institute,
Russian Academy of Sciences, 38 Vavilov Str., 119991 Moscow, Russia
[4]Laboratory for Solar Energy and New Energy Sources,
Bulgarian Academy of Sciences, 72 Tzarigradsko Chaussee, 1784 Sofia, Bulgaria

E-mail: lazarova@issp.bas.bg

Abstract. Thin zinc oxide films (ZnO) were developed as a matrix-free platform for surface assisted laser desorption-ionization (SALDI) time-of-flight mass spectrometry. The ZnO films were deposited by RF magnetron sputtering of ZnO ceramic targets in Ar atmospheres on monocrystalline silicon. The generation under UV (355 nm) laser irradiation of positive ions of atenolol, reserpine and gramicidin S from the ZnO layers deposited was studied. All analytes tested were detected as protonated molecules with no or very structure-specific fragmentation. The mass spectra obtained showed low levels of chemical background noise. All ZnO films studied exhibited high stability and good reproducibility. The detection limits for test analytes are in the 10 femtomol range.

1. Introduction

Surface assisted laser desorption ionization (SALDI) is a novel method of ionization of organic and bioorganic compounds which has been under intensive development over the last decade. In SALDI, the ionized species are formed in a gas phase from analytes deposited on the surface of a specially designed SALDI-active substrate which is irradiated by a pulsed laser. This process does not require the entrainment action of an added matrix compound for desorption [1]. A variety of inorganic materials in the form of powders or monolithic plates have been studied as a possible platform for SALDI; it was found that only a few selected substrates have the ability to strongly promote ion formation. One of the well-known examples of such a substrate is porous silicon obtained by electrochemical etching [2]. In particular, porous silicon is used for small-molecule detection in DIOS (desorption/ionization on silicon), a variant of SALDI [3]. It was found recently that thin nanocrystalline and/or amorphous silicon films prepared by RF magnetron sputtering are very

[5] To whom any correspondence should be addressed.

effective SALDI substrates [4]. When compared to porous silicon, deposited films have a number of advantages, such as easy and controllable method of fabrication, greater uniformity over large areas, higher purity and good reproducibility. However, the amorphous Si substrates were found to be unstable in ambient air.

The present work is focused on examining thin zinc oxide films (ZnO) as a platform for SALDI time-of-flight mass spectrometry.

2. Experimental

We used a laboratory-built, linear, turbopumped time-of-flight mass spectrometer with a 0.70 m flight tube. The ion acceleration voltage was 16.0 kV; the ions were detected by a 40-mm diameter chevron multichannel plate (MCP) assembly. The detection system was adjusted so as to obtain a direct reading of the ions number hitting the detector. The signal from the MCP was recorded by an ADC digitizer with 1 Gs/s sampling rate and ability to store mass spectra at a rate of up to 1 kHz. A diode-pumped YAG laser with near-diffraction limited beam quality 0.5 ns pulse duration was operated at 300 Hz. A frequency-tripling crystal yielded 355 nm laser pulses of about 100 μJ. The laser pulse energy was varied in the range of 10-30 μJ using an attenuator.

In order to utilize the analyte more efficiently, the laser focus was scanned over a 1×2 mm^2 sample deposition area using a computer-controlled dual-mirror scanner. The laser focus moved approximately 0.08 mm between successive laser shots; about 300 individual mass spectra were obtained from different points on the probed area of the SALDI substrate and were accumulated by the ADC digitizer during one “scanning cycle”.

Thin films of ZnO were deposited by RF 13.56 MHz magnetron sputtering of ZnO ceramic targets in Ar atmosphere (0.5 Pa) on monocrystalline silicon (for SALDI and X-ray diffraction (XRD) measurements) and glass substrates (for optical measurement) held at 500ºC. The distance between the target and the substrates was 7.5 cm. The vacuum chamber was evacuated to a residual gas pressure of 2×10^{-4} Pa. Sputtering was carried out at RF power of 180 W. The thickness of the films deposited measured by a “Talystep” profilometer was about 0.5 μm.

The XRD spectra were obtained using a DRON 3 spectrometer with CuK_α radiation: λ $CuK_{\alpha1}$ = 1.540560 Å and λ $CuK_{\alpha2}$ = 1.544426 Å (intensity half that of λ $CuK_{\alpha1}$).

The SALDI substrates used in the present work were approximately 5×5 mm^2 and were attached to the instrument’s stainless steel insertion probe.

All chemicals were obtained commercially and used without further purification. Atenolol (Mr 266.3 g/mol), reserpine (Mr 608.7 g/mol) and gramicidin S (Mr 1140.7 g/mol) used as test analytes were dissolved in acetonitrile-water at a concentration of 50 to 1000 ng/ml. A volume in amount of 1-4 μl of analyte solution was spotted on the surface of a ZnO film, air-dried and then analyzed.

3. Results and discussion

3.1. ZnO films characterization

The XRD spectrum of the ZnO films deposited was measured. The reflection corresponding to the (002) plane is observed which demonstrates the preferential orientation of the grains in the ZnO films. The grains size was calculated using the Debye-Scherrer formula and was about 160 nm. The transmission of all samples was about 92-93% at wavelengths higher than 600 nm, assuming a value of 100% for the transmission of the glass substrate. The absorption coefficient, α, was calculated. The formula for direct allowed transitions was used to obtain the optical gap, E_g, whose value was 3.27 eV, typical for ZnO.

3.2. Mass spectra

Atenolol, reserpine and gramicidin S were selected as test analytes because of their high proton affinities, differences in molecular weights and their widespread use as drugs, so that we could obtain a good indication of the usefulness of the ZnO assisted laser desorption ionization method.

The preliminary study of the ZnO films threshold behavior, i.e., the ion currents dependence on the laser fluence showed that 40 mJ/cm^2 of laser intensity gives the best performance for all analytes studied. Decreasing the laser fluence leads to a dramatic decrease of the SALDI-signal. As the laser fluence is increase, zinc containing peaks appear in the mass spectra, indicating the destruction of ZnO films. Therefore, all following experiments were carried out at 40 mJ/cm^2 of laser fluence. Figure 1 shows a typical SALDI mass spectrum of 400 pg of atenolol deposited on the ZnO surface from a 200 ng/ml solution. It is seen that the base peak in the spectrum in figure 1 is associated with the protonated atenolol molecule at m/z=267. No fragment ions are observed in the spectrum. The minor background peaks throughout the mass range result from surface contamination, possibly from vacuum pump oil. K^+ at m/z=39 and Na^+ at m/z=23 ions were also observed in the low-mass region (not shown). In general, however, the mass spectra obtained are “clean” with low chemical background. This property the of nc-ZnO surface minimizes the need for sample cleanup.

Figure 2 shows a SALDI mass spectrum of 1 ng (0.9 pmole) of gramicidine S measured from a 500 ng/ml concentration solution of this peptide. The base peak in the mass spectrum is the protonated molecule at m/z = 1141, but there are also sodium $[M+Na]^+$ and potassium $[M+K]^+$ cation adduct peaks. Similarly to the case of atenolol, no fragment ions are observed in the mass spectrum of gramicidine S. These observations prove that ZnO based SALDI is a soft ionization technique. However, as it was shown earlier [4] SALDI mechanism involves the stage of thermally induced desorption of ions. This can result in at least partial decomposition of thermally labile compounds.

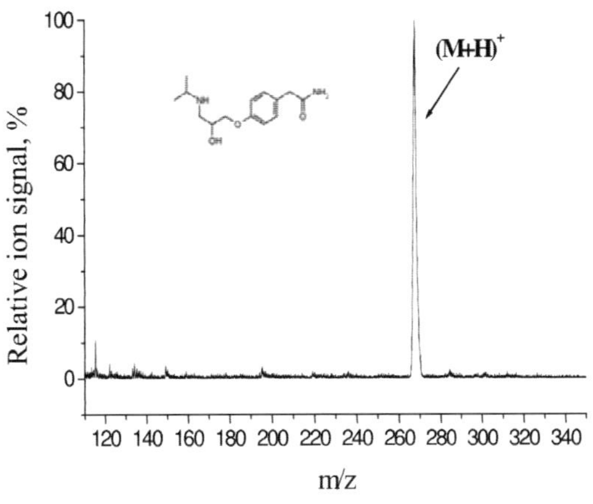

Figure 1. SALDI mass-spectrum of atenolol.

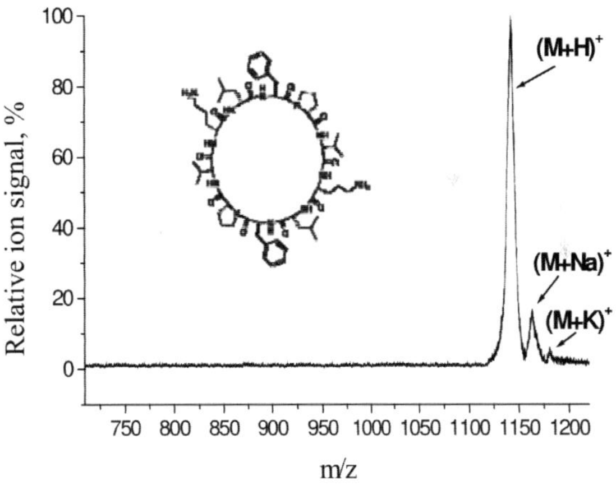

Figure 2. SALDI mass-spectrum of gramicidin S.

Reserpine exhibits such behavior. This is illustrated in figure 3, which shows the mass spectrum of 400 pg of reserpine deposited on the ZnO surface from a 200 ng/ml solution. The mass spectrum presented contains a protonated molecule at m/z=609 and a few structure-specific fragment ions. The major fragment ion at m/z = 195 is formed by heterolytic cleavage of the bond between oxygen of the ester group and carbon of the ring accompanied by hydrogen transfer to the oxygen and elimination of the 3,4,5-trimethoxybenzoic acid. Another fragment ion at m/z=397 can be interpreted as a result of the C–O bond cleavage of inside the ester group, as is shown in figure 3.

3.3. Analytical parameters

Based on the actual amount of analyte deposited onto the ZnO surface, it was found that the average number of ions detected per femtomole of atenolol, reserpine and gramicidin S were 50 ions/fmol, 45 ions/fmol and 5 ions/fmol, respectively. With about 100 ions required for reliable compound detection, the detection limits obtained were in the range of 2-20 femtomoles.

Journal of Physics: Conference Series **223** (2010) 012038 doi:10.1088/1742-6596/223/1/012038

A relatively high mass spectrometric selectivity was obtained, as a relatively high resolution, m/Δm ~800, was implemented in a simple linear TOF mass spectrometer due to a short desorbing laser pulse and to a well defined ionization zone located on the active ZnO surface.

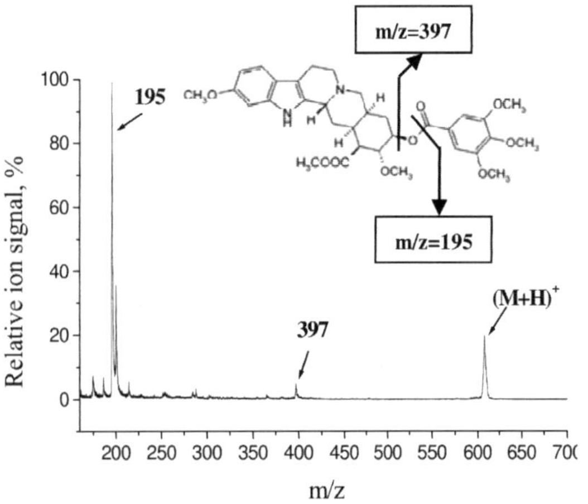

Figure 3. SALDI mass-spectrum of reserpine.

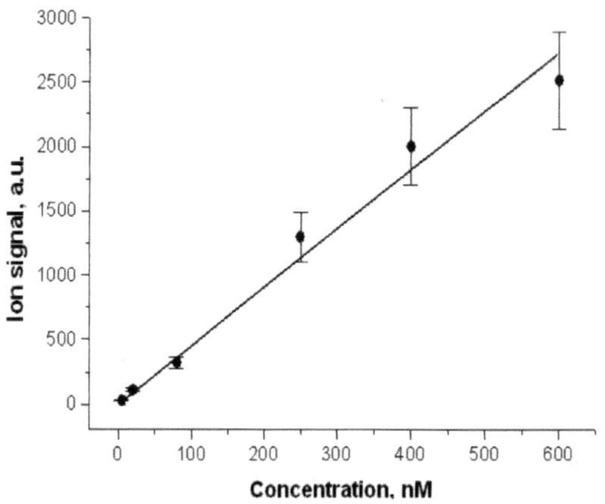

Figure 4. Calibration curve for reserpine.

The SALDI ions signal was measured for different amounts of analytes over a range of 2 orders of magnitude by changing the concentration of the solutions deposited. A typical dependence of the analytical signal on the analyte concentration is illustrated in figure.4 for reserpine. Very similar calibration curves were obtained for atenolol and gramicidin S. It was found that the linearity was better than 0.98 for all three compounds studied. The reproducibility was monitored by measuring the analytical signal using the same experimental conditions but with the different ZnO film samples. Twenty replicate analyses were carried out over a one-week period for each of the compounds studied. The relative standard deviation was less than 15% for all three compounds. This result clearly demonstrates the applicability of ZnO films for quantitative analysis of low molecular weight drugs. However, internal standards are needed for better quantification using the droplet analyte deposition method. In contrast to silicon based SALDI-active substrates, such as porous, electrochemically etched and amorphous silicon, ZnO based SALDI substrates are very stable under atmospheric conditions. There was no significant decrease in the ionization efficiency after a few months storage in air. The substrates were not "cleaned" in any way, except by laser irradiation during the experiments.

4. Conclusions

Thin ZnO films, prepared by an RF method, were investigated as the platform for analysis of organic and bioorganic molecules by SALDI-TOF mass-spectrometry. The suitability of the films prepared for drug detection in the mass range 250-1150 m/z was examined. Atenolol, reserpine and gramicidin S were studied as test analytes.

It was found that the ZnO films are very stable and show good sensitivity in the range of analyte concentrations examined. The mass spectra obtained are "clean" with almost no chemical background and with no or very specific analyte ions fragmentation. The detection limits are in the 2 - 20 femtomol range. Good reproducibility under the same experimental conditions was found for all analytes.

The high purity, the ease of reproducible fabrication, the good analytical parameters, and the ability to analyze samples rapidly make thin ZnO films a potentially powerful tool for the detection of small bioorganic and organic molecules especially in the field of drug analysis by the SALDI method.

Acknowledgement
The present work was supported by a bilateral project (2009-2011) between GEOKHI, RAS and IFTT, BAS and by the Bulgarian National Science Fund – projects DO02-207-2008 and NT-3-03/2006.

References
[1] Alimpiev S S, Nikiforov S M, Karavanskii V A and Sunner J 2001 *J. Chem. Phys.* **115** 1891-1901
[2] Bisi O, Ossicini S and Pavesi L 2000 *Surf. Sci. Rep.* **38** 1-126
[3] Wei J, Buriak J M and Siuzdak G 1999 *Nature* **399** 243-5
[4] Alimpiev S S, Grechnikov A A, Sunner J, Karavanskii V A, Zhabin S N, Simanovsky Ya O and Nikiforov S M 2008 *J. Chem. Phys.* **128** 014711-19

Journal of Physics: Conference Series **223** (2010) 012039 doi:10.1088/1742-6596/223/1/012039

Optical and vibrational spectra analysis of CVD - mixed oxide films: optimization of the films electrochromic performance

T Ivanova[1], K A Gesheva[1,4], M Abrashev[2], P Sharlandjiev[3] and D Nazarova[3]

[1]Central Laboratory of Solar Energy and New Energy Sources,
Bulgarian Academy of Sciences, 72 Tzarigradsko Chaussee, 1784 Sofia, Bulgaria
[2]Faculty of Physics, St. Kl. Ohridski University of Sofia,
5 J. Bourchier Blvd., 1164 Sofia, Bulgaria
[3]Central Laboratory of Optical Storage and Processing of Information,
Bulgarian Academy of Sciences, Acad. G. Bonchev Str., Bl. 101, 1113 Sofia, Bulgaria

E-mail: kagesh@phys.bas.bg

Abstract. Mixed oxide films based on Mo and W were successfully prepared by atmospheric pressure CVD at the low substrate temperature of 200°C. High amount of oxygen was used to ensure a high degree of oxidation resulting in more stoichiometric oxide films. The structural transformations under different thermal treatments were studied by Raman spectroscopy and FTIR spectroscopy analysis. The films were characterized electrochemically by cyclic voltammetry using different electrolytes, scan rates, etc. The mixed oxide films exhibited a strong electrochromic (EC) effect.

1. Introduction

Electrochromic materials have been studied intensively due to their ability to change reversibly their optical transmittance upon application of a voltage [1, 2]. WO_3 has been the most widely investigated material due to its superior electrochromic characteristics. It has also been studied in view of applications as gas sensors and photosensitive electrodes [3, 4]. Mixed W/Mo based oxide systems are interesting due to the possibility to combine the advantages of the single components' properties. Electrochromic MoO_3 has an absorption band close to the maximum of human eye sensitivity. Various technologies have been reported for their preparation, such as vacuum evaporation, sol-gel and CVD techniques [5,6]. Due to the presence of electronic W^{5+}, W^{6+} states, and the corresponding lower energy states of Mo, W/Mo based oxide systems exhibit stronger optical absorption. Electrochromic films are functional element in "smart windows" envisaged as energy control windows in buildings and automobiles. They control the energy of the solar flux entering the building, thus saving air-conditioning cost in summer and heating cost in winter [7].

This paper presents the investigation of low-temperature chemical-vapor growth of W/Mo oxide films. Their structure and chemical bonding are studied by Raman and FTIR spectroscopy. The optical transmittance is investigated by UV-VIS spectrophotometry. The optical modulation and color efficiency are determined from cyclic voltammetry data.

---

[4] To whom any correspondence should be addressed.

2. Experimental

Mixed oxide films based on Mo and W were successfully prepared by means of an atmospheric pressure CVD method and CVD equipment that were described previously [8]. The substrates were heated to 200 °C. The precursor used was a physical mixture of $Mo(CO)_6$ and $W(CO)_6$ in a ratio 1:6 in favor of $W(CO)_6$; the sublimation temperature was 90 °C. The carbonyls vapors were carried by an Ar flow from the sublimator to the reactor. This Ar flow rate is related to the oxygen flow rate (reactive gas) at a given ratio, which is a measure for the oxygen amount. The ratios studied were 1/16 and 1/32. The samples were annealed at 500°C for 1 hour. The film thickness was 280 nm for WO_3-MoO_3 films prepared at a gas flow rate of 1/32, and 200 nm, for films at flow rate of 1/16. The IR measurements were carried out on Shimadzu FTIR Prestige-21 equipment. The Raman study was performed using a SPEX 1403 Raman spectrometer, Ar^+ laser line 488 nm, laser power 54 mW. The optical transmission and reflection were measured by a Cary 5E Spectrophotometer at normal incidence in the range 350 - 2500 nm. The cyclic voltammetry experiments were performed in a standard three-electrode cell arrangement.

3. Results and discussions

The Raman spectra were recorded in the spectral range 100 – 1200 cm^{-1}. The peak located at 520 cm^{-1} and the broad band at 960 cm^{-1} are due to vibration modes of the Si substrates. The Raman spectra of as deposited MoO_3-WO_3 films (figure 1) reveal crystallized film produced at a higher oxygen amount (gas flow ratio of 1/32). The well-defined Raman lines are evidence of crystallization. Our previous studies [8] of MoO_3-WO_3 films deposited from different precursor mixture 1:4 showed that the films remain amorphous after annealing at 400 °C. The Raman spectrum of an as deposited film obtained at a lower oxygen amount (figure 1a, curve 1) is typical for an amorphous sample, the weak broad band at 783 cm^{-1} being attributed to the stretching mode of monoclinic MoO_3 [9]. The spectrum of the film deposited at a higher amount of oxygen (1/32) contains four Raman lines at 261, 326, 711 and 809 cm^{-1}. These lines are assigned to crystalline monoclinic WO_3 as they match well the fundamental W-O-W bands of monoclinic WO_3 (273, 327, 715, 804 cm^{-1}) [10].

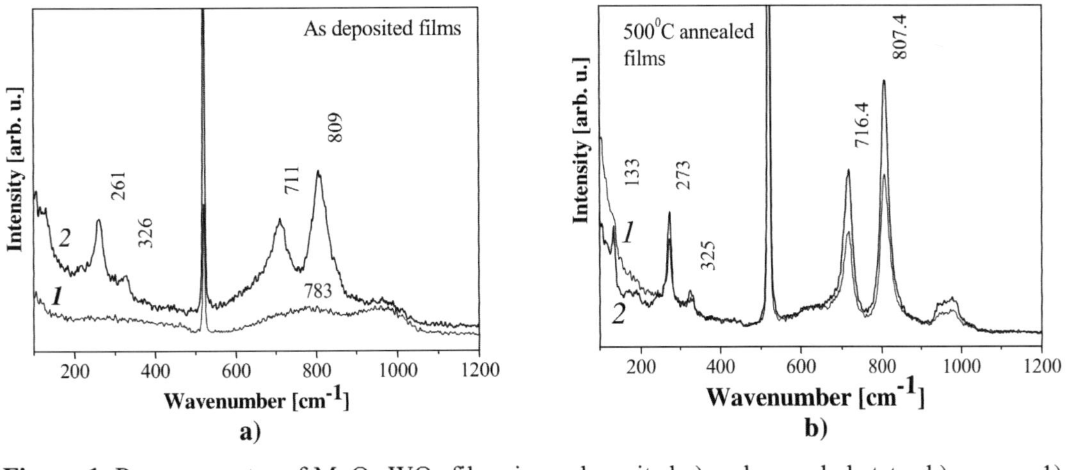

Figure 1. Raman spectra of MoO_3-WO_3 films in as-deposited a) and annealed states b): curves 1) presents spectra of films produced at the lower gas ratio 1/16 and curve 2) at ratio 1/32, i.e., a higher amount of oxygen.

The broad Raman bands are an indication for the presence of an amorphous fraction into the crystallized film. A possible explanation of the crystallization in the as-deposited film is its bigger thickness (280 nm) compared to the other film (200 nm). Thicker films crystallize during the film growth. Another contribution could be the faster growth-rate of WO_3 component in the presence of Mo precursor. We obtained similar results in some of our previous studies of MoO_3-WO_3 films [8].

Annealing of MoO_3-WO_3 at 500 °C in air for 1 hour (figure 1b) leads to full crystallization of the films, and stronger Raman lines are observed for the film deposited at a higher oxygen content. Intense Raman lines at 237.7, 325, 716.4 and 807.4 cm^{-1} are seen; are assigned to monoclinic WO_3. In the spectrum of the MoO_3-WO_3 film deposited at the higher oxygen amount, a stronger peak at 133 cm^{-1} and weak peaks at 170 and 190 cm^{-1} are present. The weak lines are related to deformation modes of monoclinic β-MoO_3 and the peak at 237 cm^{-1} is also assigned to MoO_3. This analysis leads to the conclusion that the CVD MoO_3-WO_3 films crystallize in monoclinic WO_3 with small inclusions of MoO_3.

The FTIR spectra presented in figure 2 were recorded for CVD Mo-W oxide films prepared at the two gas flow ratios, 1/16 and 1/32. The as-deposited samples reveal broad absorption bands at different locations. For the lower oxygen content (1/16), the band covers the spectral range 600-800 cm^{-1} centered at 705 cm^{-1}; for the film prepared with higher oxygen content (1/32), the band is centered at 653.7 cm^{-1}. Weak IR bands are observed at 359, 418 cm^{-1} for 1/16, and at 453 cm^{-1} for 1/32. Our IR study of CVD MoO_3 and WO_3 films [8] showed main bands at 790 cm^{-1} for MoO_3 and at 690 cm^{-1} for WO_3 for as-deposited samples. A shift of the bands is observed due to vibration modes characteristic for MoO_3 and WO_3. The broad absorption band at 705 cm^{-1} is assigned to W-O-W bridging modes and Mo-O vibrations [11]. The IR bands at 653.7 cm^{-1} is attributed to stretching W-O vibrations of monoclinic WO_3.

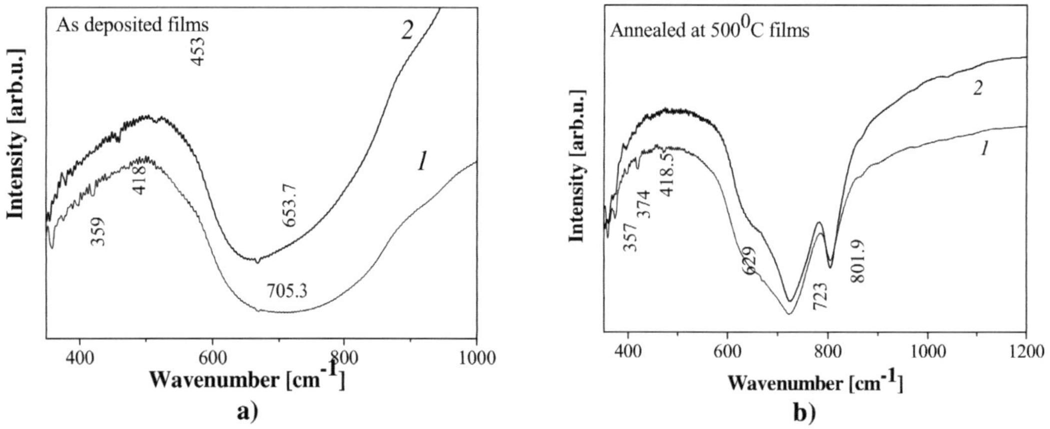

Figure 2. FTIR spectra of MoO_3-WO_3 films, a) as deposited and b) annealed at 500°C: curve 1 - for the lower oxygen amount 1/16, curve 2 - for 1/32.

After annealing at 500 °C, the spectrum of the MoO_3-WO_3 film prepared at 1/16 shows bands at 357, 374 (weak), 418.5, 475, 802 and a strong band at 723 cm^{-1}. The IR spectrum of the 1/32-film is similar, with weak bands at 395.6 and 639.9 cm^{-1}, a sharp line at 801.9 cm^{-1}, and a strong one at 723 cm^{-1}. The maxima at 723 cm^{-1} and 801.9 cm^{-1} belong to monoclinic WO_3 [12]. The line at 723 cm^{-1} is shifted compared to that at 706 cm^{-1}, this shift being related to the presence of Mo (VI) species [13]. The FTIR study confirms the presence of monoclinic WO_3 and reveal the influence of the MoO_3 fraction on the spectra.

The UV-VIS spectra of as-deposited MoO_3-WO_3 films on glass substrates show transparency, as high as 80% for films deposited at higher amount of oxygen, and 70% for the lower oxygen content samples.

The higher transmittance is related to an oxidized film grown in the presence of a high amount of oxygen. The optical transmittance of MoO_3-WO_3 films decreases by 20%, when conductive glass is used as a substrate (figure 3). An unexpected tendency is observed for the films prepared at the lower oxygen content (1/16), namely, the transmittance decreases after annealing. We observed such behavior in CVD MoO_3-WO_3 films deposited at similar process parameters with precursor mixture

$Mo(CO)_6$:$W(CO)_6$ = 1:4 [14]. A possible explanation is a mass transfer process that would lead to structural rearrangement, but is not completed as it needs a longer time.

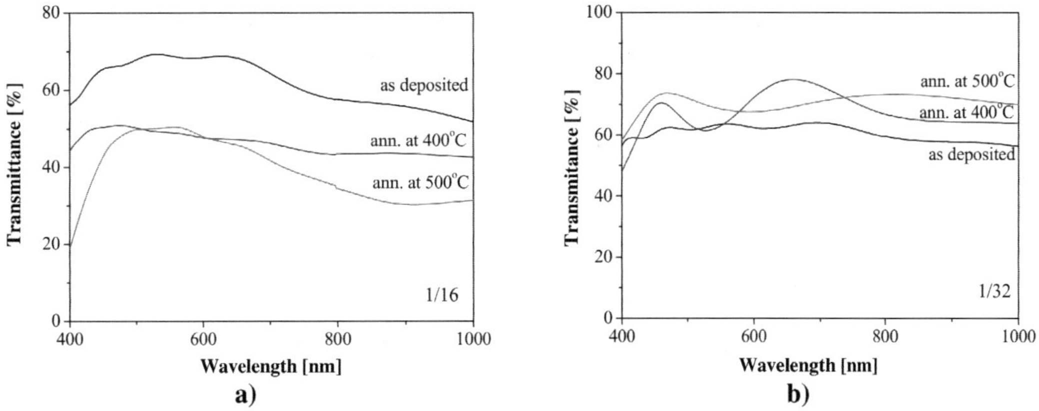

Figure 3. Transmittance of CVD annealed MoO_3-WO_3 films obtained at the two gas flow rates ratios, a) 1/16 and b) 1/32. The substrates are conductive glass.

Films prepared at higher amount of oxygen (1/32) increase their transmittance after annealing. The same was observed for the single oxide components. The MoO_3-WO_3 films show coloring in 1 M $LiClO_4$ + PC and a good reversibility. The optical modulation and color efficiency were determined from voltametric data. Figure 4 presents the values obtained for as-deposited films. The optical modulation is 30 - 40 % for MoO_3-WO_3 films produced at the lower oxygen amount (1/16) with a maximum at λ = 550 nm. MoO_3-WO_3 films prepared at 1/32 show higher values of the color efficiency, namely, CE = 140.8 cm^2/C (550 nm). Our earlier study of CVD MoO_3-WO_3 films deposited from precursor Mo $(CO)_6$:$W(CO)_6$ = 1:4 showed ΔT = 45,2 % (550 nm) and CE = 196.5 cm^2/C (550 nm) for as-deposited samples [15]. These superior films were amorphous in structure. The results show that CE and ΔT are influenced by the Mo component in the mixed oxide structure.

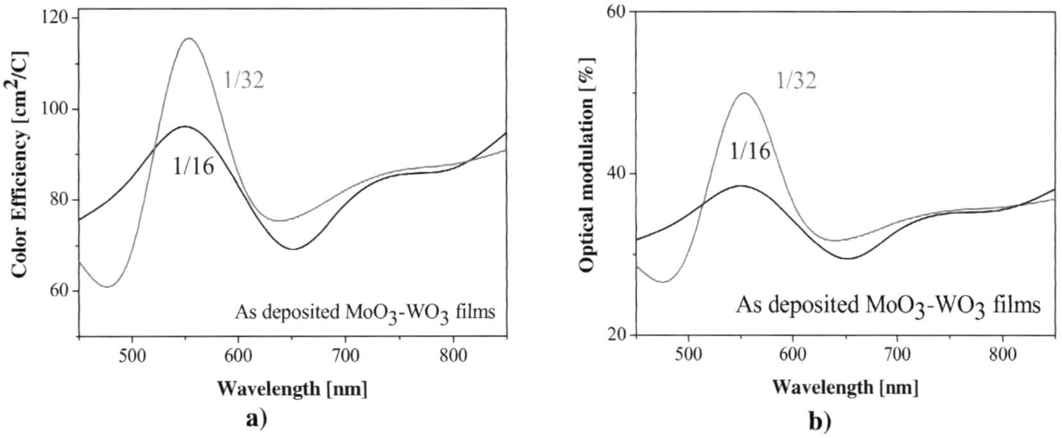

Figure 4. Color efficiency a) and optical modulation b) of CVD MoO_3-WO_3 films in as deposited state. The electrolyte used is 1 M $LiClO_4$ + PC.

Annealing at 500 °C leads to lower values of the optical modulation and color efficiency, namely, the films prepared at 1/16 showed ΔT =28 - 43 % and CE values of 40 - 100 cm^2/C. We assume that this is due to crystallization of the films, as it is known that amorphous structures favor the electrochromic effect.

4. Conclusions

The structure of CVD MoO_3-WO_3 films is found to be strongly influenced by the oxygen content. WO_3 is the predominant crystalline phase in the mixed films structure. The transmittance is 80% when the films are deposited on glass substrates. The as-deposited at higher amount of oxygen mixed Mo/W based oxide films exhibit higher values of color efficiency and optical modulation.

Acknowledgements
The work was supported by the Bulgarian Ministry of Education and Science under the National Science Fund Project DO1-377/NTE-04/16.06.2006, module 2.

References

[1] Deb S K 2008 *Solar Energy Mater. Solar Cells* **92** 245
[2] Patil P 2008 *Solid State Ionics* **179** 314
[3] Papaefthimiou S, Leftheriotis G and Yianoulis P 1999 *Thin Solid Films* **343-344** 183
[4] Galatsis K; Li Y X, Wlodarski W and Kalantar-Zadeh K 2001 *Sensors Actuat.* B **77** 478
[5] Lee E S and Tavil A 2007 *Building and Environment* **42** 2439
[6] Ivanova T, Gesheva K, Popkirov G, Ganchev M and Tzvetkova E 2005 *Mater. Sci. Eng.* B **119** 232
[7] Klisinska A, Mamede A S, Caigneaux E M 2008 *Thin Solid Films* **516** 2904
[8] Zhou L, Ren Q, Zhou X, Tang J, Chen Zh and Yu Ch 2008 *Microsc. Mesoscopic Mater.* **109** 248
[9] Krasovec U, Vuk A and Orel B 2001 *Electrochimica Acta* **46** 1921
[10] Rougier A, Portemer F, Quédé A and El Marssi M 1999 *Appl. Surf. Sci.* **153** 1
[11] Anwar M, Khan G A, Hogarth C A and Theocharis C R 1989 *J. Mater. Sci.* **24** 4409
[12] Gesheva K, Szekeres A and Ivanova T 2003 *Solar Energy Mater. Solar Cells* **76** 563
[13] Gesheva K A, Ivanova T, Marsen B, Zollo G and Kalitzova M 2008 *J. Cryst. Growth* **310** 2103

Auger-electron spectroscopy investigation of thin Ag-As-S-Se films

R Todorov, J Tasseva[1], G Spasov and K Petkov

Acad. J. Malinowski Central Laboratory of Photoprocesses,
Bulgarian Academy of Sciences, Acad. G. Bonchev Str., Bl. 109, 1113 Sofia, Bulgaria

E-mail: jordanka@clf.bas.bg

Abstract. The photoinduced changes in the refractive index and optical band-gap of thin $As_{32}S_{34}Se_{34}$ films photodoped with silver were studied using spectrophotometric methods. The compositional profile of the films was revealed by means of Auger-electron spectroscopy.

1. Introduction

Photodoping of metals into chalcogenide glasses has been of continued interest [1-4] since the first report on the phenomenon in 1924 [5] due to the possibility of its practical application in grey-scale lithography for MEMS [6], photovoltaic cells [7], etc. Industrial applications require that the phodoped layers possess reproducible properties – optical, electrical, mechanical. Reasonably, the reproducibility of the optical properties demands optical homogeneity of the films, which is closely related to the uniform distribution of the metal into the chalcogenide layer. Few papers dealing with the metal distribution in photodoped chalcogenide films have been published [1, 8-9].

We report the results from our study on the optical properties of very thin (0.12 μm) $As_{32}S_{34}Se_{34}$ films photodoped with silver. Investigations of such thin photodoped chalcogenide layers have not been reported earlier. The claim for compositional homogeneity (i.e. uniform elemental distribution in the depth of the layers) of our films was supported by Auger-electron spectroscopy (AES) profiles.

2. Experimental

Powdered bulk glass of composition $As_{32}S_{34}Se_{34}$ was placed into a Mo crucible wherefrom thermally evaporated at room temperature under vacuum of 3×10^{-3} Pa. Thin chalcogenide films were deposited in the same vacuum cycle onto bare optical BK-7 glass and Si substrates and onto preliminarily sputtered Ag layers. Optical transmission and reflection measurements in the spectral range 400-2000 nm were carried out by a UV–VIS–NIR spectrophotometer (Cary 05E, USA) both prior to and following illumination of the mono - (chalcogenide) and double layered (chalcogenide/Ag) structures. The samples were exposed to a halogen lamp (20 mW/cm^2) for 30 min. The illumination of the double layered structure was conducted through the chalcogenide side until the silver layer (with thickness 17 nm) was consumed.

The optical constants – linear refractive index, *n*, and extinction coefficient, *k* – and the thickness of the films, *d*, were determined. by spectrophotometric measurements Since the thickness of the layers was 120 nm, the number of interference extrema in the transmittance spectra was insufficient to

[1] To whom any correspondence should be addressed.

allow application of the envelope method [10]. Hence, double TR methods [11], as described elsewhere [12], were applied. The program used to calculate n can determine it to an accuracy better than 3 % in the transparent region (where $d/\lambda < 0.3$) for an experimental inaccuracy of ± 0.1 % and ± 0.5 % in transmittance and reflectance, respectively [13].

The absorption coefficient, α, of the films was calculated from the relation $\alpha = 4\pi k/\lambda$, where λ is the wavelength. The optical band gap, E_g^{opt}, for direct transitions was derived by Tauc's law [14]:

$$(\alpha\hbar\omega)^{1/2} = B(\hbar\omega - E_g^{opt}), \quad (1)$$

where $\hbar\omega$ is the photon energy and B, a factor inversely proportional to the localized state bands.

The Auger analysis was implemented by a Riber's ASC 2000 microprobe at normal incidence of the primary electron beam with energy 3 keV. The elemental composition of the sample's surface was determined from the registered differentiated Auger-spectra. Ar^+ ion sputtering was used to measure the elemental depth distribution. A detailed description of the analysis was published in [9].

3. Results and discussion

Figure 1 shows the transmittance and reflectance spectra of a double layered $As_{32}S_{34}Se_{34}/Ag$ structure and a Ag-photodoped $As_{32}S_{34}Se_{34}$ layer.

As seen in figure 1a, prior to illumination we have a double layered structure for which the transmittance (curve 3) is below 50 % and the reflectance from the glass substrate side (i.e. from the metal layer side) is higher that that from the chalcogenide film side. Upon light exposure, photodiffusion of silver is initiated and eventually a homogenous Ag-doped chalcogenide layer is produced. The transmittance of such a layer (curve 3, figure 1b) in the interference maximum is close toh that of the substrate used and the reflectance in the spectral region of material's transparency is identical from both the film and substrate sides.

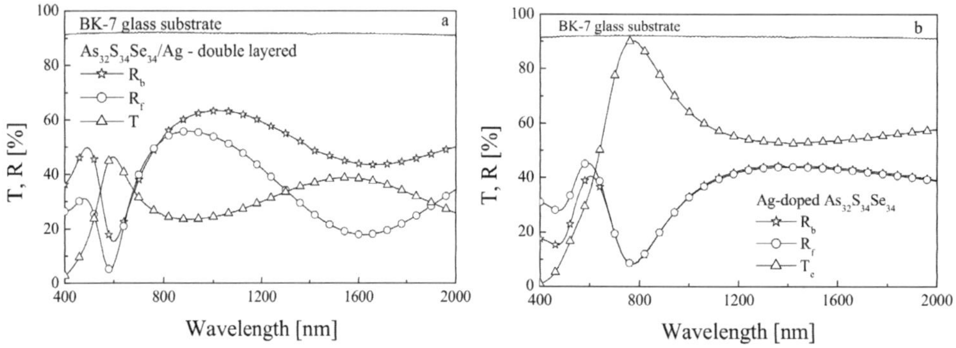

Figure 1. Transmittance (T) and reflectance (R) spectra of a double layered $As_{32}S_{34}Se_{34}/Ag$ structure prior to (a) and following irradiation (b); the subscript "b" denotes the reflectance from the glass substrate side and "f", the one from the side of the chalcogenide film.

Following the above described procedure for determination of the optical properties, the refractive index and extinction coefficient were calculated and, thence, the optical band-gap was derived. Figure 2 presents the refractive index dispersion for $As_{32}S_{34}Se_{34}$ – undoped and silver doped. Table 1 provides the exact values of d, n and E_g^{opt}.

It is seen both in figure 2 and table 1 that the refractive index of the layers of the composition studied is almost not influenced by the exposure In fact, the relative photoinduced refractive index increase (1.6 %) may be within the accuracy of the method applied for calculation of n. The presence of silver, on the other hand, could increase the refractive index by 11 %. The values of n calculated for the thin films are close to those previously published for 1 μm-layers [15]. Silver dissolution could also be related to the expansion of the film – the thickness of the photodoped layer is more than 6 % higher than that of the silver-free one. The reduction of the optical band-gap following silver photodissolution is up to 15 % and could be related to an increase in the density of the localized states. Our results are in agreement with those obtained for 0.8 μm films from the system $Ag/As_{33}S_{33.5}Se_{33.5}$ (for silver concentration of 25 at.%) published in [4].

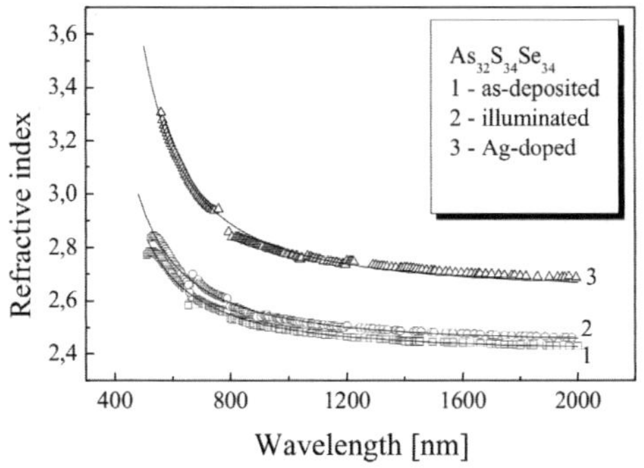

Figure 2. Refractive index dispersion of thin $As_{32}S_{34}Se_{34}$ as-deposited undoped (1), undoped exposed (2) and Ag-doped films (3); the values derived from the experimental data (represented by symbols) were fitted by the Sellmayer's equation (solid lines).

Table 1. Optical properties of thin Ag-As-S-Se films; the relative increase of the refractive index in comparison to that of the unexposed, undoped layer of the relevant composition is given in the parentheses.

Composition		d, nm	n at 1.55 μm	E_g^{opt}, eV
$As_{32}S_{34}Se_{34}$	un	125	2.44	2.0
	exp	126	2.48 (1.6 %)	2.0
	Ag	133	2.71 (11 %)	1.7

Figure 3 presents Auger profiles of the undoped and silver-doped As-S-Se layers. The AES profiles reveal uniform distribution of all the elements in the depth of both undoped and silver doped chalcogenide films (figure 3a, c). It should be mentioned, however, that the methods of Auger profiling for the layer of $As_{32}S_{34}Se_{34}$ and Ag-$As_{32}S_{34}Se_{34}$ differ from one another. It was found that the analyzing electron beam induced additional silver diffusion into the sample of $As_{32}S_{34}Se_{34}$ composition. It is seen in figure 3b that for electron bombardment time of 10 minutes the Auger signal originating from silver increases by a factor of two compared to that at the beginning of the analysis. This is known as "electron-induced chemical modification", which was reported for Ag-Ge(As)-S(Se) glasses [16]. The process could be explained by migration of positively charged silver ions from the non-irradiated parts of the sample in direction to the negatively charged irradiated region. Following these considerations, it becomes clear that the time of electron-beam interaction with matter should be reduced to less than 5 minutes and, consequently, the conventional approach of simultaneously

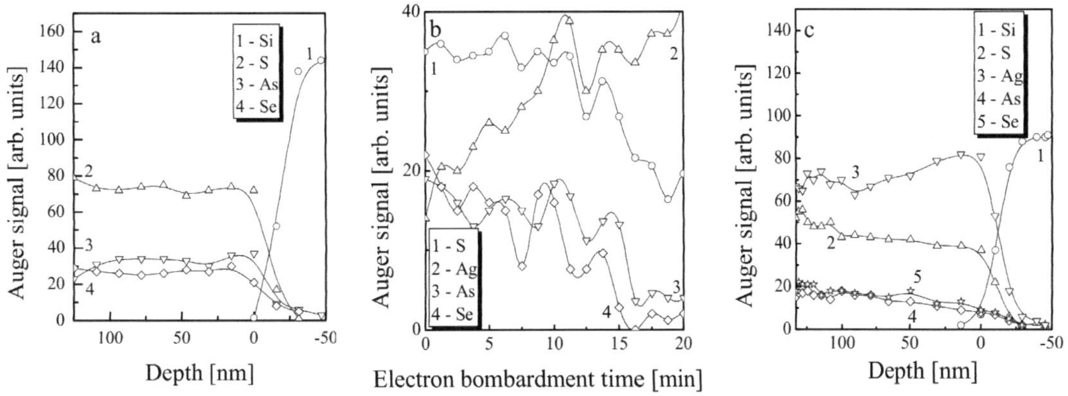

Figure 3. Auger profiles of thin undoped (a) and silver doped $As_{32}S_{34}Se_{34}$ films obtained by conventional AES analysis (a, b) and crater-edge profiling (c).

analyzing and etching the sample (which requires time of approximately 30 min) appears to be inapplicable to silver doped films. Instead, crater-edge profiling (figure 4) was utilized. The latter approach consists in etching a crater in the studied sample and successively scanning the fabricated edge by an electron beam. Thus, the irradiation duration of each spot to be scanned is reduced to less than 1.5 min and accumulation of silver is prevented. On the other hand, a distortion of the Auger results could arise from preferential sputtering of any of the elements. Following ion bombardment, an "altered layer" is produced at the surface of the film with different composition than that in the depth. Since the thickness of the altered layer is much smaller than the thickness of the whole film [17], our conclusion for homogeneity remains correct.

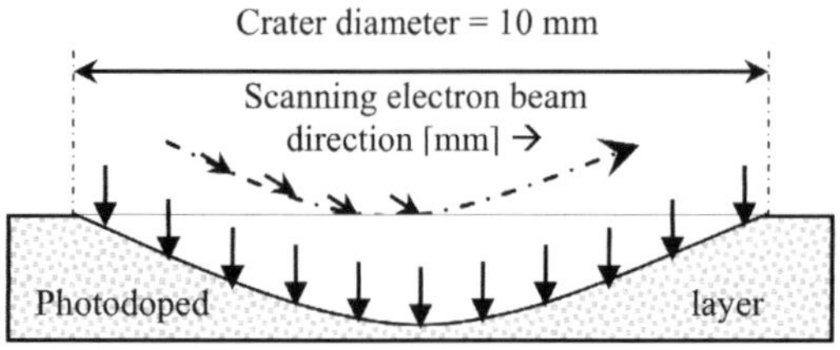

Figure 4. Crater-edge profiling scheme.

4. Conclusions

Thin films (~ 120 nm) of composition $As_{32}S_{34}Se_{34}$ were produced and photoinduced diffusion of silver was performed. The reflectance spectra reveal no perceptible inhomogeneity of the photodoped layer. It is shown that the refractive index of the latier is up to 11 % higher than that of the undoped films. Auger-electron spectroscopy profiles support our assumption that silver has been uniformly distributed in the chalcogenide films.

References

[1] Kalyva M, Siokou A, Yannopoulos S N, Wágner T, Orava J and Frumar M 2009 *J. Non-Cryst. Solids* **355** 1844
[2] Kovalskiy A, Jain H and Mitkova M 2009 *J. Non-Cryst. Solids* **355** 1924
[3] Yeo J B, Yun S D, Kim T W and Lee H Y 2008 *J. Non-Cryst. Solids* **354** 5343
[4] Krbal M, Wágner T, Vlček Mil, Vlček Mir and Frumar M 2006 *J. Non-Cryst. Solids* **352** 2662
[5] Freundlich H and Nathansohn M 1921 *Koll. Zeitschr.* **29** 16
[6] Kovalskiy A, Cech J, Tan C L, Heffner W R, Miller E, Waits C M, Dubey M, Churaman W, Vlček M and Jain H 2009 *Proc. SPIE* **7273** p 72734A
[7] Yoshida N, Harada H and Tanaka K 1997 *Solid State Ionics* **95** 323
[8] Tomova R, Spasov G, Stoycheva-Topalova R and Buroff A 2000 *J. Non-Cryst. Solids* **266-269** 985
[9] Tomova R, Spasov G, Stoycheva-Topalova R and Buroff A 2004 *J. Non-Cryst. Solids* **338-340** 582
[10] Swanepoel R 1983 *J. Phys.* E *Sci. Instrum.* **16** 1214
[11] Abeles F and Theye M 1996 *Surf. Sci.* **5** 325
[12] Tasseva J, Todorov R and Petkov K 2005 *Nanoscience & Nanotechnologies* **5** ed E Balabanova and I Dragieva (Sofia: Heron Press) p 81
[13] Konstantinov I, Babeva Tz and Kitova S 1998 *Appl. Opt.* **37** 4260
[14] Tauc J 1974 *Amorphious and liquid semiconuctors* (New York: Plenum Press)
[15] Tasseva J, Todorov R and Petkov K 2009 *J. Optoelectron. Adv. Mater.* **11** 1257
[16] Yoshida N and Tanaka K 1997 *J. Non-Cryst. Solids* **210** 119
[17] Pamler W and Röll K 1983 *Surf. Interface Anal.* **5** 105

Modulated optical reflectance method for analysis of magnetoelectric nanomaterials

V Pencheva[1], **S Penchev, I Nedkov, T Kutzarova and V Naboko**

Emil Djakov Institute of Electronics,
Bulgarian Academy of Sciences, 72 Tzarigradsko Chaussee, 1764 Sofia, Bulgaria

E-mail: vasilka@ie.bas.bg

Abstract. A recently developed technique of modulated optical reflectance and its applicability to structural and defectoscopic analysis with high spatial resolution of the ferromagnetic LSMO materials is reported. The variation of the optical reflectance is measured within the active area in the laser focus on the surface of the thin film subjected to photothermal modulation. The measurement described is indicative of the magnetoelectric properties of the sample films and is found to be proportional to the thermal variations of the conductance and free carrier density.

1. Introduction

Ferromagnetic manganites of rare earth elements with Ca, Ba, Sr and Pb, in particular, LSMO compounds with general formula $La_{1-x}Sr_xMnO_3$, have been the subject of intensive studies in recent years. They are promising for the development of a new range of electronic magnetoresistive devices owing to the effect of colossal magnetoresistance of metal-insulator transition at temperatures higher than the Curie point. Among the nondestructive methods to characterize these materials, the optical methods are distinguished by the capability to measure various physical parameters with high sensitivity without being affected by high electromagnetic fields. Layer thickness, optical indices and thermal properties determining the physical and chemical structure are assessed by phase-sensitive interferometric methods, such as ellipsometry and laser heterodyne measurement of photothermal displacement [1-3]. However, these are typically sophisticated and susceptible to random changes of the optical path causing problematic vibration and depolarization noises. Moreover, in such systems the optical properties of the film are dependent on the properties of the substrate. An appropriate alternative is to employ an analytical scheme based on the periodic modulation of the optical reflectance (MOR) induced by photothermal variations of the conductive properties of the probed surface. MOR is characterized by an additional advantage as it is applied only to the conductive films, and is thus not affected by the properties of the dielectric substrate. Being a fast, contactless and selective technique of high spatial resolution, MOR is employed successfully for characterization of YBCO thin films and yields tractable results. The intensity of the modulated reflectance of the sample surface reveals the structural inhomogeneities and the conductivity mechanism of the films. The present investigation is intended to validate the method for structural and defectoscopic analysis of LSMO manganites.

[1] To whom any correspondence should be addressed.

2. Theoretical analysis

The origin of modulated reflectance has been discussed in earlier studies of YBCO materials [4-7] where it has been attributed to carrier density modulation; this model explains well the experimental results. It was found theoretically and experimentally that MOR depends linearly on the optical conductivity scaling as a function of the carrier density. For free carriers, Drude model expresses the real part of the conductivity at optical frequency ω as a function of the carrier density and scattering rate through the dependence on the plasma frequency ω_p:

$$\sigma(\omega) = \frac{\sigma_0}{1+\omega^2\tau^2}; \; \sigma_0 = \frac{1}{\rho} = \tau\varepsilon_0\omega_p^2 = \frac{Nq^2\tau}{m^*}; \quad (1)$$

$$\omega_p^2 = \frac{Nq^2}{m^*\varepsilon_0}; \; \tau = \frac{1}{\Gamma}, \quad (2)$$

where σ_0 is the electrical conductivity at constant current; ρ is the electrical resistivity; N is the carrier density; q is the electric charge; ε_0 is the dielectric constant; τ is the free carrier lifetime; Γ is Drude scattering rate and m^* is the carrier effective mass.

A detailed consideration is given in ref. [5] where Drude equations are used to find the thermal derivative of reflectivity. The MOR coefficient scales with ω_p^2, which is, in turn, proportional to the carrier concentration. The MOR signal intensity is found to pass through a peak for sensing radiation wavelength with photon energy ~2eV, close to the photon energy of He-Ne laser radiation at 632,8 nm (1,956eV). The modulating laser wavelength does not induce any change of the probe surface except photothermal modulation at the sensing wavelength. Automodulation by the optical reflectivity of the heating wavelength is a negligible fraction of MOR signal. The photothermal modulation efficiency is determined by the sensing radiation energy characteristics that are higher towards the higher-frequency spectrum. The choice of another wavelength is to be compensated by higher power characteristics of the laser radiation that would increase the MOR signal intensity.

The complex physics of conductivity of the LSMO manganite is not fully understood yet, which applies as well to the strong MOR signal detected from these ferromagnetic nanostructures. Several mechanisms can contribute to the change in the optical properties of the manganites. A key solution of the problem is that manganites of various compositions reveal similar quasi-Drude extremal dependence on the plasma frequency and, respectively, on the free carrier density.

For a given wavelength, the following basic expression of optical reflectance can be written according to the Fresnel equations as a function of the complex refractive index [8]:

$$R(\omega) = \left|\frac{\tilde{n}(\omega)-1}{\tilde{n}(\omega)+1}\right|^2; \; \tilde{n} = n + ik. \quad (3)$$

The complex refractive index is related to the dielectric response $\tilde{\varepsilon} = \varepsilon' + j\varepsilon''$ as:

$$\varepsilon' = n^2 - k^2 = 1 + \frac{\omega_p^2\tau^2}{\omega^2\tau^2+1}; \; \varepsilon'' = 2nk = \frac{\omega_p^2\tau}{\omega(\omega^2\tau^2+1)}. \quad (4)$$

Journal of Physics: Conference Series **223** (2010) 012041 doi:10.1088/1742-6596/223/1/012041

The MOR signal proportionality to the photothermal change of carrier concentration is derived assuming the theoretical consideration given originally in [9]. The temperature dependence of the reflectance is reproduced below for reference:

$$\frac{dR}{R} = \mathrm{Re}\left[\frac{4\Delta N}{(\tilde{n}-1)(\tilde{n}+1)} \times \left(\frac{\partial n}{\partial N} + i\frac{\partial k}{\partial N}\right)\right], \tag{5}$$

where $\Delta N = \frac{\partial N}{\partial T}\Delta T$ is the change of carrier density. In the case of, YBCO equation (4) is simplified considering $n >> k$, $\varepsilon' >> \varepsilon''$, $\omega\tau >> 1$. For LSMO ferromagnetic manganites, we use the calculated parameters in [9] assuming $n >> k$, $\varepsilon' >> \varepsilon''$ in the temperature range considered (250-400K); a simplified expression for the frequency-dependent thermal derivative of the reflectance can be written after approximations:

$$\frac{dR}{R} \approx \frac{4\Delta N}{(n-1)(n+1)}\frac{\partial n}{\partial N}; \quad \frac{\partial n}{\partial N} = \frac{1}{2n}\frac{\omega_p^2\tau^2}{N}\frac{\omega^2\tau^2-1}{\left(\omega^2\tau^2+1\right)^2}. \tag{6}$$

The relative amplitude of the measured MOR signal $\Delta R/R$ is finally derived as:

$$\frac{\Delta R}{R} = K\frac{1}{N}\frac{\partial N}{\partial T}\Delta T \tag{7}$$

$$K = \frac{2}{\sqrt{1+\frac{\omega_p^2\tau^2}{\omega^2\tau^2+1}}}\frac{\omega^2\tau^2-1}{\omega^2\tau^2+1}, \tag{8}$$

where $\Delta R/R$ is proportional by the coefficient K (dielectric response factor) to the relative change in the carrier density. We assume a thermal character of the modulated reflectance, which excludes generation of free carriers by the pumping laser radiation.

3. Experiment

The MOR scheme is a modified version of the laser dual-wavelength setup employed in different optical remote sensing methods. It was introduced initially for characterization of HTS YBCO films [4, 6] and was developed further by us by implementing a flexible optical tract of fiber-optics elements [7]. It combines two laser beams of different wavelengths and power precisely aligned on the sampled surface (figure1). Their spectral difference should not be great so that the same focusing optics could be used determining the focal length and the focal spot size. MOR is based on thermal interaction and efficient spectral separation of the two wavelengths is necessary to eliminate false crosstalk signal. The active region on the sample surface is confined by the aligned focal spots of laser radiation of both wavelengths, which determines the spatial resolution of ~10 μm. The probe laser beam is directed collinearly with the heating laser beam and separated in the detector tract via a fiber-optics monochromator and high-finesse interference-filter (figure 1). The probe laser beam is coupled into the optical fibre aperture by the zero order diffraction maximum and the back-reflected beam is coupled into the photodetector through the selective first order diffraction maximum of the diffraction grating. Ihe laser source employed for photothermal modulation is a powerful laser diode of 100 mW (780 nm wavelength) square-wave modulated at a frequency of 1 kHz to modulate locally the temperature of the sample film. The driving laser pulses serve simultaneously to synchronize the lock-in amplifier with the repetition rate, which determines the frequency band of its filter. The probe beam is modulated solely by the variation of the reflection. The spectral dependence of the photothermal modulation of the probed surface is expressed in a range of maximal intensity which is determined by Drude effect [6]. The sophisticated alignment of the two laser beams with the necessary micrometer precision is replaced in our scheme by the less demanding input of the laser power into an optical fibre of larger aperture facilitating the tuning procedure. In addition, the optical fiber employed is multimode Corning 100/140 μm set reducing the phase noise due to interference in the optical tract, reaching essential modulation of the signal in uncompensated systems.

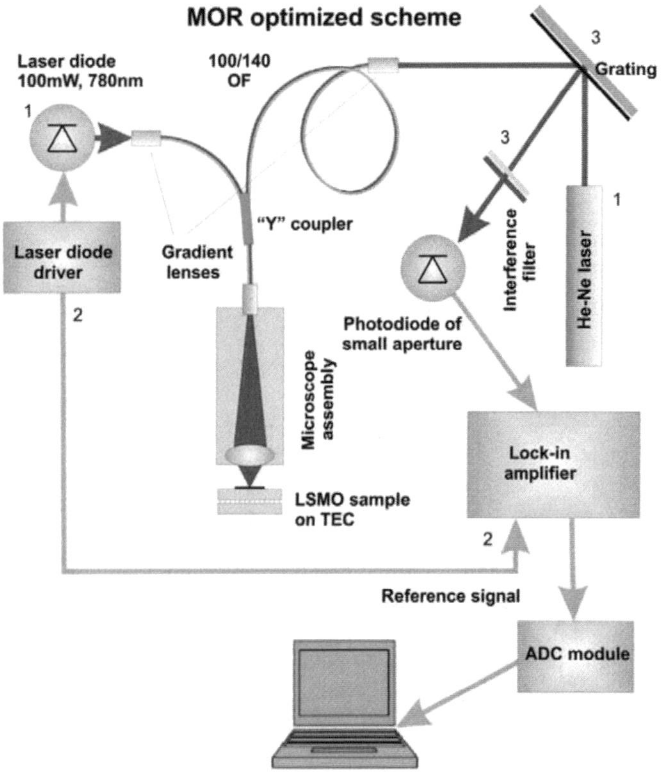

Figure 1. Outline of the experimental scheme for modulated optical reflectance of the LSMO film: (1) optical dual-wavelength setup employing two lasers of different power characteristics (a probe and a modulator); (2) synchronous modulation and detection electronics; (3) selective filter and fibre-optics monochromator of the probe (632,8nm) wavelength.

The ferromagnetic film investigated was a monocrystalline $La_{0,7}Sr_{0,3}MnO_3$ structure of 80 nm thickness grown in-situ on a (100) $LaAlO_3$ substrate. The annealing temperatures of the substrate and the film are $700C^0$ and $600C^0$ respectively [10, 11]. The applications of LSMO manganites are related to their colossal resistance of the ferromagnetic–paramagnetic phase transition which occurs at temperatures higher than the Curie point - 367K in the case of bulk materials. The film resistivity is evaluated by its integral resistance measured by a digital multimeter (Tektronix TM 506) using the four-contact method simultaneously with the MOR signal. The Peltier TEC with the sample is mounted onto a horizontal X-Y stage allowing the scans to be recorded via the ADC module in the computer.

The absolute reflectance R_0 of the sample at wavelength of 632,8 nm and temperature of 300 K is measured and the value obtained of 0,13 is in good agreement with the values reported in [8, 12] as a function of the temperature for single crystals of $La_{0,7}Sr_{0,3}MnO_3$. The temperature profiles show a relative increase of R_0 with a maximum value around the Curie point and a subsequent decrease in the paramagnetic phase. The difference ΔR with the reflectance of the substrate is negligible- 0,0035 at 300 K. The magnitudes of MOR (a few percent of R_0) obtained in the experiment are significant and we discuss below the possible origin of this high response.

Journal of Physics: Conference Series **223** (2010) 012041 doi:10.1088/1742-6596/223/1/012041

4. Results and discussion

The sample LSMO film was investigated by the MOR technique described to reveal the surface quality and the thermal properties affecting the resistivity and the magnetic phase transitions. We completed line scans across the central interface between the deposited film and the adjacent substrate plane in a confined region of 1100 μm / 3700 μm size (figure 2) which exhibit good homogeneity. The spatial resolution of the system, which is limited by the laser focal spot, is determined in this case by the steps between the measurement points of the scans - 50×100μm in both directions as indicated in the plot. It is demonstrated that the method is promising for surface imaging of sufficiently high resolution to reveal larger assemblies of imperfections.

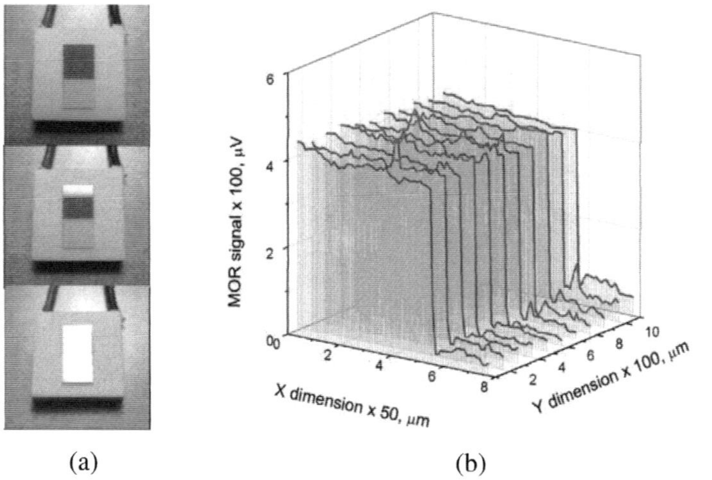

Figure 2. View of the sample film- substrate interface compared to 3D MOR signal: (a) film-substrate interface exposed to transient and reflected light; (b) MOR signal measured at the lock-in output in z-direction; scans in x-direction by 50 μm step; scans in y-direction by 100 μm step.

In the figure, the optical reflectance of the sampled region irradiated by a white light source and the relevant MOR signal are compared. It is obvious that there is lack of contrast of the unmodulated optical reflectance of the film and the substrate due to the close similarity of the optical properties of the two materials. MOR scans provide better indication of the structural differences due to the thermal modulation that separates the reflectance of the temperature dependant LSMO film from the unaffected dielectric substrate (figure 2(b)). This behavior is considered also characteristic of the explicit spatial distribution of the magnetic resistivity ρ along with the general profiles of the LSMO film plotted in the next figure.

We investigated the thermal properties of the sample up to the Curie temperature of the ferromagnetic phase transition of the LSMO sample as another important application of MOR technique. In terms of the discussed proportionality of the MOR response to conductivity, one can write the following relation for their thermal derivatives reverting to the expression of resistivity (1):

$$\frac{d}{dT}\left(\frac{\Delta R}{R}\right) \sim \frac{d\sigma}{dT};\ \frac{d\sigma}{dT} = \frac{d}{dT}\left(\frac{1}{\rho}\right) = \frac{1}{\rho}\left(\frac{1}{\rho}\frac{d\rho}{dT}\right) = \frac{1}{\rho}\beta, \tag{9}$$

where $\beta = \frac{1}{\rho}\frac{d\rho}{dT}$ is the relative thermal derivative of resistivity [8].

The experimental series of measurements in the temperature range of 260-350 K is visualized by an animated graph (Fig. 3). Considering the above, the corresponding experimental data of the thermal derivative of the MOR signal plotted in the figure against the calculated values of $\frac{\beta}{\rho}$ are compared to reveal the thermal nature of the MOR signal. The maximal value of the profiles at ~335 K is typical for LSMO compounds. The temperature shift of the MOR maximum compared to the film resistivity derivative is due to the local temperature rise of ~40K caused by the photothermal modulation in the laser focal spot. One can note the very good superposition (see animation) of the profiles of the thermal derivatives of the MOR signal and the film resistivity in the temperature range investigated. We have corrected the conclusions in [8] using expression (9) instead of the relative thermal derivative of the resistivity as a logical assumption. At the present stage of the experiment, this would remove, in our opinion, the discrepancy of the profiles at temperatures above 350K.

Animated by fax & picture viewer or internet explorer

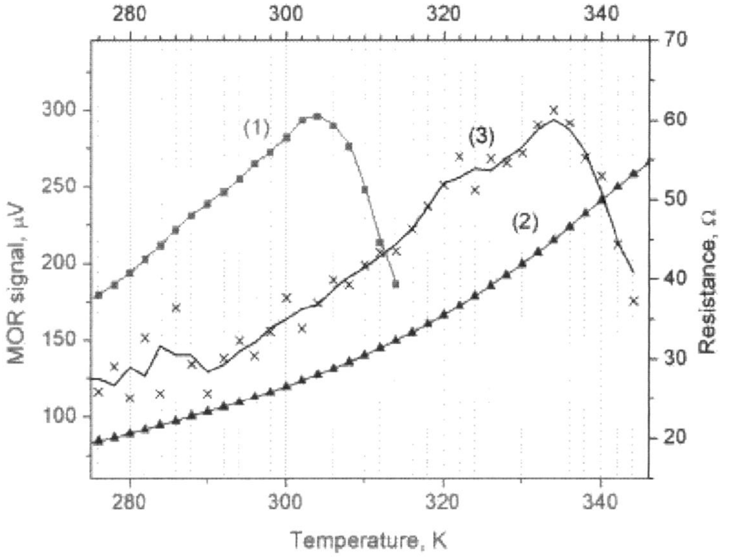

Figure 3. MOR signal revealed by the functional dependence vs. temperature. left axis (1) lock-in output; right axis- (2) resistance of the integral LSMO sample film; (3) thermal derivative of the film resistance given in arbitrary units.

5. Conclusion

In conclusion, we have demonstrated the applicability of the method of photothermal modulation of the reflectance for contactless measurements of the physical properties of highly resistive LSMO ferromagnetic nanolayers at different temperatures. Further studies will be directed to the thermal interaction close to the phase transitions of samples of these materials subjected to magnetic fields.

Acknowledgments

The work is supported by the Bulgarian National Science Fund under Thematic Project: "New magnetic and magnetoelectric materials for the next generation electronic devices"

References

[1] Sumie S, Takamatsu H, Noshimoto Y, Horiuchi T, Nakayama H, Kanata T and Nishino T 1992 *Jpn. J. Appl. Phys.* **31** 3775
[2] Tanabe T, Penchev S and Toyoda K 1994 *Lasers in Engineering* **3** 59
[3] Pencheva V, Penchev S, Naboko V, Donchev T, Kolev S and Kutzarova T 2006 *Plasma Processes and Polymers* **3** 2 253
[4] Almond D P, Barker A K, Bento A C, Singh S K, Appleyard N J, Jackson T J and Palmer S B 1995 *Supercond. Sci. Technol.* **8** 667
[5] Holcomb M J, Perry C L, Collman J P and Little W A 1996 *Phys. Rev.*B **53** 6734

[6] Almond D P, Nokrach P, Stokes E W, Porch A, Foulds S A L, Wellhöfer F, Powell J R and Abell J S 2000 *J. Appl. Phys* **87** 12 8628
[7] Pencheva V, Penchev S, Naboko V, Naboko S and Donchev T 2004 *Vacuum* **76** 253
[8] Mechin V L, Flament S, Perry A, Almond D P and Chakalov R A 2005 *J. Appl. Phys.* **98** 10 1
[9] Opsal J, Taylor M W, Smith W L and Rosencwaig A 1987 *J. Appl. Phys* **61** 1 240
[10] Nurgaliev T, Demidov V V, Petrzhik A M, Ovsyannikov G A, Miteva S and Blagoev B 2008 *J. Optoelectr.Adv. Mater.* **10** 2 273
[11] Nurgaliev T 2008 *J. Magnetism and Magnetic Materials* **320** 304
[12] Okimoto Y, Katsufuji T, Ishikawa T, Arima T and Tokura Y 1997 *Phys. Rev* B **55** 4206

Method for nondestructive testing of the film coating behavior of surface acoustic wave (SAW) sensors

M A Taslakov[1,3] and I D Avramov[2]

[1]Emil Djakov Institute of Electronics,
Bulgarian Academy of Sciences, 72 Tzarigradsko Chaussee, 1784 Sofia, Bulgaria
[2]Georgi Nadjakov Institute of Solid State Physics,
Bulgarian Academy of Sciences, 72 Tzarigradsko Chaussee, 1784 Sofia, Bulgaria

E-mail: matasel@gmail.com

Abstract. This paper presents a practical non-destructive method for studying the film coating behavior of SAW devices by using a water soluble dielectric film (manitol) deposited on the SAW device surface by resistive evaporation. After measuring the electrical parameters of the film coated SAW device, the film can easily be removed from its surface by water rinsing without causing any damage to it. The SAW device can then be used over and over again in a large number of film depositions. The method was tested on a 1 GHz surface transverse wave (STW) resonator coated with manitol of varying thickness. After each coating and evaluation, the STW device was successfully recovered without significant performance degradation. Data is presented on the electrical changes of the STW device as a result of depositing manitol coatings of various thicknesses.

1. Introduction

Surface acoustic wave (SAW) devices coated with thin chemosensitive polymer films are widely used as sensors for the detection of small amounts of various gases in the ambient atmosphere [1-3], as well for studying properties of liquids [4]. The most important part of the SAW sensor design is the optimization of the polymer film thickness on the device surface for maximum frequency sensitivity and dynamic range [5]. The practical film thickness optimization is a tedious and expensive process since it requires a large number of SAW devices which become unusable after each film deposition [6, 7]. This process is illustrated in figure 1, which compares the frequency and phase responses of a 700 MHz surface transverse wave (STW) sensor resonator before and after deposition of a 190 nm solid hexamethyldisiloxane (HMDSO) film which causes strong distortion of the device characteristics and overall performance degradation, making the SAW resonator unusable after only one single film deposition. To reduce the number of SAW devices used in the thickness optimization process, in this paper we propose a new practical method for studying the film coating behavior of SAW devices. Instead of depositing solid dielectric films like Parylene C which eventually causes the destruction of the tested device [7], we use a water soluble dielectric film (manitol) which is deposited onto the SAW device surface in a resistive evaporation process. After the measurement, the manitol film can easily be removed from the substrate surface by water rinsing thus completely recovering the SAW device

[3] To whom any correspondence should be addressed.

without performance degradation. The same SAW device can thus be used over and over again for a large number of manitol depositions with different film thicknesses.

2. Experimental setup

The experimental setup for manitol deposition is shown schematically in figure 2. The SAW device is placed in the vacuum chamber of a resistive evaporator above the tantalum crucible containing manitol crystals. By electrically heating the crucible, the manitol is evaporated at a controlled rate and a thin manitol film is deposited on the SAW device surface. Its thickness is monitored by a quartz crystal microbalance (QCM). After the film deposition, the device is taken out of the chamber and electrically characterized using a vector network analyzer. Finally, the SAW device is rinsed in deionized water, dried in dry nitrogen and is ready for the next deposition.

3. Experimental results

Figure 3 compares the narrowband frequency and phase responses of the 1 GHz STW resonator used in the manitol coating experiments. Figure 3a) shows the resonator data prior to the first deposition and figure 3b) illustrates its behavior after 5 deposition cycles and subsequent manitol removal. There is no significant change in the performance. Precise data on the insertion loss, loaded Q and center frequency prior to and after the depositions are summarized in table 1. As a result of the depositions, the device center frequency is shifted downwards by about 80 ppm and the loaded Q is degraded by 14 %. Surprisingly, the insertion loss decreases insignificantly (0.5 dB). We attribute this improvement to the cleaning of the initially contaminated device surface in the process of water rinsing.

4. Discussion

Our experiments show that manitol, due to its water solubility, seems to be appropriate for mass sensitivity studies on SAW based resonant sensors without wasting too many test devices. Typically, this material is used as a low adhesion separation layer in precise copying of fine optical surface profiles. In our opinion, however, it could also be successfully used in initial experiments on SAW based microbalances which we expect to be much

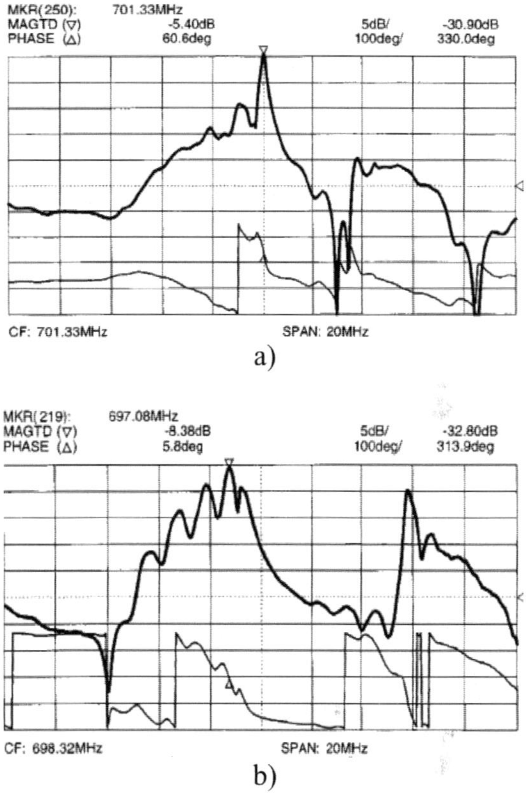

Figure 1. Frequency (upper curves) and phase (lower curves) responses of a 700 MHz STW sensor resonator (a) before and (b) after 190 nm HMDSO deposition.

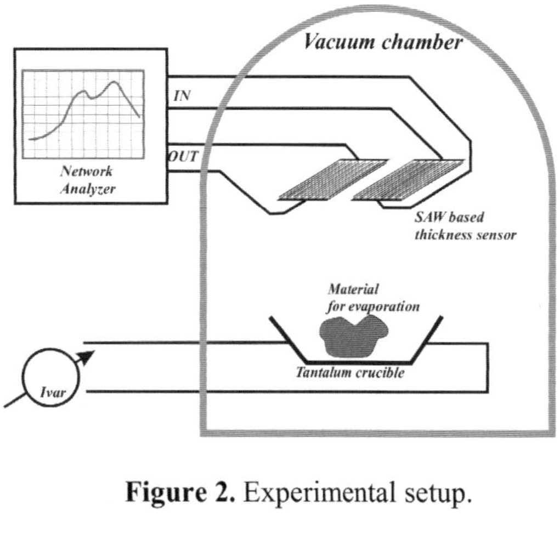

Figure 2. Experimental setup.

more sensitive than QCM, as well as for studying nonlinear slip effects at the interface between the SAW device surface and low-adhesion layers. We plan future research in this field.

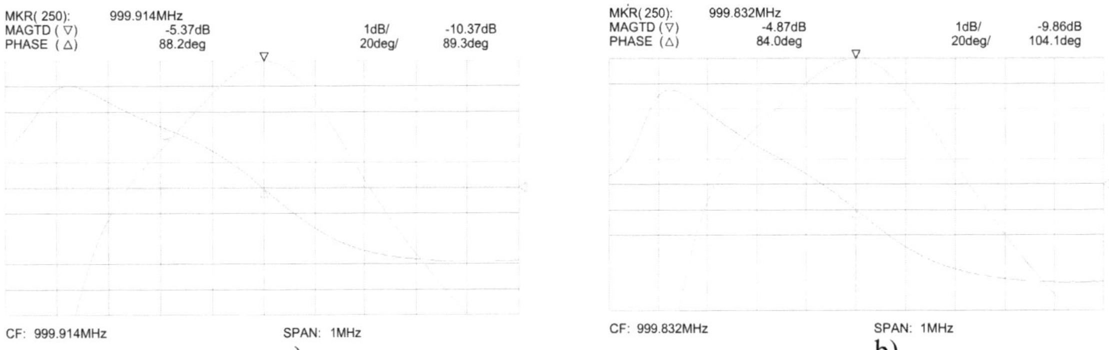

a) b)

Figure 3. Comparison of the frequency (upper curves) and phase (lower curves) responses of the test resonator (a) prior to and (b) after 5 manitol depositions.

Table 1. Electrical performance of the test STW resonator prior to and after 5 depositions.

Device data	Initial device data	After 5 depositions
Center frequency [MHz]	999.914	999.832
Loaded Q	2990	2560
Insertion loss [dB]	5.37	4.87

5. Conclusions

We have demonstrated a successful nondestructive method for mass sensitivity studies on SAW-based resonant sensor devices. By controlled resistive evaporation of manitol, which is water soluble, the SAW device can be fully recovered after the deposited film is removed by water rinsing. This allows for the SAW device to be used for multiple depositions thus saving a large number of test devices.

References

[1] Wohltjen H and Dessy R 1979 Surface acoustic wave probe for chemical analysis Parts I–III *Anal. Chem.* **5** 1458–75
[2] Frye G C and Martin S J 1991 Dual output acoustic wave sensor for molecular identification *Proc. IEEE Transducers* pp pp 566–9
[3] Grate J W, Martin J S and White R M 1993 Acoustic wave microsensors *Anal. Chem.* **65**, 940A–8A, 987A–96A
[4] Shiokawa S and Kondoh J 1999 Surface acoustic wave sensor for liquid phase detection *Proc. 1999 IEEE Ultrasonics Symp.* (Lake Tahoe NV USA) pp 446-52
[5] Avramov I D, Rapp M, Kurosawa S, Krawczak P and Radeva E 2002 Gas sensitivity comparison of polymer coated SAW and STW resonators operating at the same acoustic wave length *IEEE Sensors J.* **2**/3 150-9
[6] McGill R A, Chrisey D B, Mlsna T E, Stepnowski J L, Chung R and Cotal H 1997 Performance optimization of surface acoustic wave chemical sensors *Proc. IEEE Int. Freq. Contr. Symp.* (Orlando FL USA) pp 140–6
[7] Avramov I D, Laenge K, Rupp S, Rapp B and Rapp M 2007 Polymer coating behavior of Rayleigh SAW resonators with gold electrode structure for gas sensor applications *IEEE Transactions on Ultrasonics, Ferroelectrics and Frequency Control* **54**/1 157-66

Density-functional study of the crystallographic structure of chromium nitride films

V Antonov[1,3] and I Iordanova[2]

[1]Institute for Nuclear Research and Nuclear Energy,
Bulgarian Academy of Sciences, 72 Tzarigradsko Chaussee, 1784 Sofia, Bulgaria
[2]Department of Solid State Physics and Microelectronics, Faculty of Physics,
St. Kl. Ohridski University of Sofia, 5 J. Bouchier Blvd., 1164 Sofia, Bulgaria

E-mail: vladiant@inrne.bas.bg

Abstract. First principles calculations within the framework of the density functional theory with ultrasoft pseudopotentials and plane-wave wavefunctions basis set were performed to investigate the electronic and crystallographic structure of chromium nitride. Different types of antiferromagnetic orderings were considered in the calculations. Three crystallographic planes, namely {100}, {110} and {111}, most commonly observed during the growth process of CrN layers, were investigated assuming $AFM1_{[110]}$ magnetic ordering. The corresponding surface energies and magnetic structures were simulated.

1. Introduction

Over the last decades, hard and corrosion resistant coatings have frequently been used to protect machine parts and tools and to extend their service life. Transition metal nitrides and among them chromium nitride (CrN) are between the most common materials for these applications. CrN is a hard coating material with excellent wear, corrosion and oxidation resistance. The coatings with potential for industrial applications are mainly deposited by magnetron sputtering. In view of optimizing the technologies and extending the practical applications, the CrN films have been widely analyzed by the existing and by newly developed modern techniques.

The Density Functional Theory (DFT) [1] has recently been implemented in relatively simple computational codes which may help in the characterization of structure formation processes on an atomic level and of important coating parameters. These codes have widely been used for the first-principles calculations of surface reconstruction and surface parameters which may significantly influence the properties [2].

In the present study a method based on the quantum mechanical DFT wasw developed and applied to the evaluation of surface energies of CrN films as a function of their crystallographic orientations.

2. Theoretical background

The atomic arrangement in perfect crystals can be described by a periodically repeated unit cell, represented in the DFT by periodic boundary conditions. However, in many physical systems, the perfect periodicity is violated and the system is either approximately periodic or periodic only in one

[3] To whom any correspondence should be addressed.

or two directions. Examples of such systems include surfaces, point defects in crystals, substitutional alloys, heterostructures, liquid crystals, etc. These systems may be successfully simulated by a periodically repeated fictitious supercell with a shape and size depending on the system studied. In the case of solid surfaces, a crystal slab alternated with a slab of empty space is used. Both slabs are large enough to ensure that the bulk behavior is recovered inside the crystal slab. The surface behavior should be unaffected by the presence of the periodical boundary conditions. Due to the absence of an abrupt termination, the supercell approach is usually more convenient than the simulation of an extended system by taking a finite piece of the material. During the study of the slab, a structural relaxation should be performed in order to reflect the effects of surface reconstruction due to the abrupt change of the density on the surface. This process corresponds to a convergence towards the atomic configuration with a minimum energy. The lowest-energy structure produced with the aid of DFT will be the equilibrium structure at zero temperature and zero pressure.

In our earlier analyses of CrN with DFT [3], we found that in order to obtain a stable crystallographic structure, antiferromagnetic ordering along the [110] crystallographic direction should be introduced. The NaCl type crystallographic structure of CrN observed at room temperature can be successfully reproduced by a number of antiferromagnetic layers with the same spin orientation, which alternate their spin every next layer [3]. This spin arrangement is known as $AFM1_{[110]}$ and was used in the present study.

3. Computational procedure

The calculations were performed using ultrasoft pseudopotentials and a plane wave basis as implemented in the Quantum Espresso (QE) integrated suite of computer codes [4]. The generalized gradient approximation (GGA) of Perdew, Burke and M. Ernzerhof (PBE) [5] was used. The applied pseudopotentials with QE were ultrasoft, generated by the Vanderbilt scheme [6] and taken from the QE distribution [4]. The wave functions at each k-point are represented by the numerical coefficients of a finite set of plane waves, determined by a kinetic energy cut-off. The k-point sets were generated automatically, following the Monkhorst - Pack (MP) scheme [7]. To decrease the number of k-points needed to sample the Brillouin zone, the crystallographic symmetry of the samples was considered. The computation of interatomic forces was performed by the Hellmann Feynman theorem [8]. The initial calculations started with convergence studies to determine the kinetic energy cut-off and the size of the MP grid. Iterations using self-consistent field (SCF) cycles were performed until the difference of the total energy was less than 10^{-5} eV. The forces in the unit cell and the lattice vectors were relaxed using ground state structural optimization via the Broyden algorithm as implemented in the QE. Calculations were performed until the interatomic force became less than 7×10^{-3} eV/Å and the cell pressure (for the lattice constant calculations) was less than 0.05 kbar.

The film surfaces were simulated with a supercell containing CrN layers and vacuum space using the calculated relaxed lattice constant. The characterization of the surface properties started with convergence studies with respect to the slab and vacuum thickness. The latter was calculated from the convergence studies performed until the error became less than 0.1% for the cases when {100}, {110} or {111} crystallographic planes are lying in the surfaces. During the calculations, the angle between the [110] direction, along which the spins are aligned, and the directions of the surface normals [100], [110] and [111] respectively, were taken into account. For the case of the {100} lying on the surface, the possible angles are 45^0 (supercell system noted as {100}a) and 90^0 (supercell system noted as {100}b). When the crystallographic orientation of the surface is {110} the angles are 0^0 (supercell system noted as {110}a), 90^0 (supercell system noted as {110}b) and 60^0 (supercell system noted as {110}c). For the case of (111) the possible angles are 90^0 (supercell system noted as {111}a) and 32.27^0 (supercell system noted as {111}b). The numerical experiments were performed both for spin orientation fixed along [110] and with non-collinear relaxation to estimate the changes of the magnetic structure due to surface effects. Because of the requirements for the accommodation of the magnetic structure, the calculations for all surface orientations were limited to six layers only.

4. Results and discussion

The convergence studies showed that the required cut-off energy for CrN is 748 eV. To reflect the metallic band structure of CrN, electron level smearing via the scheme of Methfessel-Paxton (MP) [9] with smearing width of 0.27eV was introduced. The MP grid with size 6×6×6 is sufficient for determination of the lattice constant with a relative error of less than 1%. The calculated value of the CrN lattice constant is 4.140 Å, which is equal to the experimental data taken from the X-Ray JCPDS No 11-0065 data file. In order to reflect the unit cell size and the vacuum layer, the MP grid used for the surfaces with different crystallographic orientations were as follows: for {100}a – 14×1×14 and {100}b – 1×10×10; for {110}a – 14×10×1, {110}b – 1×10×14 and {110}c – 7×1×10; for {111}a – 7×7×1, and for {111}b – 7×7×1. The thickness of the vacuum slab was calculated from the convergence studies performed until the error was less than 0.1% and thus it was defined to be 12 Å. In the two layers in the middle of the CrN slab the atoms are kept fixed, while the others was allowed to relax free until the force acting on them was minimized below the given threshold. The analysis of the atom positions after surface relaxation was performed considering the fact that the {100} and {110} planes contain both Cr and N atoms, while the {111} planes are built of Cr or N atoms only [10]. The surface energy of the layers was calculated by subtracting the energy of the slab from the energy of the bulk (supercell without vacuum slab) for the same number of atoms and dividing the result by 2 to reflect the fact that each supercell has two surfaces of the same kind.

The results from the non-collinear calculations showed no change in the spin direction, so it can be concluded that for all the cases the spins stay collinear with the [110] direction. This was used in further calculations assuming a collinear magnetization. The results obtained are given in the table 1.

Table 1. Relative change of the surface atoms positions along the surface normal (in % of the lattice parameter) and surface energy values (in J/m^2).

	{100}a	{100}b	{110}a	{110}b	{110}c	{111}a	{111}b
Cr	2.96	0.92	2.92	2.92	3.12	1.71	1.70
N	-1.04*	-0.2	0.85	0.84	0.43	2.99	3.07
Surface energy:							
relaxed	1.09	1.26	2.16	2.16	2.17	2.52	2.52
non relaxed	1.30	1.32	2.53	2.53	2.49	3.88	3.88

* The minus sign indicates outward displacement.

The results show that for both relaxed and non relaxed structures the surface energies in CrN are maximum for the {111} crystallographic orientation of the surface while for the {100} orientation they are minimal. For the two types of the {111} surface orientations there is no difference in the surface energy values but it is not so for the different types of {100} and {110} orientations. From a thermodynamic point of view, it can be expected that the orientations with a lower energy will be more stable.

The relaxation leading to a decrease of the surface energy is accompanied with an inward and outward displacement of the surface atoms. According to the results in the Table 1, the inward displacement is predominating while the outward displacement is observed only for the N atoms placed in {100} planes. The displacements are the highest for the Cr atoms lying in the {110} and for N atoms lying in the {111} surface planes. These results correlate with the higher degree of relaxation of the surface energy observed in these orientations in comparison with {100}. Thus, for the discussed {111} surface orientations, the decrease of the surface energy due to relaxation is 35%, while for the {110} and {100} it varies from 13% to15% and from 5% to16%, respectively. The results discussed are in a qualitative agreement with the data in [2] obtained by similar analyses performed on TiN surfaces.

Journal of Physics: Conference Series **223** (2010) 012043 doi:10.1088/1742-6596/223/1/012043

5. Conclusions

Computer simulations based on the DFT method made it possible to analyze CrN solid surfaces with {100}, {110} and {100} crystallographic orientations.

Successful simulation of the crystallographic structure of CrN films needs antiferromagnetic ordering of the spins along [110] direction.

The surface energies depend on the type of crystallographic planes and their orientation with respect to the direction of antiferromagnetic ordering. They were evaluated and the relaxation processes accompanied with surface atoms displacements and surface energy decrease were analyzed: the relaxation is the most pronounced for CrN surfaces with {111} orientation which has the highest initial energy and the lowest one for the {100} surfaces with the minimal initial surface energy. After the relaxation, the energy of the {111} surfaces is preserved maximal, while the {100} stays minimal.

Acknowledegement

Vladislav Vihrov Antonov acknowledges support from the European Operational program HRD through contract BGO051PO001/07/3.3-02/53 with the Bulgarian Ministry of Education and Science. The authors would also like to express their gratitude for the CPU time granted on the PHYSON computer cluster operating under NSF grant VU-205-2006 project Modeling and analysis of physical systems, the Scientific funds of Sofia University "St Kliment Ohridski" (proj.149/2009) and the National Scientific Fund at the Ministry of Education and Science, Bulgaria (project DO-224/2008).

References

[1] Kohn W 1999 *Rev. Mod. Phys.* **71** 1253
[2] Marlo M and Milman V 2000 *Phys. Rev.* B **62** 2899
[3] Antonov V and Iordanova I 2009 *AIP Conf. Proc.* **1203** 1149
[4] Giannozzi P et al http://www.quantum-espresso.org
[5] Perdew J P, Burke K and Ernzerhof M 1996 *Phys. Rev. Lett.* **77** 3865
[6] Vanderbilt D 1990 *Phys. Rev.* B **41** 7892
[7] Monkhorst H and Pack J 1976 *Phys. Rev.* B **13** 5188
[8] Feynman R 1939 *Phys. Rev* **56** 340
[9] Methfessel M and Paxton A T 1989 *Phys. Rev.* B **40** 3616
[10] Iordanova I, Kelly P J, Mirchev R and Antonov V 2007 *Vacuum* **81** 830

Journal of Physics: Conference Series **223** (2010) 012044 doi:10.1088/1742-6596/223/1/012044

Study of the long-range proximity effect in LSMO/YBCO bilayers

V Štrbík[1,3], Š Beňačka[1], E Mateev[2], B Blagoev[2], T Nurgaliev[2] and S Miteva[2]

[1]Institute of Electrical Engineering,
Slovak Academy of Sciences, 9 Dúbravská cesta, 841 04 Bratislava, Slovak Republic
[2]Emil Djakov Institute of Electronics,
Bulgarian Academy of Sciences, 72 Tzarigradsko Chaussee, 1784 Sofia, Bulgaria

E-mail: elekstrb@savba.sk

Abstract. Superconductor-ferromagnet thin film heterostructures in lateral geometry were prepared to study the long-range proximity effect. The high temperature superconductor $YBa_2Cu_3O_x$ (YBCO) as superconductive electrode and $La_{0.67}Sr_{0.33}MnO_3$ (LSMO) as ferromagnetic separation were used. The length of the ferromagnetic layer between two YBCO microstrip electrodes was about 3 μm. The temperature dependence of the resistance of these structures was measured. The results obtained showed that the separation of the superconducting YBCO microstrips is too large for observation of a long-range proximity effect; further optimization of the structures is, therefore, needed.

1. Introduction

The interplay between superconductive and spin-polarized materials has potential applications in spin cryoelectronics. The perovskite half metals, which are completely spin polarized in one spin direction, are particularly interesting because of their ability to form high-quality heterostructures with high-Tc cuprate superconductors. Electron transport through contact of superconductor (S) with normal (N) or ferromagnet (F) metal is mediated by Andreev reflection, the process in which an electron with energy less than the energy gap of S is reflected as a hole with phase coherence preserved over some distance from the interface [1]. There are two essential features of the proximity effect in FS structures, which make it different from that in SN structures. In SN structures the penetration depth of Cooper pairs (CP) into the N metal is determined by the N metal coherence length $\xi_N=(\hbar D_N/2\pi k_B T)^{1/2}$, where D_N is the diffusion coefficient of N, T is temperature, and $\hbar$ and k_B are the Planck and Boltzmann constants. In case of a SF bilayer, the penetration depth of CP into a ferromagnet is much shorter $\xi_F = (\hbar D_F/2\pi E_{ex})^{1/2}$ provided the magnetic exchange energy $E_{ex} >> k_B T$, or the Thouless energy $E_T=D/L^2$ (L is the length of F) are rather large, which is fulfilled in the case of perovskite (half metal) materials. Despite the short coherence length ξ_F in the F material, there is a possibility of a long-range proximity effect (LRPE) in FS structures if inhomogeneous magnetization is present in the vicinity of the SF interface. It has been shown [2] that, besides the ordinary singlet spin structure of CP (↑↓), the inhomogeneity (e.g. a domain wall) generates a triplet spin structure with amplitude comparable to the singlet one containing phase correlations between electrons with the same spin projections of CP (↑↑).

[3] To whom any correspondence should be addressed.

The penetration depth ξ^*_F of a triplet CP, in the case of LRPE, should be in the order of ξ_N. Thus, the research attention has also been focused on investigating Josephson effects in SFS structures because of the much larger variety of Josephson junction properties realized in combinations of these materials [3].

The magnetic inhomogeneity (domain wall, spin active interface) necessary for the LRPE to arise must be localized, in the case of serial SF geometry, immediately at the SF interface, which is experimentally very difficult to realize; the triplet amplitude is otherwise extremely small. It was found recently [4] that the triplet Andreev reflection amplitude may, in the so called lateral geometry (figure 1), be enhanced by the factor l_d/d in comparison with the serial geometry, where l_d is the domain-wall width and d is the thickness of the F film. In addition, in lateral geometry the LRPE should be independent from the domain-wall location with respect to the SF interface, which is a very significant simplification of the experimental situation. The present contribution is an experimental attempt to verify the theoretical result in [4].

2. Experimental

RF and DC magnetron sputtering systems were used to grow F (LSMO) and S (YBCO) bilayer heterostructures. The polycrystalline $La_{0.7}Sr_{0.3}MnO_3$ thin films were directly deposited by RF off-axis single magnetron sputtering on $LaAlO_3$ single crystal substrates. During the sputtering in $Ar:O_2$ (1:1) atmosphere with total pressure of 5.3 Pa, magnetron RF power of 30 W and substrate temperature of 780°C were used. The thickness of the LSMO layer was about 40 nm [5]. DC off-axis double magnetron sputtering was used for deposition of YBCO films on the top of the LSMO layer. The film deposition was carried out at a substrate temperature of T_{dep} = 780°C and the optimal annealing temperature was T_A = 530°C. The thickness of the YBCO layer was 70 nm [6].

Patterning of the structures was carried out using optical photolithography and wet etching in a 0.1% solution of HNO_3. The lateral cross-section of the structure is shown in figure 1; figure 2 is a potograph of a real SFS sample with three YBCO microstrips with widths of 15, 10 and 5 μm and separation L of 3 μm. The electrical current flows from one part of the YBCO strips to the other through the ferromagnetic LSMO layer.

Before patterning, the bilayer SF heterostructure was investigated by the dc four-point technique and by a contactless inductive method. After patterning the lateral geometry of the structure, the dc technique was used to study the transport properties. X-ray diffraction (XRD) was used to characterize the growth of both films.

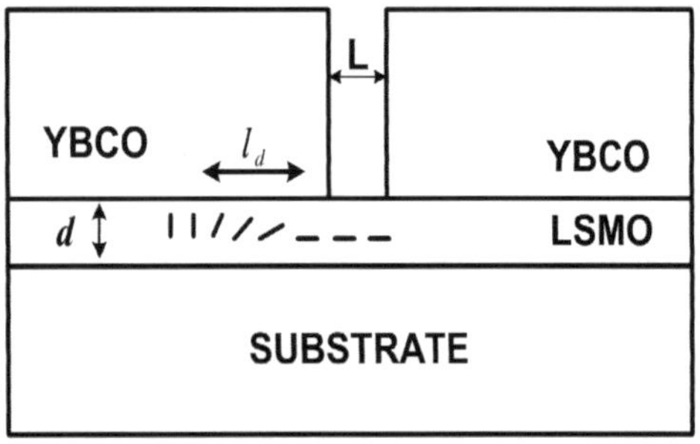

Figure 1. Cross-section of the sample used to study the long-range proximity effect. The domain-wall width l_d and the YBCO separation L by a LSMO layer of thickness d are shown.

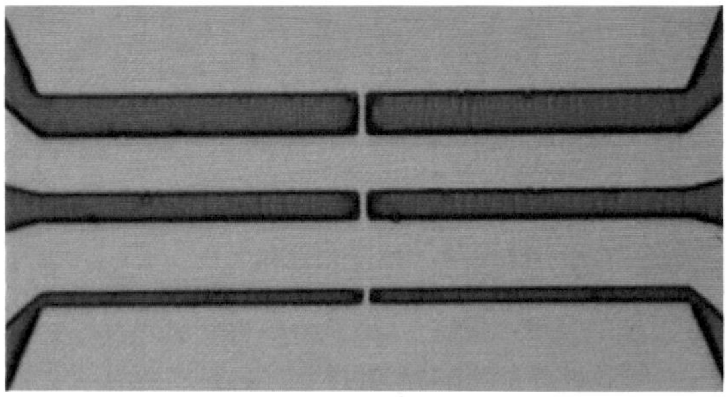

Figure 2. Experimental lateral structure consisting of three YBCO strips (wide 15, 10, and 5 μm) with 3 μm long separation.

3. Results and discussion

The XRD spectrum of the FS heterostructure (figure 3) shows preferential *c*-axis growth of both LSMO and YBCO films on the $LaAlO_3$ single-crystal substrate. The corresponding LSMO lattice parameter is $a = 0.391$ nm which matches very well the YBCO lattice parameters.

Resistance vs temperature dependences of both an as-prepared YBCO film and a lateral proximity structure (5 μm wide strip) are shown in figure 4. While the zero-resistance critical temperature T_{C0} of the as-prepared YBCO film is 86 K (figure 4, squares), the T_{C0} of the bilayer structure is below 77 K. The transition width of this structure is significantly larger and a low resistance tail appears. This is confirmed by the similar results obtained by inductive measurement (figure 5). The non-zero resistance of the lateral microstrip geometry was constant below 80 K, indicating absence of LRPE, apparently due to the large YBCO microstrips separation ($L \cong 3$ μm).

The Cooper pairs transparency between the S and F films is too small, and/or the superconductivity is suppressed (figures 4 and 5), probably due to the ex-situ deposition of the YBCO/LSMO bilayers.

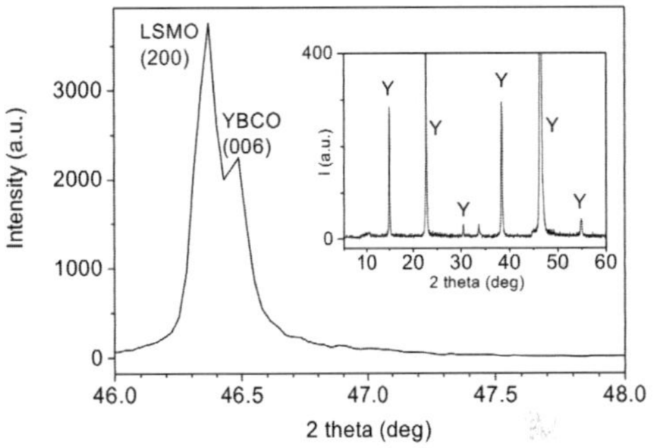

Figure 3. Peaks in XRD diagram showing the reflections of both materials. Other peaks belonging to the YBCO film (Y) are shown in the inset.

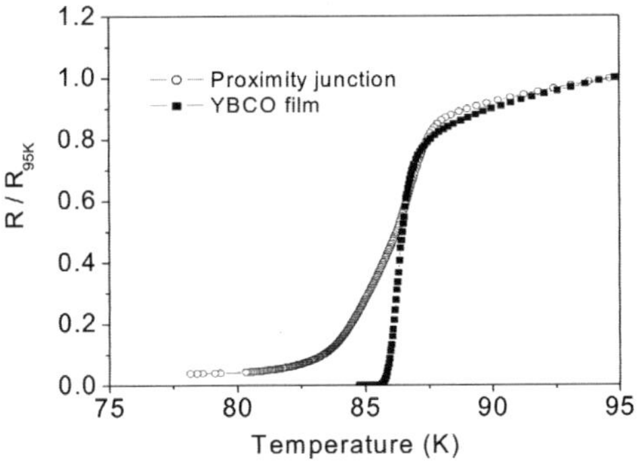

Figure 4. Superconducting transition curves for a YBCO film (squares) and a lateral YBCO/LSMO bilayer strip (circles).

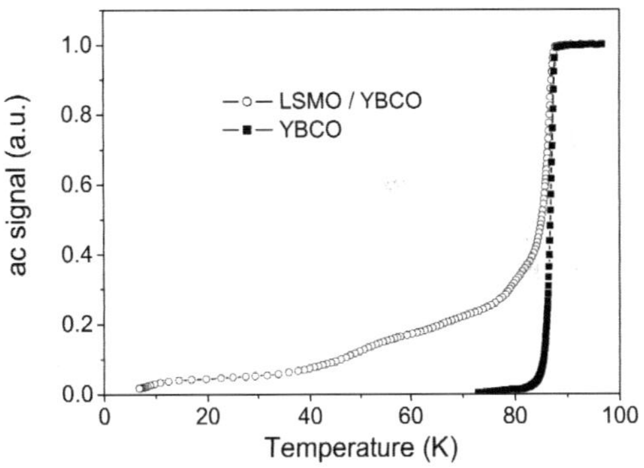

Figure 5. Inductive measurements of a single YBCO layer (squares) and a LSMO/YBCO bilayer (circles). Some degradation (probably at the FS interface) is evident.

4. Conclusions

In the case of a successful implementation of bilayer structures, and to confirm the theoretical analyses [4], i.e. the presence of LRPE in the LSMO separating YBCO microstrips, the resistance R should be zero at temperature $T < T^*_C$, where T^*_C is the onset of a constant resistance of the sample. The resistance of our samples $R \approx 9\ \Omega$ was constant below about 80K which allows us to draw several conclusions concerning the samples and the effect investigated:

- the length $L \cong 3$μm of the LSMO between YBCO thin film microstrips is very large in comparison to the penetration depth ξ^*_F of triplet correlations;
- no magnetic inhomogeneity in the LSMO thin film is present near the lateral SF interface;
- the Cooper pairs transparency between the S and F films is too small, and/or the superconductivity is suppressed, probably due to the ex-situ deposition of the YBCO/LSMO bilayers;
- the quality of the very thin ferromagnetic LSMO layer deteriorated during the samples preparation.

In order to perform successful future studies, new samples are in preparation to eliminate the above mentioned problems using both electron-beam lithography and optimized ion-beam etching in view of achieving superconducting films separation $L < 1$ μm and larger transparency of the SF interfaces.

Acknowledgements
This work was supported by the Slovak Grant Agency VEGA under project No. 2/7125/27 and by the Ministry of Education and Science of Republic of Bulgaria under project F1503/05.

References

[1] Bergeret F S, Volkov A F and Efetov K B 2005 *Rev. Mod. Phys.* **77** 1321
[2] Eschrig M and Lofwander T 2008 *Nature Phys.* **4** 138
[3] Haltermann K and Valls O T 2009 *Phys. Rev.* B **80** 104502
[4] Kupferschmidt J N and Brouwer P W 2009 *Phys. Rev.* B **80** 214537
[5] Blagoev B, Mateev E, Strbik V and Nurgaliev T 2009 *Abstract Book 16th Int. Summer School VEIT 2009* (Sunny Beach Bulgaria)
[6] Nurgaliev T, Mateev E, Miteva S, Blagoev B, Strbik V, Uspenskaya L S, Benacka S and Chromik S 2008 *Proc. 10th Workshop Nanoscience & Nanotechnology* (Sofia, Bulgaria)

Antireflection coatings for GaAs solar cell applications

Z I Alexieva[1,3], Z S Nenova[1], V S Bakardjieva[1], M M Milanova[2] and Hr M Dikov[1]

[1]Central Laboratory of Solar Energy and New Energy Sources, Bulgarian Academy of Sciences, 72 Tzarigradsko Chaussee, 1784 Sofia, Bulgaria
[2]Central Laboratory of Applied Physics, Bulgarian Academy of Sciences, 59 St Petersrburg Blvd, 4000 Plovdiv, Bulgaria

E-mail: alexieva@phys.bas.bg

Abstract A double-layer structure of Al_2O_3 over ZrO_2 film is studied. Minimization of the average weighted reflectance is carried out to optimize the thickness of the two layers in the antireflection coating. An optimal value of 2.17% for the weighted average reflection is estimated. The optimal thicknesses of the layers are 49 nm for the bottom and 45 nm for the top layer. Low temperature spin coating technique is used to deposit ZrO_2 and Al_2O_3 films from sol gel solutions on polished silicon wafers, GaAs multilayer heterostructures and AlGaAs/GaAs solar cells. The density of the short-circuit photocurrent increases from 25 $mA.cm^{-2}$ for solar cells without an antireflection coating to 36 $mA.cm^{-2}$ for those with a double layer coating.

1. Introduction

Fabrication of high efficiency solar cells (SC) requires a reduction of the incident light radiation loss through decreasing the solar cell surface reflection. For this purpose, different types of transparent thin film coatings are deposited on the solar cell with appropriate refractive indices and thicknesses – antireflection coatings (ARC). As a result, the reflection losses are minimized, the number of incident photons actually reaching the active areas of the SC is maximized, and the short-circuit current is increased (J_{sc}). Unlike many other optoelectronic devices, solar cells operate in a broad range of wavelengths, which means that they need a broadband ARC.

The choice of a suitable layer for ARC is very important for the production of high efficiency solar cells. However, it has been demonstrated that single layer coatings are not sufficiently effective - they lead to reflectivity reduction in a narrow spectral range only [1]. A further reflectivity decrease could be achieved using a double-layer ARC with high and low refractive indices respectively. In addition, the optical performance of these coatings is less sensitive to variations in the layer thickness [2].

A widely used double layer ARC for heteroface AlGaAs/GaAs solar cells is the MgF_2/ZnS structure with refractive indices 1.38 and 2.32 respectively and a transmission range 400 –1400 nm. Suitable materials with high refractive indices (2 - 2.4) are also ZrO_2, HfO_2, TiO_2. These layers are stable and durable and are used in combination with low refractive index layers, such as SiO_2 or Al_2O_3. Materials with higher refractive index are preferred for the bottom layer of the ARC and materials with lower refractive index, for the upper layer [3].

[3] To whom any correspondence should be addressed.

In this study we consider an alternative set of coatings with Al_2O_3/ZrO_2 films transparent in the range 400 – 900 nm [4]. The layers are deposited from sol solutions by the spin coating – an inexpensive technique with fairly reproducible results.

Optimization was carried out with respect to the thicknesses of the two layers in a double-layer Al_2O_3/ZrO_2 ARC on a GaAs substrate covered with a thin $Al_{0.8}Ga_{0.2}As$ a window layer. Subsequently, layers with optimal thicknesses were deposited on AlGaAs/GaAs solar cells. A relatively good agreement was obtained between the calculated and the experimental reflection data. The short-circuit current of the solar cells with the optimized ARC increased as compared to that of cells without ARC.

2. Experimental

2.1. Solar cell fabrication

The solar cell was fabricated on an A^3B^5 multilayer heterostructure grown by liquid phase epitaxy. It consisted of a low resistivity n^+-GaAs:Si(100) substrate; an n-AlGaAs buffer layer (back surface field - BSF); an n - GaAs: Te base; a p - GaAs:Mg emitter; a p-AlGaAs:Mg window; a p^+- GaAs:Mg contact layer; front and back Sn/Ni/Al multi-metal contacts; an antireflection coating Al_2O_3/ZrO_2. Figure 1 presents a cross-section of a AlGaAs/GaAs heteroface solar cell with a cap layer for contact formation.

The characteristics of the solar cell were evaluated before and after ARC deposition. The external quantum efficiency (EQE), the reflection from the surface of the solar cell, the short-circuit current (J_{sc}), and the open circuit voltage (V_{oc}) were measured. The EQE measurements were carried out in the spectral range of 340 - 920 nm. The photoresponse in absolute units was evaluated by comparing the photocurrent magnitude of a calibrated reference cell (NASA Lewis Research Center) and the structure under test at monochromatic illumination. The J_{sc} and V_{oc} were measured under exposure by a halogen lamp with intensity of 100 mW/cm^2.

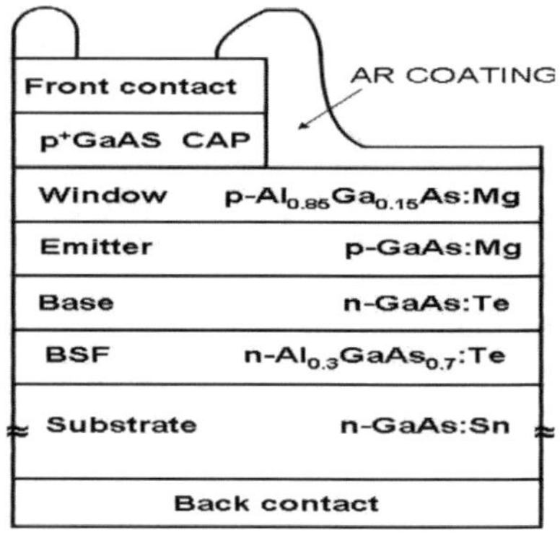

Figure 1. Cross-sectional view of a single heterojunction solar cell based on a AlGaAs/GaAs multilayer heterostructure.

2.2. ARC deposition

The layers were deposited from sol solutions by spin coating for 20sec. The spinning rate was varied in the range 6000 - 8000 rpm depending on the solution used. Al_2O_3 was deposited from an aluminum sol (aluminum butilate), followed by low temperature annealing at 400°C for 15 min in N_2 ambient to removing the organic components [5]. ZrO_2 was deposited from tri–isopropoxide [6]. Substrate heating during the coating process improved the quality of the layers deposited.

Single and double layers of ZrO_2, Al_2O_3 and Al_2O_3/ZrO_2 were deposited on three types of substrates: 1 - polished silicon wafers, 2 - GaAs substrates with a multilayer heterostructure and 3 – solar cells on the basis of these structures. The layers on polished silicon wafers were used to determine the refractive indices using a He-Ne laser ellipsometer at 632.8 nm (LEF-3M, produced by the Russian Academy of Sciences, Novosibirsk). The reflectance of all sample types was measured by a Shimadzu UV-VIS-IR 3600 Spectrophotometer in the range 400 - 900 nm with an attachment for specular reflectance measurement at 5° angle of incidence.

2.3. Calculation of the reflectivity; minimization of the average weighted reflection

We modelled the reflectivity at normal incidence the structure $Al_xGa_{1-x}As$ layer on a GaAs substrate with a double-layer of Al_2O_3/ZrO_2. The thickness of the window layer (30 nm) and its aluminium content (0.8) were determined by anodic oxidation and Raman spectroscopy [7]. The complex refractive indices of GaAs and $Al_{0.8}Ga_{0.2}As$ for the range 400 – 900 nm were taken from Aspnes et al. [8]

Journal of Physics: Conference Series **223** (2010) 012045 doi:10.1088/1742-6596/223/1/012045

and spline interpolated. The transfer matrix method of Abeles [9] was employed to determine the reflectance of the system, as this method can be used for analysis of multi-layer systems with varying thickness, refractive indices (n) and extinction coefficients (k) on a substrate.

Since the reflectivity varies with the wavelength, we used the average weighted reflectance to assess the efficiency of the system modeled [1]:

$$Rw = \frac{\int_{\lambda\,min}^{\lambda\,max} R(\lambda)Nph(\lambda)d\lambda}{\int_{\lambda\,min}^{\lambda\,max} Nph(\lambda)d\lambda},$$

where R(λ) is the wavelength-dependent reflection and N_{ph} is the photo flux of the AM1.5 (air mass 1.5) solar spectrum as a function of the wavelength. The data for the AM1.5 spectrum were taken from the site of the American Society for Testing and Materials (ASTM).

Since the refractive indices of ZrO_2 and Al_2O_3 were determined experimentally and could not be varied, the optimal thickness of the two layers in the double layer ARC was sought by minimizing the weighted average reflectance for two-layer: $Al_2O_3/ZrO_2/Al_{0.8}Ga_{0.2}As/GaAs$. The minimization was carried out using the Simplex search method.

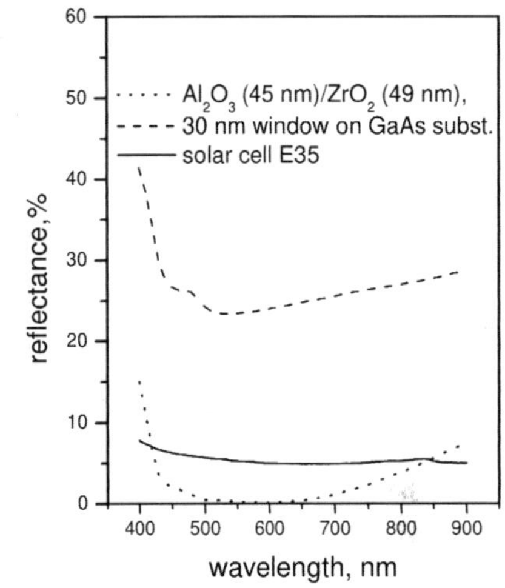

Figure 2. Modeled reflectance for the optimal thicknesses of the double ARC, a structure without ARC and an experimental SC structure.

3. Results and discussion

The minimization of the average weighted reflectance for the double layer structure on a 30 nm thick $Al_{0.8}Ga_{0.2}As$ window layer on GaAs resulted in 49 nm thickness of the lower (ZrO_2) layer and 45 nm for the top (Al_2O_3) layer. The value of the minimized average weighted reflectance was 2.17 %. Figure 2 shows the modeled reflectance of the structure $Al_2O_3/ZrO_2/Al_{0.8}Ga_{0.2}As/GaAs$ for these optimal layer thicknesses. The modeled reflectance of the GaAs substrate with an $Al_{0.8}Ga_{0.2}As$ window is also presented. The experimental reflectivity of such a structure is given as well.

The density of the short circuit photocurrent (J_{sc}) was 25 $mA.cm^{-2}$ for solar cells without ARC. After the deposition of a double layer antireflective coating Al_2O_3/ZrO_2, densities of 36 $mA.cm^{-2}$ were achieved.

The external quantum efficiencies of solar cells with and without an antireflection coating are presented in figure 3. The reflectivity spectrum of the solar cell without ARC is given as well.

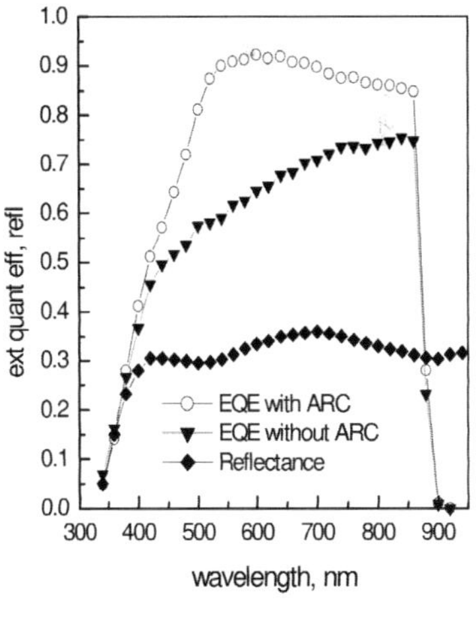

Figure 3. External quantum efficiency with and without ARC and reflectance (R) of SC.

4. Conclusion

A double layer antireflection coating was deposited on the front side a AlGaAs/GaAs solar cell by spin coating – a low cost method with good reproducibility. The optimal thickness of the two layers was determined by minimizing the average weighted reflection. A theoretical value of 2.81% was obtained together with thicknesses of the top (Al_2O_3) and the bottom (ZrO_2) layers of 45 nm and 49 nm, respectively. The density of the short-circuit photocurrent (J_{sc}) increased from 25 $mA.cm^{-2}$ for solar cells without ARC to 36 $mA.cm^{-2}$ for those with a double layer ARC.

Acknowledgments
This work was supported by the National Science Fund at the Ministry of Education and Science of Bulgaria under Grants D 03 369/06 and TN 1504.

References

[1] Richards B S 2003 *Solar Energy Materials & Solar Cells* **79** 369
[2] Sttehlke S, Bastide S, Guillet J and Levy-Clement C 2000 *Mat. Sci. Eng.* B **69** 81
[3] Algora C and Alcaraz M F 1990 *IEEE Trans. Electron Devices* **44** 1499
[4] Vitanov P, Harizanova A, Ivanova T and Dimitrova T 2010 *Materials Sci and Engin. IOP Conf. Series* **8** 012027
[5] Vitanov P, Harizanova A, Ivanova T and Dimitrova T 2009 *Thin Solid Films* **517** 6327
[6] Vitanov P, Tsanev A, Stefanov P, Harizanova A and Ivanova T 2007 *J. Optoelect. Adv. Mater.* **9** 256
[7] Milanova M, Kakanakov R D, Arnaudov B, Evtimova S, Vitanov P, Goranova E, Ivanov P, Alexieva Z, Shvarts M, Timoshina K and Vlasov A 2009 *Proc. Workshop Nanosci. & Nanotechnol.* ed E Balabanova and I Dragieva (November 2008 Sofia Bulgaria) **8** 124
[8] Aspnes D E, Keiso S M, Logan R A and Bhat R 1986 *J. Appl. Phys.* **60** 754
[9] Abeles F 1950 *Annals de Physique* **5** 596

AUTHOR INDEX

AUTHOR INDEX

Institute of Physics
Dirac House, Temple Back
Bristol BS1 6BE UK

CURRAN ASSOCIATES INC
proceedings
.com